Design of
Machine Members

Design of Machine Members

VENTON LEVY DOUGHTIE

Professor of Mechanical Engineering
University of Texas

ALEX. VALLANCE

Late Chief Designer, Reed Roller Bit Co.,
and Associate Professor of Mechanical Engineering
University of Texas

With Some Revisions by
LEONARDT F. KREISLE

Associate Professor of Mechanical Engineering
University of Texas

Fourth Edition

McGraw-Hill Book Company

New York   San Francisco   Toronto   London

Preface

This fourth edition of a textbook first published in 1938 has been extensively revised and rewritten to provide the student and the practicing engineer with current design information based on technological advances in the field. The basic premise of the book, however, remains unchanged; it is that design problems are best approached through a combination of theory, modern knowledge of materials, awareness of the limitations and practicality of various production methods, and a fair measure of common sense.

As in previous editions, the book begins with a general account of design problems and procedures; materials and their engineering properties are reviewed; theories of failure, loading, and resulting stresses and strains are presented; and the design of specific machine elements, such as shafts, bearings, gears, brakes, clutches, fasteners, cylinders, tubes, chains, and belts, are discussed in various chapters. The discussion of materials is essentially new, as is the chapter on stresses and strains, which considers modern data on stress concentration, fatigue, endurance, contact stresses, theories of failure, and dynamic loading. The discussions of both materials and stresses have been included for ready reference or review and may be omitted from a course of study. Each of the chapters on various machine elements has been written for independent study and with as little reference to other chapters as practical.

Among other significant changes, the chapter on sliding-element bearings has been completely revised and includes the new load numbers, effects of surface roughness, and other information resulting from recent experimental and analytical investigations. Factor of safety, margin of safety, allowable stress, and design loads have been clarified by the definition of a design factor. Dynamic and stress concentration have been considered in shaft and gear design, and a completely new chapter on seals, packings, gaskets, and shields has been added.

A majority of the illustrations are new, and a wealth of new and revised practical problems have been included to provide the student with realistic design experience. Some information derived from handbooks, standards, and manufacturing manuals has been included to aid the student in the solution of problems and to illustrate the type of information available. The original source is indicated for those desiring more complete information. The student with some knowledge of kinematics, statics, dynamics, materials, and strength of materials should not have great difficulty in mastering the subject matter.

The author wishes to acknowledge the many suggestions he has received for the improvement of this book from users of previous editions, the use of information from technical societies and manufacturers, and the encouragement received from the publisher and friends in the writing of this edition. The author is particularly indebted to Dr. L. F. Kreisle, his colleague and former student, for suggestions, encouragement, revision of some of the chapters, and review of the entire manuscript.

Venton Levy Doughtie

Contents

1

Design Philosophy

1-1 Design. Engineering design is the culmination of an engineering education. Without design, industry would have no new or improved products, basic information resulting from research would not be put to the use of man, and progress would halt. Engineering design is found in all professions and specialities—medicine, chemistry, architecture, and so on. Design is a dynamic field and includes various types of designers. In industry one finds product designers, appearance designers, apparatus designers, industrial designers, machine designers, systems designers, tool designers, electrical circuit designers, structural designers, creative designers, and others. What is designed? Anything and everything from a miniature ball bearing 0.1 in. in outside diameter to a complete power or manufacturing plant covering many acres, a spacecraft, a submarine, an airplane, or an artificial timer for the human heart. It may be a small simple piece such as a breech-bolt that goes to make up a complicated subassembly, or the jet engine starter used in jet aircraft. In systems engineering, the operation of a group of components is studied as if the group were one large device. Analytical investigations, tests, and study of environmental effects are carried out as if for a single huge component.

1

The design goal for any one of the small components is subject to the common good of the system as a whole.

Engineering is the art and science by which the properties of matter and the sources of power in nature are made useful in structures, machines, and fabricated parts. A *machine* is a combination of resistant parts so arranged as to cause the forces of nature to produce definite work with constrained motion. To *create* is to bring into existence something new. To *design* is to contrive, fashion, or invent according to a plan.

Creative design, creative engineering, and professional creativity are popular terms which embrace the entire field of design. Industrial concerns have developed courses for training of designers. Books have been written and courses taught in colleges and industry to train people for creativity. No longer can it be said that a designer is born and not made by study and training. The student of design and the young designer would do well to study the techniques used and suggested for attacking the problem of designing devices for doing a job.

A design requires the answers to at least the following questions:

1. What device, gadget, mechanism, or combination of these should be used?

2. What components or elements should be used—mechanical, pneumatic, or electrical?

3. What material should be used?

4. What loads are present?

5. How will the size be determined?

6. How will size, space, weight, manufacture, operation, reliability, maintainability, and cost affect the design?

7. What will be the user's reaction?

The above questions and discussion indicate the truthfulness of the first sentence of this article. An engineering designer need be well versed in mathematics and the basic sciences. He needs mechanics, materials, electricity, and circuits, and production know-how. His education is never complete. The qualities of a designer may be summed up as follows: an active and inquisitive mind; an inner desire that is satisfied only by a job well done; a broad background of information and fundamental knowledge; ability to present ideas and solutions to problems effectively; a creative approach to any problem; and the ability to get along with people.

It is apparent that design is engineering and covers a broad field. This book is concerned with one phase of design—design of machine members, commonly called machine design. It is hoped that this text will provide guidance to students in applying, in a practical way, the knowledge learned previously in their curriculums and be helpful in preparing them better to take their places in the engineering world.

1-2   The Designer.   As the designer develops his ideas into finished plans for a machine, he must bring into play an extensive knowledge of subjects that may be roughly classified as:

1. Technical factors
2. Experience factors
3. Human factors

The *technical information* necessary to design a machine varies with the type and field of application; and no one designer can become expert in all types of design. However, an understanding of mechanisms, mechanics, structures, materials, mechanical processes, fluid mechanics, thermodynamics, electrical circuits, and similar technical subjects is essential to every type of design.   Since the stresses and the deflections in all parts of a machine cannot always be accurately determined, the designer is forced to rely upon experimental data. The ability to analyze and use experimental data is a very important characteristic of the designer.

Design experience and knowledge of existing designs are essential to a thorough understanding of machine design, but previous designs should not be considered in any way that would hinder the designer's creative ability. The young designer will find unlimited fields for original ideas. The designer needs not only technical experience but also experience in the conduct of business and a knowledge of commerce and economics.

The *human element* is receiving more and more attention in the design room. There has been considerable study leading to the simplification of the operating controls and to a reduction of the physical effort necessary to operate the machine. Automation has acquired a dominant influence in the industrial world. Safety devices are being built into the machines, thus reducing accidents to a minimum. Pleasing appearance and sense appeal are now being seriously considered as aids to the sales force. To improve his knowledge of the human element, the designer should acquire the habit of discussing problems with engineers, technicians, salesmen, operators, mechanics, and even unskilled labor, since the different viewpoints will always help in developing well-balanced designs.

The designer uses mathematics, graphical methods, experimental stress analysis, dimensional analysis, statistical analysis, models, laboratory testing, and thought. His sources of information are libraries, patents, vendor catalogs, competitors' and noncompetitors' products, technical and nontechnical journals, learned society meetings, associates, and nature.

1-3   The Designer's Problems.   The factors to be considered by a designer are often complex and in some cases are intangible and so must be estimated or assumed. A machine member with constant or variable forces acting upon it is shaped to give the required motion, and is sized to cause the resulting stresses, deformation, and perhaps wear to remain within definite limits under a particular environment influenced by temperature, corrosion, nuclear effects, and other ambient conditions.

The motion required is dependent upon the function of the machine part. In some cases the positions of the part must be known throughout the entire cycle, whereas in other cases it is necessary to design the part only so that it will be at certain places at certain times; attention is not given to its motion at all times. The velocity and acceleration of the part must often be considered in order to care for the resulting vibration and inertial effects.

The forces resulting from static and dynamic forces are often difficult to determine exactly, and in some cases only an estimation is possible or necessary. The accuracy necessary in determining these forces will depend upon the importance of the part, the effect of the forces upon its size, and its weight.

The method of fabrication of the part must be considered because there may be many choices. The designer must be certain first that the part can be made as he has designed it; second, that the cost is the least when all factors are considered; and, third, that the part can be assembled in the machine and is accessible for repair and replacement.

The choice of material may be limited by availability, cost, manufacturing techniques, physical properties, requirements for finish, and other considerations. The physical properties of a material vary according to its compounding, size, shape, homogeneity, and forming. The designer knows its average properties as determined by a standard test specimen, but he does not know that the machine part made of the same material will have identical properties. The use of the materials which are now being developed should be considered in both new and old applications.

The size, area, and shape of cross section depend upon the resulting stresses and deformation in the member. The latter two items may be determined by theoretical stress analyses; formulas based upon experiment; empirical relations based upon theory, experiment, or practice; or code equations which may be a compromise among all previous methods. In some cases the designer may not have available any of the above relations and must depend upon either judgment or appearance.

From the above discussion it is apparent that the design of a machine, or even a member of a machine, is not an exacting science. Perhaps this is the reason there are various designs of a machine to perform the same function. The designer must often rely upon models and other testing techniques to determine whether or not his machine will perform. His problem is not one which always has a ready answer. Usually there is more than one way of doing a thing; sometimes there is no way.

1-4 Design Procedure.   There are many ways of attacking the same problem, and no rigid rules can be laid down for the designer to follow.

In the design of a new machine, the following logical procedure may be used:

1. Make a complete statement of the purpose for which the machine is to be designed.

2. Select the groups of mechanisms which could give the desired motion or group of motions.

3. Determine the energy transmitted by, and the forces acting on, each member of the selected mechanism.

4. Select the material best suited for each member.

5. Determine the size of each member by considering the forces acting, the permissible stress on the material, and the permissible deflection or deformation.

6. Modify the members to agree with previous experience and judgment and to facilitate manufacture.

7. Make assembly and detail drawings of the machine and include complete specifications for the material and manufacturing methods.

Often situations arise in which the size depends upon space limitations, size of mating parts, the housing, wear, rigidity, fabrication, lubrication, weight, or environmental effects. The beginning designer is often prone to rely entirely upon mathematical analyses. There are many parts in a machine that need to be a certain size in order to either fit the application or cause the desired results. In any case standard sizes should be used whenever possible. Often the "calculated size" should be changed to a standard size if available. Standard-sized parts are less costly, more readily obtained, and often more easily assembled. In certain members, no exact analysis of stresses and deformations can be made, and it is then necessary for the designer to resort to his experience and that of others in order to obtain properly proportioned machine members. In such cases, experimental stress analysis may be helpful. In many cases, empirical equations (based on experimental results) must be used. Some slight modification may result in an easier manufacturing method and, hence, result in lower production costs. It is also essential that the assembly of the various parts, the replacement of worn parts, the allowance for wear, provision for lubrication, and similar items be kept constantly in mind during the design of the machine.

2

Materials

2-1 General Considerations. After the general layout of the machine has been determined and the necessary mechanisms chosen or devised, it becomes necessary that the designer select a proper material for each machine member. This involves the consideration of such factors as the engineering properties of the available materials; the weight, size, and shape of the machine member as well as the loads that it must carry; cost of the material; cost of fabricating the machine element from each material, usually with several alternative production procedures possible for each material; and any properties of the material peculiar to the use to which the member will be put. The major engineering properties of materials which usually are of importance to the designer are strength, stiffness, ductility, toughness, resilience, fatigue resistance, shock resistance, wear resistance, creep characteristics, corrosion resistance, hardness, hardenability, machinability, formability, castability, weldability, ability to be surface finished in an acceptable manner, effects of high and low temperatures upon the behavior of the material, visual appearance, frictional properties, and internal vibrational damping properties.

7

2-2   Fabrication of Materials.   The designer must keep in mind the available fabrication methods and the effects of each on the properties of the finished member.

Castings are used for members of intricate shape that would be difficult to manufacture by other methods. When cast in metal molds under pressure, the products are called die castings; they are more accurate in size and shape than are ordinary castings, and in many cases they require no machining. Centrifugal castings are made by rapidly revolving the molds during the introduction and solidification of the molten casting material.

Although castings are desirable for many parts of intricate shape, they present many problems of design. Shrinkage during the cooling period, and nonuniform cooling of irregular thick and thin sections, bring residual stresses which may cause some castings to rupture before they can be put into service, or to warp during machining operations. Important castings always should be annealed or normalized to relieve these stresses.

Large castings expose proportionately less cooling surface, cool more slowly, and have a coarser grain structure than do small castings of the same material. This usually results in less strength and ductility in the larger castings, unless they are properly heat-treated.

Hot-working of ductile materials by *rolling, forging,* and similar processes refines the grain and generally improves the properties. Rolling, *pressing,* and *extruding* processes work the metal throughout and produce nearly uniform structure in all parts of the material. In forging, the working of the material is more or less local, and the inner part is not affected unless the forging hammer is relatively heavy. After hot-working, the material should be allowed to cool slowly and evenly to avoid residual stresses. Brass, lead, and other soft materials are frequently extruded into intricate shapes that cannot be made by rolling or forging.

2-3   Tabulated Properties of Materials.   Tables of the most important engineering properties of materials commonly used in machine design are included for reference. The properties of materials vary with the mechanical and thermal treatment received during manufacture, their treatment during the usage of the machine parts, the size of the part, the ambient conditions surrounding the produced part, and other factors. The tabulated properties of materials must be considered as average values only. Higher values may be obtained by careful manufacture and heat treatment; under certain conditions lower values may occur. The veteran designer must utilize his own experience with the material under consideration in order to arrive at a suitable value for the particular engineering property of the material that he desires.

2-4   Standard Specifications for Designating Materials.   Several national organizations and governmental agencies and a few large manufacturers have established standard specifications for designating and describing the various materials which are widely used by the designer. These specifications are very helpful in assisting the designer to make intelligent selections of materials. The most commonly used sets of specifications are the AISI (American Iron and Steel Institute) and SAE (Society of Automotive Engineers), used commonly for ferrous materials, the AA (Aluminum Association) specifications for aluminum and its alloys, and the ASTM (American Society for Testing and Materials) and ASA (American Standards Association) for a wide variety of materials. The Air Force, Army, and Navy each have their own set of specifications. In addition, various joint specifications exist between the Army and Navy (AN) as well as joint specifications common to all three military forces (MIL).

2-5 AISI and SAE Designations of Steels. The American Iron and Steel Institute (AISI) specifications for steels designate materials by use of a four-digit number (five digits in a few cases) with a one- or two-letter prefix and occasionally a one-letter suffix, such as C 1045, NE 8615, and 4620 H. The Society of Automotive Engineers (SAE) system for steels is essentially the same as the AISI system with the omission of the letter prefixes and suffixes.

In both the AISI and the SAE systems, the first digit indicates the class of steel; 1 is used to indicate carbon steels, 2 for nickel steels, 3 for chrome-nickel steels, 4 for molybdenum steels, 5 for chromium steels, 6 for chrome-vanadium steels, 7 for tungsten steels, 8 for the national emergency steels (also called triple-alloy steels), and 9 for silico-manganese steels (also sometimes called national emergency steels). The second digit indicates the approximate percentage of the principal alloying element present in the steel. The remaining two (sometimes three) digits represent a number 100 times the approximate percentage of carbon present in the steel.

In the AISI system, a one- or two-letter prefix (usually one letter) is added to the above four or five digits to indicate the method of producing the steel. These designations are A for basic open-hearth alloy steels, B for acid Bessemer carbon steels, C for basic open-hearth carbon steels, D for acid open-hearth carbon steels, E for electric furnace steels, and NE for the national emergency steels. The suffix F is employed to indicate free-machining steels, and H to indicate steels with more restricted hardenability limits than are commonly obtainable.

2-6 AA Designation for Aluminum and Its Alloys. The Aluminum Association (AA) system is employed for designating aluminum and its alloys and consists of a four-digit number followed by a letter and one or two numbers, such as 2024-T4 and 3003-H12. In some cases, an X will be prefixed to indicate a special experimental material, such as X7178.

The first digit of the four-digit number indicates the major alloying element(s), the remaining three digits arbitrarily being assigned to particular materials. If this first digit is 1, an aluminum of approximate 99 per cent purity is indicated; 2 indicates that copper is the main alloying element, 3 is used when manganese is the major alloying element, 4 is used where silicon is the main alloying element, 5 refers to alloys in which magnesium is the major alloying element, 6 indicates that magnesium and silicon are the major alloying elements, 7 is used when zinc is the main alloying element, and 8 indicates alloys of special elements other than those indicated above.

A suffix letter O indicates an annealed wrought material, F indicates a material that has been fabricated without any effort exerted to control properties of the material, H is employed to indicate a material strain-hardened by cold-working, and T is used for tempered materials.

Whenever H is used, it is followed by two digits. If the first digit is 1, temper produced by cold-working is indicated; 2 refers to a material cold-worked to harder temper than finally desired and then reduced to desired temper by partial annealing; 3 is used for cold-worked materials which then are stabilized by heating for a short period of time at a slightly elevated temperature. The second digit represents the relative hardness of the material, 0 referring to annealed condition, 2 to a hardness just above annealed condition, and so on up to 8, which indicates the hardest practical temper obtainable for the material concerned.

When T is used, a single digit suffix also is employed. T2 is used for annealed castings, T3 for solution heat-treated and then cold-worked, T4 for solution heat-treated, T5 for artificially aged only, T6 for solution heat-treated and then artificially aged, T7 for solution heat-treated and then stabilized, T8 for solution heat-treated followed by cold-working and then artificial aging, and T9 for solution heat-treated followed by artificial aging and then cold-working.

2-7 Static Strengths of Materials.   To obtain engineering properties of materials, designers usually refer more to information obtained from tensile testing than to that obtained from practically all other types of testing combined. The more important properties are tensile ultimate and yield stresses, modulus of elasticity, the corresponding compressive and shearing properties, and Poisson's ratio. General properties of commonly used materials are given in Table 2-1. More detailed information for specific materials is given in tables throughout this chapter.

2-8 Fatigue Strengths of Materials.   Machine parts frequently are loaded in a dynamic fashion in which the stresses vary with time, usually periodically. These stresses may vary periodically from a given value of tensile stress to the same magnitude of compressive stress, or from a par-

Table 2-1 **General properties of commonly used materials**

Material	Young's modulus E, psi	Modulus of rigidity G, psi	Poisson's ratio μ	Specific weight w, lb/in.^{3}	Thermal coefficient of linear expansion c_t, in./(in.)(°F)
Aluminum alloys	10,000,000	3,800,000	0.33	0.095,5	0.000,013,3
Beryllium	44,000,000		0.024–0.632	0.065,8	0.000,006,9
Brass	17,000,000	5,300,000–6,000,000	0.33–0.36	0.309	0.000,010,4
Bronze	16,000,000	5,100,000–5,900,000	0.35–0.36	0.294–0.321	0.000,010,1
Cast iron	12,000,000	5,200,000–8,200,000	0.28–0.29	0.256	0.000,006
Concrete	2,000,000		0.10–0.25	0.083,3	
Copper	17,000,000	5,800,000	0.335	0.322	0.000,009,3
Lead	2,600,000		0.43	0.411	0.000,016,1
Magnesium alloys	6,500,000	2,500,000	0.281	0.062,8	0.000,014
Monel	25,000,000	9,500,000	0.315	0.321	
Nickel	30,000,000		0.31	0.310	0.000,007,4
Plastics	1,000,000			0.037,6–0.059,6	
Steels	30,000,000	10,500,000–11,500,000	0.26–0.30	0.282	0.000,006,1–0.000,007,3
Tin	6,000,000		0.33	0.265	0.000,013
Titanium	16,500,000			0.164	0.000,004,7
Wood	1,500,000			0.015,1–0.035,9	0.000,002,1–0.000,030
Zinc			0.21	0.254	0.000,009–0.000,022

ticular value of shear in one direction to the same value of shear in the opposite direction. Either of these two cases would be examples of *fully reversed stress variation*. This periodic variation may be harmonic or non-harmonic. Even though a stress variation may be periodic, it may not be fully reversed. Stress variations from one magnitude of tension to a smaller magnitude of tension, or from one magnitude of tension to a different magnitude of compression are examples of non-fully reversed stress variations. Each of these possible variations in stresses produces what is known as *fatigue loading* on the machine member.

Experimental data indicate that repeated loading and unloading of a material will cause fatigue failure at a stress much lower than the ultimate strength of the same material as determined by static testing. The number of stress repetitions necessary to produce failure depends upon both the maximum stress imposed and the nature of the stress variation. Experimental evidence indicates that there is a limiting stress below which failure essentially will not occur even though the stress may be repeated an "infinite" number of times. This limiting value of stress often is referred to as the *endurance limit* of the material. This endurance limit is s_{ea} for axial loading, s_{ef} for flexural loading, and s_{es} for shear loading of the material with harmonic stress variation.

Endurance-limit data are very difficult to determine by experimental or other means, particularly for an infinite number of fatigue cycles. For this reason, experimentally determined endurance limits generally are expressed by an average amplitude of fully reversed harmonically varying stress as determined from several test specimens in which failure occurs in a specified number of cycles. If the method of loading is not specified, fully reversed harmonically varying flexure in fatigue is assumed. The usually employed endurance-limit cycle life is 5 million for wrought ferrous materials, 10 million for cast ferrous materials, 1 million or less for magnesium alloys, up to 500 million for aluminum alloys, and 2 million for wood.

The typical relationship between the amplitude of the applied stress and the number of stress repetitions required to produce fatigue rupture is represented by an experimentally determined curve similar to that indicated in Fig. 2-1. These same data plotted logarithmically yield two approximately straight lines as indicated in Fig. 2-2. Most experimenters take the stress corresponding to the break in the logarithmic curve as the endurance limit. It was from similar curves that the 1-million- to 500-million-cycle lives listed in the previous paragraph were determined.

2-9   Factors Affecting the Polished-specimen Endurance Limit of Materials Subjected to Harmonic Stress Variation.   Originally it was thought that endurance fatigue of materials was formed by crystallization of the material. Endurance fatigue failure appears to be caused by the progressive growth across the material of a minute crack formed in the material at some point of highly localized stress. This point-of-stress raiser may be at a small flaw in the material, or at some scratch or discontinuity of the surface.

So long as the rate of stress reversal does not exceed approximately 10,000 cycles/min, the endurance limit of a material appears to remain

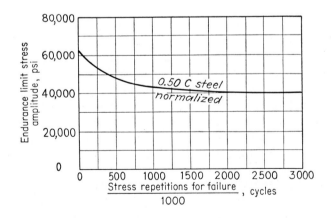

Fig. 2-1 Typical stress-repetition fatigue behavior of carbon steel.

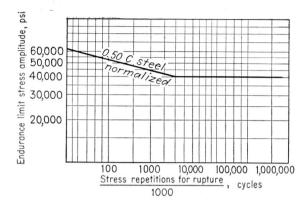

Fig. 2-2 Typical logarithmic plot of stress-repetition fatigue data for carbon steel.

essentially constant. For higher rates of stress reversal, the endurance usually rises slightly.

The endurance limit may be raised somewhat by highly polishing the surface of the test specimens, by cold-working, by burnishing, or by shot-peening. Nitriding, casehardening, or any other method for producing a thin layer of high-strength material on the outer surface also increases the endurance limit.

2-10   Typical Endurance Limits for Harmonically Varying Fully Reversed Axial, Bending, and Shear Loadings. Practically all available endurance limit data are based on specimens with highly pol-ished surfaces subjected to conditions of fully reversed harmonic stress variations. Some of these data are given in this chapter in the tables of engineering properties of various materials. Unfortunately not much data are available and the designer must resort to other procedures for obtain-ing useful endurance values. For wrought ferrous materials, the endurance limit values for fully reversed axial loading are approximately 70 per cent those for reversed bending. Data available for shear endurance limits of wrought ferrous materials indicate that s_{es} is approximately 55 per cent the corresponding s_{ef} values. Typical s_{ef} values for a few commonly used materials are given in Table 2-2.

2-11   Obtaining Actual Endurance Limits s_e. Each of the pre-viously mentioned endurance limits s_{ea}, s_{ef}, and s_{es} should be called *pol-ished-specimen endurance limits* for *harmonic stress variations*. They are based upon a more or less arbitrarily selected number of life cycles in the favorable ambient atmosphere of the testing laboratory and for test specimens with highly polished surfaces. Unfortunately for most machine

Table 2-2 Typical s_{ef} flexural endurance limits for fully reversed periodic loading of polished specimens of several commonly used materials*

Material	s_{ef}, psi
Aluminum:	
Cast alloys	6,000–11,000
Wrought alloys	8,000–18,000
Brass	7,000–20,000
Bronze:	
Phosphor	12,000
Tobin	21,000
Copper	12,000–17,000
Iron:	
Armco	24,000
Cast	6,000–18,000
Malleable	24,000
Magnesium alloys	7,000–17,000
Monel	20,000–50,000
Nitraloy	80,000
Steels:	
Cast	24,000–32,000
Plain carbon	25,000–75,000
6150 heat-treated	80,000

* L. S. Marks, "Mechanical Engineer's Handbook," 5th ed., p. 403, McGraw-Hill Book Company, New York, 1951.

parts, their surface roughnesses are not so low as those for the polished specimens, the ambient atmosphere may be more harmful than the air in the laboratory, the actual pattern of stress variation may not be harmonic, and possibly a different fatigue life is desired than the usual value for the specific material. Thus the *actual endurance limit* s_e is lower than the polished-specimen endurance limit as indicated by

$$s_e = \frac{s_{ea}, \; s_{ef}, \; \text{or} \; s_{es}}{flp} \tag{2-1}$$

where f is a surface finish and environment factor as indicated in Fig. 2-3, l is a life factor which takes into account the desired fatigue life of the member, and p is a factor which takes into account the pattern of stress variation imposed upon the member. Under essentially static loading, the surface finish appears to have little effect upon the strength of materials, hence f becomes essentially unity for this case. The life factor l can be determined only after experimentally determined endurance limit versus stress repetitions to failure data are available for the specific material for the reversed axial, flexural, or shear loading corresponding to the type of failure. Figures 2-1 and 2-2 are typical data for 0.50 per cent carbon normalized steel. The life factor l is the ratio of the stress amplitude of the endurance limit at the number of life cycles standard for a given material to the stress amplitude at the desired number of life cycles. Unless data similar to those in Figs. 2-1 and 2-2 are available, most designers assume a unity value for l. For nonharmonic stress variations, values of the stress pattern factor p are obtained as indicated in Eqs. (2-2) through (2-5). Whether s_{ea}, s_{ef}, or s_{es} is used in the numerator of Eq. (2-1) depends upon whether the combined stress causing fatigue failure corresponds to axial, flexural, or shear loading conditions.

2-12 Endurance Limits for Wrought Ferrous Materials Subjected to Not Completely Reversed Loadings. If a fatigue specimen is subjected to stresses which are not fully reversed and vary offset-harmonically on either side of a nonzero average stress (such as from one value of tension to a lower value of tension, zero, or a compressive stress of different absolute value from the highest tensile stress; from zero to a

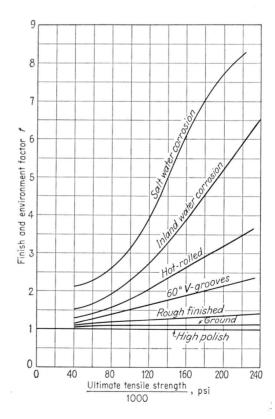

Fig. 2-3 Finish and environment factor f.

compressive stress; or from one low compressive stress to a greater magnitude compressive stress), the resulting endurance limit is different from that determined for fully reversed loading. For wrought ferrous materials subjected to offset-harmonic fatigue loading, experimental findings indicate

$$s_e = \frac{1.4s_{ea}}{fl[1 - 0.4(s_{min}/s_{max})]} = \frac{s_{ea}}{flp} \qquad (2\text{-}2)$$

where $p = \dfrac{1 - 0.4(s_{min}/s_{max})}{1.4}$

$$s_e = \frac{3s_{ef}}{fl[2 - (s_{min}/s_{max})]} = \frac{s_{ef}}{flp} \qquad (2\text{-}3)$$

where $p = \dfrac{2 - (s_{min}/s_{max})}{3}$

$$s_e = \frac{2s_{es}}{fl[1 - (s_{min}/s_{max})]} = \frac{s_{es}}{flp} \qquad (2\text{-}4)$$

where $p = \dfrac{1 - (s_{min}/s_{max})}{2}$

In the above equations, s_{max} and s_{min} respectively are the algebraic largest and smallest stress within the member per cycle, any value of compression considered as being smaller than either zero or a tensile stress. Since experimental data for offset-harmonic axial fatigue loading are very scarce and rather scattered, Eq. (2-2) should be considered as grossly approximate; the 1.4 and 0.4 factors were selected by the author on the basis of the data available to him. The above three equations obviously are incorrect for stress variations in which the ratio of s_{min} to s_{max} causes the denominator to approach zero.

2-13   Endurance Limits for Cast Iron.   The fatigue properties of cast iron are not so well known as those for steels. Unlike steel, under static loading conditions cast iron is many times stronger in compression than in tension. A similar condition exists for fatigue loading of cast iron. Tests made on cast iron at the University of Illinois indicate that the flexural endurance limit s_{ef} is approximately 35 per cent of the tensile ultimate strength s_{tu} and approximately 10 per cent of the ultimate compressive strength s_{cu}. When the stress cycle is not completely reversed and remains entirely compressive, the endurance limit is approximately four times that obtained when the stress cycle remains completely tensile.

For offset-harmonic flexure when tension is the largest absolute value of stress in the cycle, Eq. (2-3) fairly well fits experimental data. For offset-harmonic flexure when compression is the largest absolute value of stress in the cycle, the following equation is applicable:

$$s_e = \frac{6s_{ef}}{fl[1 - 5(s_{\min}/s_{\max})]} = \frac{s_{ef}}{flp} \tag{2-5}$$

where $p = \dfrac{1 - 5(s_{\min}/s_{\max})}{6}$

2-14 Endurance Limits for Nonharmonic Stress Variations.
Each of the endurance limits presented so far is based upon harmonic or offset-harmonic variation of stresses. This exact condition seldom exists in machines. Apparently almost no data are available for endurance limits under conditions of periodic stress variation other than harmonic or offset-harmonic. Since other types of stress variation are more abrupt than harmonic variation (such as square wave, triangular, random, and other forms), it appears logical that the corresponding endurance limits would be less than those determined for harmonic conditions. The harmonically determined endurance limits usually suffice for almost all fatigue conditions in the design and analysis of machines even though they are not exactly correct.

2-15 Graphical Determination of the Endurance Limit. Often it is more convenient to use a diagram in determining the endurance limit than to use formulas.

The endurance diagram, shown in Fig. 2-4 for SAE 1045 steel, can

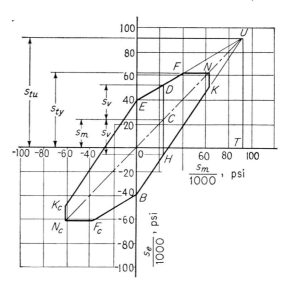

Fig. 2-4 Typical endurance diagram for SAE 1045 steel.

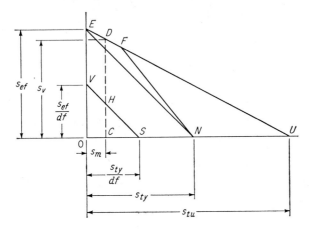

Fig. 2-5   Soderberg-Goodman diagram.

be constructed for any material when the ultimate strength, yield stress, and endurance limit for completely reversed stresses are known for the conditions of loading (flexure, torsion, or axial loading). Although flexural fatigue loading will be used as an illustration, the methods employed are applicable to torsion and axial fatigue loading as well. Analysis and experimental work indicate that, if a variable stress is superimposed upon a steady stress, the plotted results will determine a maximum- and a minimum-stress line between which safe operation can be maintained. These lines are EU and BU in Fig. 2-4 and are slightly curved, but may be assumed to be straight without appreciable error. The stress corresponding to point U is the ultimate strength, and the points E and B correspond to the endurance limit for complete stress reversals. Any point on OU, as C, represents a steady or mean stress s_m, and CD and CH represent the variable stress s_v, which may be combined with s_m. The lines EU, BU, and OU indicate stress combinations that will ultimately cause rupture, but in no case should the maximum stress exceed the yield stress. Hence the figure $EFNKB$ limits the possible stress combinations. This figure forms the endurance diagram for flexural stresses. Direct tensile stresses may be taken as 80 per cent of those shown in the diagram, and the shear stresses as 55 per cent.

As shown in Fig. 2-5, for certain purposes this diagram may be simplified by moving the point U to the horizontal axis. Points F and N will take the positions indicated, and CD will represent the variable stress that, combined with the steady stress OC, will just cause failure. For simplicity, the straight line EN can replace the line EFN as the stress limit line. This line EN often is called the Soderberg* line, and the line

---

* C. R. Soderberg, Working Stresses, *Trans. ASME*, vol. 55, Ap.M. 5546, 1933.

EU is the modified Goodman* line; hence this diagram is often called the Soderberg-Goodman diagram. By introducing the design factor df,† the line VS, or working-stress limit line, is obtained. The CH represents the variable stress that may be superimposed on a steady stress OC.

This procedure is essentially correct if the time variation of s_v relative to the mean stress s_m is harmonic. It usually is sufficiently accurate for most purposes of design for nonharmonic periodic variations.

Example 1. To illustrate this method of graphically determining endurance limits, consider a coiled valve spring. When the valve is open the stress in the spring is 35,000 psi and when closed the stress is 25,000 psi. The spring material has a yield stress of 80,000 psi in torsion and an endurance limit of 45,000 psi in reversed torsion. For this service, a design factor of 2 is judged suitable, and the allowable stresses are

$$s_{s\ all} = \frac{s_{ty}}{df} = \frac{80,000}{2} = 40,000 \text{ psi} \qquad \text{in torsion}$$

and

$$s_{e\ all} = \frac{s_{es}}{df} = \frac{45,000}{2} = 22,500 \text{ psi} \qquad \text{in reversed torsion}$$

These values are used to construct the line VS in Fig. 2-6. Lay off OC equal to 30,000 psi to represent the average or mean stress on the spring. Then CD, measured to scale, represents the permissible variable stress. The permissible variable stress scales 5,625 psi, and since the variable stress on the spring is only 5,000 psi, the imposed stresses are safe.

Example 2. To illustrate the use of the diagram in determining permissible design stresses, consider a piston rod for a reciprocating pump. The rod will be subjected to a total load having cyclic variation from 30,000 lb in compression

* J. Goodman, "Mechanics Applied to Engineering," Longmans, Green & Co., Inc., New York, 1924.

† Often also called factor of safety, factor of ignorance, margin of safety, or safety ratio. See Art. 4-3.

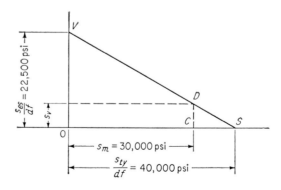

Fig. 2-6 Graphical determination of permissible variable stress s_v.

to 5,000 lb in tension. The expected life is 10 years of operation at 30 strokes/min, 24 hr/day, and 300 days/year.   Determine the rod size using SAE 2340 steel for which $s_{ty} = 62,000$ psi and $s_{ef} = 55,000$ psi. The expected life is equivalent to 12,960,000 stress repetitions and therefore fatigue effects must be considered.

Construct Fig. 2-7 from the given s_{ty} and s_{ef} data. The mean stress is unknown, but the stress range will be the same as the load range. The mean load is $-12,500$ lb and the variable load is 17,500 lb. Hence

$$\frac{F_v}{F_m} = \frac{s_v}{s_m} = \frac{17,500}{12,500} = 1.40$$

Through point O in the figure draw the line OB having the slope 1.40 and intersecting VS at D. The point D indicates the mean and variable stresses which can be used with the stress ratio 1.40. Scaling the stresses from the diagram yields

$$s_m = OC = 24,400 \text{ psi}$$
$$s_v = CD = 34,150 \text{ psi}$$

Thus
$$s_m + s_v = 58,550 \text{ psi}$$

Assuming a service factor of 1.25, a design factor of 1.50, a surface-finish factor of 1.10, and a stress-concentration factor of 2.25, the design stress is

$$s_d = \frac{58,550}{1.25 \times 1.5 \times 1.1 \times 2.25} = 12,620 \text{ psi allowable}$$

The required rod area is $30,000/12,620 = 2.376$ in.^{2}, corresponding to a diameter of 1.74 in., say $1\frac{3}{4}$ in., neglecting any column effects on the rod. See Art. 4-5 for more information on design stresses.

2-16   Effect of High Temperatures on Materials.   The selection of the proper design stress for members subjected to high temperatures is dependent upon two properties of the material. First, the strength of the material is modified by temperature, and, if the temperature is maintained for a sufficient period of time, structural changes that will further affect the strength will take place. The effect of temperature on the

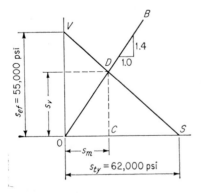

Fig. 2-7   Graphical determination of allowable mean stress s_m and variable stress s_v.

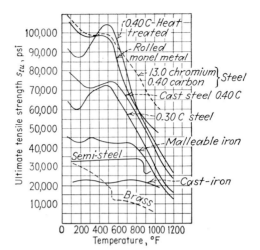

Fig. 2-8 Effect of temperature upon the ultimate tensile strength of various common metals.

strength of typical materials is shown in Fig. 2-8. Results of this kind must be used with care, since they are usually taken from tests made shortly after the material has reached the desired temperature and before structural changes due to prolonged exposure to heat have taken place.

The second item that must be considered is the change in size at high temperatures and the gradual deformation, or creep, that accompanies stress at high temperatures. *Creep* is the term used in referring to the continuous increase in the strain, or deformation, of any material subjected to stress. An examination of the curves in Fig. 2-9 shows that the rate of creep varies with the stress, temperature, and time. The rapid initial deformation produces a strain hardening of the material that tends to decrease the creep rate. The effect of continued high temperature on the structure of the metal is to temper it and increase its ductility and thus the creep rate. When the tempering effect predominates, a transition point is reached, beyond which the creep increases very rapidly. At low temperatures and stresses, the curves apparently indicate that the creep finally ceases, but

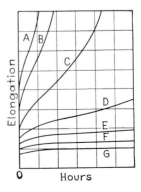

Fig. 2-9 Typical tensile creep-time curves. Tests were made at a constant temperature with the stress decreasing from *A* to *G*.

Table 2-3 Creep strength of various metals, psi, for 1,000 hours of static tensile load application for indicated creep and temperature*

Steels and high-temperature alloys

Metal	0.01 Per cent creep at				0.1 Per cent creep at					
	1000°F	1100°F	1200°F	1500°F	1000°F	1100°F	1200°F	1400°F	1500°F	1600°F
Carbon and low-alloy steels:										
Low-carbon steels (wrought, cast)	1,800	....	100	....	3,300–5,000	....	500			
Carbon-molybdenum steels (wrought, cast)	5,000–7,000	3,000	1,000	....	10,000–12,000	4,000	2,000			
Chromium-molybdenum steels, 0.5–3% (wrought, cast)	6,000–12,000	2,000–4,000	1,000–2,500	....	10,000–20,000	3,000–8,000	2,000–4,500			
Chromium steels, 4–6% (wrought, cast)	6,000–7,000	2,500–3,500	1,000–2,000	....	8,000–12,000	5,000–6,500	2,000–3,500			
Chromium steels, 6–10% (wrought, cast)	5,000–9,000	2,500–4,000	1,000–2,000	....	8,000–12,000	4,000–6,000	2,500–3,000			
Stainless steels:										
Ferritic chromium steels, 405, 430, 440 (wrought)	4,200–7,000	2,300–4,500	1,000–1,600	....	6,000–8,500	3,000–5,000	1,500–2,200			
Martensitic chromium steels, 403, 410, 416, 420 (wrought)	12,000–17,000	3,500	1,300	....	9,200	4,200	2,000			
Nickel-chromium steels, 304, 316, 321, 347 (wrought)	8,000	7,500–11,500	4,500–7,000	1,000–2,000	17,000–25,000	12,000–18,200	7,000–12,700		1,200–2,800	
Nickel-chromium steels, 309 (wrought)	....	....	4,000	500	15,900	11,600	8,000		1,000	
Nickel-chromium steels, 310, 314 (wrought)	17,000	13,000	8,000	200	17,000	13,000–14,000	9,000		1,000–2,500	
Heat-resistant cast high-temperature alloys:										
Iron-chromium alloys, HA, HC, HD (cast)	....	....	....	....	....	....	....	1,200–3,500		700–1,900
Iron-chromium alloys, HE, HF, HH, HI, HK, HL (cast)	....	....	....	....	....	....	....	3,500–7,000		2,000–4,300
Nickel-chromium alloys, HN, HT, HU, HW, HX (cast)	....	....	....	....	....	....	....	6,000–8,500		3,000–5,000
Superalloys:										
Inconel X	....	....	....	....	....	....	64,000		12,300	9,000
19-9 DL	....	....	....	....	....	....	20,000		7,100	2,400
N-155	....	....	....	....	....	....	....		10,300	....
S-816	....	....	....	....	....	....	42,000		11,500	5,800

Nonferrous metals

Metal	0.01 Per cent creep at					0.1 Per cent creep at				
	300°F	400°F	500°F	600°F	800°F	300°F	400°F	500°F	600°F	800°F
Nonferrous metals:										
Coppers (wrought, annealed)	3,000–8,000	1,500–5,000	400–2,600	....	....	....	5,000–9,000	1,000–2,000	....	....
Nonleaded brasses (wrought, annealed)	900–19,000	2,000–11,000	300–23,000	....	....	....	....	....	....	....
Bronzes (wrought, annealed)	14,000–23,000	5,000–10,000	2,000–5,000	....	....	....	....	....	....	....
Cupro-nickel (wrought, water-quenched, aged)	25,000–40,000	15,000–30,000	8,000–30,000	....	....	25,000	....	....	....	....
Aluminum 2024-T (sheet)	23,000	9,500	2,500	....	....	....	22,000	13,000	2,000	....
Aluminum 7075-T (sheet)	12,000	4,000	2,500	....	....	....	13,000	3,000	2,000	....
Titanium, commercial (sheet, annealed)	....	3,800	....	1,500	....	....	6,000	3,000	32,000	....
Ti-6Al-4V (sheet, annealed)	....	....	....	1,500	10,000	....	....	....	80,000	13,000
Ti-7Al-4Mo (bar or forging, annealed)	....	....	....	32,000	....	....	40,000	37,000	85,000	18,000

* Creep Strength of Metals, *Materials in Design Eng.*, vol. 54, no. 5, p. 29, 1961.

at the higher temperatures and stresses, the transition point is very pronounced. After a sufficient length of time has elapsed, there will probably be a transition point for any combination of temperature and stress.

Careful consideration of creep data leads to the conclusion that extrapolation beyond the time limit of the test data must be carried out with extreme caution. The designer, however, must select design stresses that will keep the creep or deformation within prescribed limits during the life of the machine, which may be 10 to 25 years or more.

The *creep strength* of a material usually is defined as the constant stress to which a material is loaded, at a specified temperature and for a specified time, that will cause a further elongation or creep of a given percentage of the original unstressed length. Usually creep strengths are

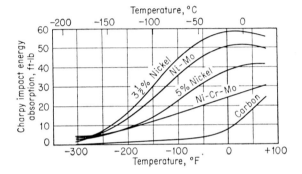

Fig. 2-10 Charpy impact resistance of annealed carbon- and nickel-alloy steels at low temperatures. (*From Nickel Steel Topics, October, 1933. International Nickel Co.*)

Steel type	Approximate steel composition	Approximate Brinell hardness as annealed
Carbon steel................	SAE 1030	
$3\frac{1}{2}\%$ nickel steel............	SAE 2315	
5% nickel steel.............	SAE 2512	140–155
Nickel-molybdenum steel.....	0.12% C, $3\frac{1}{2}\%$ Ni, 0.25% Mo	
Nickel-chromium-molybdenum steel	0.30% C, $3\frac{1}{2}\%$ Ni, 0.9% Cr, 0.25% Mo	250

specified as pounds per square inch tension for either 0.01 or 0.1 per cent creep at a particular Fahrenheit temperature, usually between 300 and 2000°F. Typical creep strengths for several metals are given in Table 2-3.

2-17   Effect of Low Temperature on Materials.   Machinery that is to operate in cold climates, in refrigerating service, and in oil dewaxing plants must show satisfactory service at low temperatures. The strength and elasticity of steels are not affected in an adverse manner, but embrittlement may become a major consideration in the selection of the proper material. Hence, the impact properties at low temperatures must be given careful consideration when selecting the design stress.

As the temperature is decreased, plain carbon and low-carbon steels reach a condition where the loss of impact resistance becomes very rapid, and then the steel becomes very brittle. Most carbon steels reach this condition within atmospheric ranges, at times only slightly below room temperatures, and thus may become dangerously brittle at service temperatures. Nickel in steel acts to offset this condition by raising the impact resistance at room temperatures, by lowering the temperature at which rapid loss of impact resistance occurs, and by lowering the rate of loss in this region. For extreme low temperatures, around −300°F, such as are encountered in liquid-air machinery, steels containing 42 per cent or more of nickel are used. These steels retain their impact resistance unimpaired to temperatures of −310°F, and probably much lower.

The effect of low temperature upon the properties of steels is depicted in Figs. 2-10 and 2-11.

2-18   Classification of Ferrous Materials.   Where iron is the principal component element, ferrous materials are classified according to

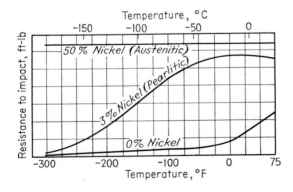

Fig. 2-11   Effects of high and low nickel contents upon the low-temperature properties of steel. (*From Nickel Steel Topics, October, 1933. International Nickel Co.*)

Table 2-4 **Properties of typical cast irons**

Material	Ultimate strength		Flexural endurance limit s_{ef}, psi	Brinell hardness number BHN	Modulus of elasticity		Tensile elongation in 2 in, %	Remarks and suggested uses
	Tension s_{tu}, psi	Compression s_{cu}, psi			Tension and compression E, psi	Shear G, psi		
Gray:								
Ordinary..........	18,000	80,000	9,000	100–150	10–12,000,000	4,000,000	0–1	General industrial castings
Good..............	24,000 16,000	100,000	12,000	100–150	12,000,000	4,800,000	0–1	Pump cylinders, etc.
High-grade........	30,000	120,000	15,000	100–150	14,000,000	5,600,000	0–1	Important castings
Malleable, SAE 32510......	50,000	120,000	25,000	100–145	23,000,000	9,200,000	10	Substitute for unimportant forgings
Nickel alloys:								
Ni–0.75, C–3.40, Si–1.75, Mn–0.55	32,000 24,000	120,000	16,000	200 175	15,000,000	6,000,000	1–2	Light machine frames
Ni–2.00, C–3.00, Si–1.10, Mn–0.80	40,000 31,000	155,000	20,000	220 200	20,000,000	8,000,000	1–2	Heavy diesel cylinders, pump, and valve bodies
Nickel-chromium alloys:								
Ni–0.75, Cr–0.30, C–3.40, Si–1.90, Mn–0.65	32,000	125,000	16,000	200	15,000,000	6,000,000	1–2	Light machine-tool table, light engine cylinders
Ni–2.75, Cr–0.80, C–3.00, Si–1.25, Mn–0.60.....	45,000	160,000	22,000	300	20,000,000	8,000,000	1–2	Heavy forming dies

Upper figures refer to arbitration test bars. Lower figures refer to the center of 4-in. round specimens.

Flexure: For cast irons in bending, the modulus of rupture may be taken as $1.75s_{tu}$ for circular sections, $1.50s_{tu}$ for rectangular sections, and $1.25s_{tu}$ for I and T sections.

Shear: The strength of cast iron in shear may be taken as $1.10s_{tu}$.

their carbon content. *Wrought irons* contain less than 0.1 per cent carbon, *steels* have a carbon content between 0.1 and 2.0 per cent inclusive, and *cast irons* have carbon exceeding 2.0 per cent.

2-19   Classification of Steels. Steels may be classified as plain carbon steels, alloy steels, and special-purpose steels. An *alloy steel* contains one or more of the following alloying elements in excess of the percentage indicated: 1.65 per cent manganese; 0.60 per cent copper or silicon; 3.99 per cent aluminum, boron, or chromium; any percentage of cobalt, molybdenum, nickel, titanium, tungsten, vanadium, zirconium, or any other element which is intentionally added to obtain a desired alloying effect. A *special-purpose steel* is a material produced to meet certain special conditions in fabrication or in use such as free machining, heat resistance, corrosion resistance, impact resistance, and tool steels. A *plain carbon steel* is one which is neither an alloy steel nor a special-purpose steel. Depending upon their carbon content, plain carbon steels frequently are divided into low-carbon steels with less than 0.3 per cent carbon, medium-carbon steels with between 0.3 and 0.5 per cent carbon inclusive, and high-carbon steels with over 0.5 per cent carbon.

Excluding cast iron, steel is the most commonly used material in machine members. The mechanical working in the manufacturing process refines the structure and produces a more uniform steel having greater strength, greater toughness, and more durability than are obtained in castings. The properties of the steels vary greatly with the carbon content and with the form in which the carbon occurs, the beneficial effects of alloying materials depending to a large extent on their action on the carbon.

Low-carbon steels are soft, very ductile, easily machined, easily welded by any process, and, since the carbon content is low, unresponsive to heat treatment. Medium-carbon steels are stronger and tougher than the low-carbon steels, machine well, and respond to heat treatment. High-carbon steels respond readily to heat treatment. In the heat-treated state, they may have very high strengths combined with hardness, but are not so ductile as the medium-carbon steels. In the higher carbon ranges, the extreme hardness is accompanied by excessive brittleness. The higher the carbon content, the more difficult it is to weld these steels.

Steel castings are more difficult to produce than iron castings and are more expensive, but are stronger and tougher. They are used for machine members of intricate shape that require high strength and impact resistance, such as locomotive frames, large internal-combustion engine frames, and well-drilling tools. Alloy-steel castings have been developed to meet the demands of industry for greater strength and reliability of cast machine members. Properties of typical carbon steels are indicated in Table 2-5.

Table 2-5 Properties of typical carbon steels

Material	Ultimate strength		Yield stress		Flexural endurance limit s_{ef}, psi	Brinell hardness number BHN	Modulus of elasticity		Elongation in 2 in., %	Remarks and suggested uses
	Tension s_u, psi	Shear s_{su}, psi	Tension and compression, s_y, psi	Shear s_{sy}, psi			Tension and compression E, psi	Shear G, psi		
Wrought iron........	48,000	50,000	27,000	30,000	25,000	100	28,000,000	11,200,000	30-40	
Cast steel:										
Soft........	60,000	42,000	27,000	16,000	26,000	110	30,000,000	12,000,000	22	General-purpose castings
Medium........	70,000	49,000	31,500	19,000	30,000	120	30,000,000	12,000,000	18	
Hard........	80,000	56,000	36,000	21,000	34,000	130	30,000,000	12,000,000	15	
SAE 1025, annealed........	67,000	41,000	34,000	20,000	29,000	120	30,000,000	12,000,000	26	Machinery and general-purpose steel
water-quenched........	78,000	55,000	41,000	24,000	43,000	159	30,000,000	12,000,000	35	
	90,000	63,000	58,000	34,000	50,000	183			27	
SAE 1045, annealed........	85,000	60,000	45,000	26,000	42,000	140	30,000,000	12,000,000	20	Large forgings, axles, shafts
water-quenched........	95,000	67,000	60,000	35,000	53,000	197	30,000,000	12,000,000	28	
	120,000	84,000	90,000	52,000	67,000	248			15	
oil-quenched........	96,000	67,000	62,000	35,000	53,000	192	30,000,000	12,000,000	22	
	115,000	80,000	80,000	45,000	65,000	235			16	
SAE 1095, annealed........	110,000	75,000	55,000	33,000	52,000	200	30,000,000	12,000,000	20	Springs, cutting instruments
oil-quenched........	130,000	85,000	66,000	39,000	68,000	300	30,000,000	11,500,000	16	
	188,000	120,000	130,000	75,000	100,000	380			10	

Upper figures: steel quenched and drawn to 1300°F.
Lower figures: steel quenched and drawn to 800°F.
Values for intermediate drawing temperatures may be approximated by direct interpolation.

2-20   Alloy Steels. Alloys are used to effect increased strength, increased elastic ratio, increased hardness in the annealed steels without loss of ductility, more uniform structure, better machinability, and better resistance to fatigue and corrosion. The proper combination of these properties depends not only on the chemical composition, but largely on the heat treatment, without which the alloy steels are not greatly superior to the plain carbon steels. With proper alloying and heat-treating, ultimate strengths of over 300,000 psi and yield stresses of over 250,000 psi are obtainable. Many of the alloy steels are expensive; therefore their use is limited.

Nickel increases the strength without sacrificing the ductility of the carbon steels. It also tends to retard the grain growth, which allows a larger range of heating and longer carburizing periods without damage due to coarse grain structure. *Chromium* is added principally to increase corrosion and oxidation resistance, to increase hardenability, to raise strength at high temperatures, and to resist abrasion and wear when high carbon content exists. *Vanadium* is used to toughen and strengthen the steel, to reduce the grain size, and to act as a cleanser and degasifier. It has the desirable effect of increasing the life of tools, springs, and other members subjected to high temperatures. *Molybdenum* acts very much like chromium but is more powerful in its action. It also increases the depth of hardening after heat treatment. *Nickel-molybdenum* and *nickel-chromium-molybdenum* steels retain the good features of the nickel-chromium steels and in addition have better machining qualities. These steels respond uniformly to heat treatment and offer the highest strengths obtainable in commercial steels. *Manganese* is a deoxidizer and contributes to strength and high toughness. For this reason, manganese steels find wide use in gears, splines, high-torque shafts, and similar applications. *Silicon* raises the critical temperature for heat treatment of steel but also increases its susceptibility to decarburization and graphitization. When employed together with chromium, nickel, and tungsten, silicon promotes resistance to high-temperature oxidation. *Copper* increases the atmospheric corrosion resistance and strength of steel. *Aluminum* serves as an excellent deoxidizing agent, and promotes the nitriding of steels. *Boron* is employed almost exclusively to increase the hardenability of steels.

Selected properties of the more commonly used alloy steels are tabulated in Table 2-6.

2-21   Heat Treatment of Metals. Heat treatment is the process of controlled heating and cooling of metals or alloys in solid state for the purpose of changing their structural arrangement and to obtain certain desirable properties. The higher-carbon steels and alloy steels are especially responsive to heat treatment.

Annealing consists of heating the metal to a temperature slightly above the critical temperature and then cooling slowly, usually in the furnace, to produce an even grain structure, reduce the hardness, and increase the ductility, usually at a reduction of strength.

Normalizing is a form of annealing (cooling in air) used to remove the effects of any previous heat treatment and to produce a uniform grain structure before other heat treatments are applied to develop particular properties in the metal.

Quenching, or rapid cooling from above the critical temperature by immersion in cold water or other cooling medium, is a hardening treatment. The degree of hardness depends on the amount of carbon present, and on the rate of cooling, which can be varied by using such cooling mediums as ice water, cool water, oil, hot oil, molten lead, etc. The hardening treatment raises the strength of the metal and increases the wear resistance, but makes the metal brittle and of low ductility.

Tempering, or drawing, consists of reheating the quenched metal below critical temperature to restore some of the ductility and reduce the brittleness. Increased toughness is obtained at the expense of high strength. The loss of hardness depends on the temperature and drawing time.

Casehardening or *carburizing* is a process of hardening the outer portion of the metal by prolonged heating free from contact with air while the metal is packed in carbon in the form of bone char, leather scraps, or charcoal. The outer metal absorbs carbon, and when the hot metal is quenched, this high-carbon steel hardens, whereas the low-carbon steel of the core remains soft and ductile. In gas-carburizing, the metal is heated in an atmosphere of gas controlled so that the metal absorbs carbon from the gas but will not be oxidized on the surface.

Cyaniding is casehardening with powdered potassium cyanide or potassium ferrocyanide mixed with potassium bichromate substituted for the carbon. For a very thin case, immersion in hot liquid cyanide is sufficient. Cyaniding produces a thin but very hard case in a very short time.

Nitriding is a surface hardening accomplished by heating certain steel alloys (Nitralloys) immersed in ammonia fumes.

The principal treatments used with cast iron are *aging*, to relieve castings of cooling and shrinkage stresses; *baking*, to remove brittleness resulting from pickling in acid during the cleaning process; *annealing*, to reduce the hardness and permit machining; *toughening*, to increase the strength of white iron; *quenching* by rapid cooling, to obtain sufficient hardness to resist wear and indentation; *drawing*, to restore the strength of quenched castings; and *carburizing*, to increase wear and impact resistance.

Table 2-6 Properties of typical alloy steels

Material	Ultimate strength Tension s_u, psi	Ultimate strength Shear s_{su}, psi	Yield stress Tension s_y, psi	Yield stress Shear s_{sy}, psi	Flexural endurance limit s_{ef}, psi	Brinell hardness number BHN	Modulus of elasticity Tension and compression E, psi	Modulus of elasticity Shear G, psi	Elongation in 2 in., %	Remarks and suggested uses
Nickel:										
SAE 2320, water-quenched..........	77,000	55,000	50,000	30,000	49,000	143	30,000,000	12,000,000	30	Casehardening stock for heavy parts. Not desirable for thin sections
	140,000	98,000	110,000	65,000	68,000	277			18	
oil quenched..........	75,000	54,000	48,000	29,000	46,000	140	30,000,000	12,000,000	30	Gears
	130,000	90,000	100,000	60,000	50,000	262			18	
SAE 2340, water-quenched..........	95,000	60,000	65,000	38,000	53,000	183	30,000,000	12,000,000	30	Forgings, axles
	175,000	110,000	150,000	87,000	75,000	340			16	
oil-quenched..........	93,000	59,000	62,000	36,000	55,000	183	30,000,000	12,000,000	17	Gears
	165,000	105,000	148,000	85,000	75,000	330			14	
Nickel-chromium:										
SAE 3120, water-quenched..........	86,000	60,000	57,000	34,000	55,000	174	30,000,000	12,000,000	34	Heavy sections requiring medium depth casehardening
	140,000	98,000	115,000	68,000	58,000	269			15	
oil-quenched..........	77,000	55,000	48,000	29,000	46,000	163	30,000,000	12,000,000	30	
	120,000	82,000	96,000	57,000	48,000	241			18	
SAE 3220, water-quenched..........	87,000	60,000	60,000	36,000	58,000	187	30,000,000	12,000,000	33	Massive sections requiring deep case-hardening
	165,000	115,000	143,000	83,000	72,000	331			16	
oil-quenched..........	55,000	56,000	53,000	31,000	51,000	174	30,000,000	12,000,000	30	
	80,000	107,000	130,000	95,000	65,000	311			20	
SAE 3340, oil-quenched..........	110,000	73,000	87,000	52,000	84,000	229	30,000,000	12,000,000	23	Rollers, sprockets, gears
	200,000	130,000	180,000	105,000	90,000	388			14	

Table 2-6 Properties of typical alloy steels (Continued)

Material	Ultimate strength Tension, psi s_{tu}	Ultimate strength Shear s_{su}, psi	Yield stress Tension s_{ty}, psi	Yield stress Shear s_{sy}, psi	Flexural endurance limit s_{ef}, psi	Brinell hardness number BHN	Modulus of elasticity Tension and compression E, psi	Modulus of elasticity Shear G, psi	Elongation in 2 in., %	Remarks and suggested uses
Chromium-molybdenum: SAE 4140, oil-quenched...........	120,000 190,000	90,000 150,000	100,000 165,000	55,000 91,000	60,000 95,000	240 380	30,000,000	12,000,000	25 12	
Chromium-vanadium: SAE 6145, oil-quenched...........	105,000 230,000	75,000 180,000	98,000 210,000	54,000 115,000	94,000 105,000	220 425	30,000,000	12,000,000	25 12	
Silicon-manganese: SAE 9260, oil-quenched...........	158,000	120,000	100,000	60,000	62,000	240	30,000,000	12,000,000	16	Springs, gears
Stainless steel, 0.12 C, 18.0 Cr, 8.0 Ni..	90,000 200,000	60,000 150,000	35,000 175,000	20,000 100,000	40,000 90,000	135 380	30,000,000	12,000,000	60 5	Hot-rolled Cold-worked

The endurance limit for reversed shear may be taken as 55 per cent of the endurance limit in reversed bending.
Tabulated values are minimum values to be expected with round bars up to 1¼ in. diameter. Smaller sections and careful heat treatment will show higher values.
Upper figures: steel quenched and drawn to 1300°F.
Lower figures: steel quenched and drawn to 800°F.

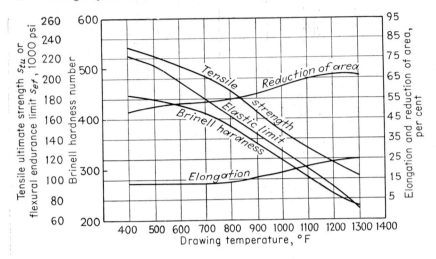

Fig. 2-12 Typical effects of heat treatment upon the physical properties of nickel-chromium 3330 oil-quenched steel.

Chemical composition, %		Approximate critical point		Suggested heat treatment
Carbon	0.25–0.35	AC_1	1345°F	1. Heat to 1425–1475°F and
Manganese	0.30–0.60	AC_2	1345°F	quench in water
Phosphorus	0.04 Max	AC_3	1355°F	2. Draw at a temperature to
Sulfur	0.04 Max			give the desired hardness
Nickel	3.25–3.75			
Chromium	1.25–1.75			

If bars larger than $1\frac{1}{4}$ in. diameter are being heat-treated, reduce the tensile strength and elastic limit approximately in accordance with the following table:

Size (diameter, in.)...	$2\frac{1}{2}$	3	$3\frac{1}{2}$	4	$4\frac{1}{2}$	5	$5\frac{1}{2}$	6
Reduction, %........	8.5	10.0	12.0	13.5	14.0	14.5	15.0	15.5

The effect of heat treatment upon the physical properties of steel is illustrated in Fig. 2-12. Typical relationship between carbon content of steels and maximum hardness is depicted in Fig. 2-13. The relationship between ultimate tensile strength and hardness of steels is tabulated in Table 2-7.

2-22 Copper. Commercial copper is a tough, ductile, and malleable metal containing less than 5 per cent of such impurities as tin, lead, nickel, bismuth, arsenic, and antimony. The strength is low compared to that of steel, and its properties depend largely on the mechanical treatment to which it has been subjected. Cold-working makes it stronger, and somewhat brittle. Pure copper is not extensively used in machine

Table 2-7 **Typical relation between tensile strength, Brinell hardness, and Rockwell hardness readings for steels***

Brinell hardness BHN	Rockwell C No. R_c	Ultimate tensile strength s_{tu}, psi	Brinell hardness BHN	Rockwell C No. R_c	Ultimate tensile strength s_{tu}, psi
97	. . . .	50,000	243	23.5	120,000
107	. . . .	55,000	266	27	130,000
116	. . . .	60,000	289	30.5	140,000
127	. . . .	65,000	311	33	150,000
138	. . . .	70,000	331	35	160,000
149	. . . .	75,000	352	37	170,000
159	2	80,000	371	39	180,000
170	6	85,000	408	43	200,000
181	8.5	90,000	444	46	220,000
192	12	95,000	482	49.5	240,000
203	15.5	100,000	522	52.5	260,000
213	17.5	105,000	563	55.5	280,000
223	20	110,000	605	58	300,000
233	21.5	115.000	635	61	320,000

* These values are approximately correct for carbon steels and low-carbon nickel, chrome-nickel, and nickel-chrome-molybdenum steels, such as SAE 2330, 4130, and 4340.

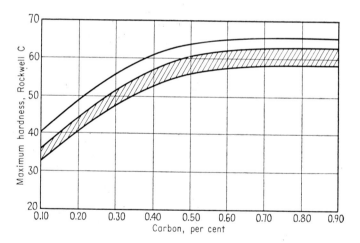

Fig. 2-13 Typical relation between carbon content and maximum martensitic hardness (upper curve), and maximum hardness usually attained (hatched band). (*Reprinted by permission from "Modern Metallurgy for Engineers" by Frank T. Sisco, 2d ed., 1948, Pitman Publishing Corporation, New York.*) The upper curve is based on data collected by J. L. Burns, T. L. Moore, and R. S. Archer, *Trans. Am. Soc. Metals*, vol. 26, 1938, pp. 1–36. The hatched band is based primarily on data given by A. L. Boegehold, *SAE Trans.*, vol. 52, pp. 472–485, 1944.

design except for castings, condenser tubes, water pipe, and sheet-metal parts where resistance to corrosion by weather and water is important, or in places where its high heat conductivity is an advantage.

The alloys of copper form an important group of materials with a wide variety of properties. Some have high strengths, some are excellent bearing materials, some retain their strength at high temperatures, and others are valuable for their corrosion resistance. Three groups of copper alloys used in machine design are the brasses, the bronzes, and Monel metal. Copper is also used as an alloy in steel to increase its resistance to corrosion.

2-23   Copper-Zinc Alloys. *Brass* is an alloy of copper and zinc, containing 45 to 90 per cent copper and small amounts of iron, lead, and tin as impurities. These alloys are highly resistant to corrosion; they machine easily and make good bearing materials.

Commercial bronze is a brass with 90 per cent copper, having excellent cold-working properties. It is readily forged, upset, drawn, or spun.

2-24   Copper-Tin Alloys. *Bronze* is an alloy of copper, tin, and a small amount of phosphorus. Bronzes with descriptive metallic names, such as aluminum bronze and manganese bronze, are not true bronzes and have no relation to the copper-tin alloys.

The bronzes are more costly than the brasses and hence are used only when the cheaper alloys do not prove to be satisfactory. They are used in the highest quality thermostatic bellows and other parts that require resistance to severe stretching together with good tensile and elastic strength.

2-25   Nickel. Wrought nickel as used in industry has mechanical properties similar to those of mild steel. It is one of the toughest of metallic materials; it retains its strength to a good degree at elevated temperatures and its ductility, toughness, and strength at subzero temperatures. It can be hot-forged and hot-rolled, cold-worked, cast, and welded. It has good corrosion resistance and is used as a lining for many types of containers.

Inconel contains 79.5 per cent nickel, 0.2 per cent copper, 13 per cent chromium, 6.5 per cent iron, and a small amount of silicon and manganese. It has the corrosion-resistant properties of Monel but has better resistance to sulfur at the higher temperatures. It retains its strength at extremely high temperature and can be used at 2100°F. Its creep properties are very good. It is nonmagnetic at all temperatures above −40°F.

Invar, an alloy iron with 30 per cent nickel, has a very low coefficient of heat expansion, making it useful for measuring instruments.

Table 2-8 Properties of typical copper-zinc alloys (brasses)

Material	Ultimate tensile strength s_u, psi	Tensile yield stress s_y, psi	Flexural endurance limit s_{ef}, psi	Brinell hardness BHN	Modulus of elasticity in tension and compression E, psi	Elongation in 2 in. %	Remarks and suggested uses
Commercial bronze.							90 Cu, 10 Zn
Rolled, hard.	65,000	63,000	18,000	107	15,000,000	18	Sheet rod, wire, tubes, hardware, screws, rivets
Rolled, soft.	35,000	11,000	12,000	52	15,000,000	56	
Forged, cold.	40,000-65,000	25,000-61,000	12,000-16,000	62-102	15,000,000	55-20	
Red brass.							85 Cu, 15 Zn
Rolled, hard.	75,000	72,000	20,000	126	15,000,000	18	Sheet, wire, shapes, tubes, radiators, plumbing
Rolled, soft.	37,000	14,000	14,000	54	15,000,000	55	
Forged, cold.	42,000-62,000	22,000-54,000	14,000-18,000	63-120	15,000,000	47-20	
Low brass.							80 Cu, 20 Zn
Rolled, hard.	75,000	59,000	22,000	130	15,000,000	18	Drawn, stamped, bellows, flexible hose
Rolled, soft.	44,000	12,000	15,000	56	15,000,000	65	
Forged, cold.	47,000-80,000	20,000-65,000		63-133	15,000,000	30-15	
Spring brass.							75 Cu, 25 Zn
Hard.	84,000	64,000	21,000	107*	14,000,000	5	
Soft.	45,000	17,000	17,000	57*	18,000,000	58	
Cartridge brass.							70 Cu, 30 Zn
Rolled, hard.	100,000	75,000	22,000	154	15,000,000	14	
Rolled, soft.	48,000	30,000	17,000	70	15,000,000	55	
Deep-drawing brass.							68 Cu, 32 Zn
Strip, hard.	85,000	79,000	21,000	106*	15,000,000	3	
Strip, soft.	45,000	11,000	17,000	13*	15,000,000	55	
Muntz metal.							60 Cu, 40 Zn
Rolled, hard.	80,000	66,000	25,000	151	15,000,000	20	Heat-exchanger tubes
Rolled, soft.	52,000	22,000	21,000	82	15,000,000	48	
Tobin bronze.							60 Cu, 39.25 Zn, 0.75 Sn
Hard.	63,000	35,000	21,000	165	15,000,000	35	Marine parts and trim
Soft.	56,000	22,000		90	15,000,000	45	
Manganese bronze.							58 Cu, 40 Zn
Hard.	75,000	45,000	20,000	110	15,000,000	20	Cast, rolled, drawn, propellers, bearings
Soft.	60,000	30,000	16,000	90	15,000,000	30	

* Rockwell hardness, F scale.

Table 2-9 Properties of typical copper-tin alloys (bronzes)

Material	Ultimate strength in tension s_{tu}, psi	Yield stress in tension s_{ty}, psi	Flexural endurance limit s_{ef}, psi	Hardness Rockwell B	Modulus of elasticity		Elongation in 2 in., %	Remarks
					Tension and compression E, psi	Shear G, psi		
Wrought bronzes:								
Signal bronze..........	......	......	...	...	....	...	...	98.2 Cu, 1.8 Sn, .05 P
Hard..........	105,000	......	...	...	....	...	...	Wire
Soft..........	45,000	......	...	...	....	...	...	Wire
Phosphor bronze:								
Grade A..........	......	......	...	...	....	...	...	94.9 Cu, 5 Sn, 0.10 P
Hard..........	80,000	65,000	...	86	....	...	8	
Soft..........	50,000	20,000	...	28	....	...	50	
Grade B..........	......	......	...	...	....	...	...	92.9 Cu, 4 Sn, 0.10 P, 3 Pb
Hard..........	62,000	52,000	...	70	....	...	25	Leaded, free cutting
Soft..........	25,000	......	...	...	....	...	...	
Grade C..........	......	......	...	...	....	...	...	91.9 Cu, 8 Sn, 0.10 P
Hard..........	93,000	68,000	...	95	....	...	10	
Soft..........	60,000	25,000	...	50	....	...	60	
Grade D..........	......	......	...	...	....	...	...	89.9 Cu, 10 Sn, 0.10 P
Hard..........	100,000	73,000	...	98	....	...	12	
Soft..........	65,000	24,000	...	55	....	...	65	
Casting bronze:								
Tin bronze..........	40,000	18,000	...	...	10,000,000	...	20	88 Cu, 10 Sn, 2 Zn
Navy bronze..........	34,000	16,000	...	...	10,000,000	...	20	88 Cu, 6 Sn, 4 Zn, 1.5 Pb, 0.5 Ni
Leaded tin bronze.......	25,000	12,000	...	...	8,500,000	...	8	80 Cu, 10 Sn, 10 Pb
Leaded red brass........	30,000	14,000	...	...	....	...	20	85 Cu, 5 Sn, 5 Zn, 5 Pb
High yellow brass........	35,000	12,000	...	...	....	...	25	71 Cu, 1 Sn, 25 Zn, 3 Pb

Table 2-10 **Properties of typical copper alloys**

Material	Ultimate strength			Yield stress		Flexural endurance limit s_{ef}, psi	Rockwell hardness	Modulus of elasticity		Elongation in 2 in., %	Remarks and suggested uses
	Tension s_u, psi	Compression s_{uc}, psi	Shear s_{su}, psi	Tension s_{ty}, psi	Compression s_{cy}, psi			Tension and compression E, psi	Shear G, psi		
Aluminum bronze	55,000 / 92,000			22,000 / 65,000		10,000 / 20,000	35 B / 92 B	17,000,000	6,500,000	65 / 7	95 Cu, 5 Al, sheet
Aluminum bronze	65,000 / 105,000			25,000 / 65,000		14,000 / 20,000	50 B / 96 B	17,000,000	6,500,000	60 / 7	92 Cu, 8 Al, sheet
Aluminum bronze	65,000 / 80,000			25,000 / 50,000		14,000 / 18,000		17,000,000	6,500,000	65 / 30	92 Cu, 8 Al, rod
Aluminum bronze: Sand cast	70,000	124,000		32,000	22,000		131*	15,000,000		18	86.5 Cu, 10.2 Al, 3.3 Fe Bushings, gears, landing gear, valve guides, pump bodies
Centrifugal cast	77,000			35,000			137*	15,000,000		15	86.5 Cu, 10.2 Al 3.3 Fe Bushings, gears, landing gear, valve guides, pump bodies
Forged	87,000			35,000			150*	15,500,000		25	Shock absorber pistons, bearings, gears
Everdure: Tubes	40,000 / 65,000			15,000 / 50,000			55 B / 75*			50 / 8	96 Cu, 1.5 Si, 0.3 Mn
Sheet	40,000 / 70,000						55 F / 80 B			50 / 15	
Rod				15,000 / 55,000							
Duronze III: Annealed	85,000										91 Cu, 7 Al, 2 Si
Drawn	95,000										
Beryllium copper:											2–2.25 Be, 0.25–0.50 Ni, + Cu
Rod, annealed	70,000							18,000,000		45	
Rod, hard-drawn	93,000	128,500			84,000		194*	17,000,000	6,900,000	4.3	
Rod, drawn and heat-treated	173,000	206,000			155,000		335*	18,400,000	7,100,000	4	
Sheet, annealed	70,000		55,000	31,000			65 B	18,000,000		2	
Sheet, No. 4 hard	118,000		59,000	105,000		35,000	102 B	17,000,000			
Sheet, hard and heat-treated	193,000		100,000	138,000			114 B	18,400,000			
Cupronickel, sheet	48,000 / 74,000			20,000 / 65,000			79*			40 / 5	80 Cu, 20 Ni

Upper figures represent annealed condition; lower figures, hard-drawn condition.
* Brinell hardness.

37

Table 2-11   **Properties of typical nickel alloys**

Material	Ultimate strength in tension s_{tu}, psi	Yield stress in tension s_{ty}, psi	Flexural endurance limit s_{ef}, psi	Brinell hardness BHN	Modulus of elasticity		Elongation in 2 in., %
					Tension and compression E, psi	Shear G, psi	
Nickel:							
Annealed....................	60,000– 80,000	15,000– 30,000	......	90–120	30,000,000	11,000,000	50–35
As drawn....................	65,000–115,000	40,000– 90,000	......	125–230	30,000,000	11,000,000	35–15
Hot-rolled..................	65,000– 80,000	15,000– 30,000	......	90–120	30,000,000	11,000,000	50–35
Forged......................	65,000–105,000	20,000– 80,000	......	90–200	30,000,000	11,000,000	40–25
Monel:							
Annealed....................	70,000– 85,000	25,000– 40,000	30,000	110–140	26,000,000	9,500,000	50–35
As drawn....................	85,000–125,000	60,000–120,000	......	160–250	26,000,000	9,500,000	35–15
Hot-rolled..................	80,000– 95,000	40,000– 65,000	......	140–185	26,000,000	9,500,000	45–30
Forged......................	80,000–110,000	40,000– 85,000	......	140–225	26,000,000	9,500,000	40–20
K Monel:							
Annealed....................	90,000–110,000	40,000– 60,000	37,500	140–180	26,000,000	9,500,000	45–35
Annealed, age-hardened........	130,000–150,000	90,000–110,000	......	240–260	26,000,000	9,500,000	30–20
As drawn....................	100,000–125,000	70,000–100,000	42,000	175–250	26,000,000	9,500,000	35–15
Drawn, age-hardened..........	140,000–170,000	100,000–130,000	41,000	260–320	26,000,000	9,500,000	30–15
Hot-rolled..................	90,000–120,000	40,000– 85,000	41,000	140–225	26,000,000	9,500,000	45–30
Rolled, age-hardened...........	135,000–160,000	100,000–120,000	44,500	260–300	26,000,000	9,500,000	30–20
Forged......................	90,000–120,000	40,000– 90,000	......	140–240	26,000,000	9,500,000	40–25
Age-hardened................	135,000–165,000	100,000–125,000	......	260–310	26,000,000	9,500,000	30–20
Inconel:							
Annealed....................	80,000–100,000	25,000– 50,000	30,000	120–170	31,000,000	11,000,000	50–35
As drawn....................	95,000–150,000	75,000–125,000	41,000	180–290	31,000,000	11,000,000	30–15
Hot-rolled..................	85,000–120,000	35,000– 90,000	38,500	140–210	31,000,000	11,000,000	45–30
Forged......................	85,000–120,000	35,000– 90,000	......	140–210	31,000,000	11,000,000	45–20

Yield stress is taken at 0.2 per cent offset.
Endurance limit at 1,000,000,000 cycles.

2-26   Aluminum Alloys.   Aluminum is one of the lightest metals used in machine construction. Commercial aluminum is soft and ductile, about one-third as heavy as steel, and approximately one-fourth as strong as structural steel. It can be strengthened by cold-working. At room temperatures, it has good resistance to the corrosive action of many chemicals.

Aluminum as commonly used is alloyed with copper, silicon, manganese, magnesium, iron, zinc, and nickel. These alloys retain the lightness of aluminum, machine better, are harder, and in the annealed or cast state have strengths up to twice that of commercial aluminum; certain of the alloys have strengths comparable to the structural steels. There are two general classes of alloys, those which can be hardened and strengthened only by cold-working and those which can be heat-treated to obtain the optimum mechanical properties.

The corrosion-resistant properties of aluminum are chiefly due to the formation of a tough adherent oxide on the surface; this film resists

Table 2-12 Typical properties of aluminum alloys

Material, Aluminum Association Code No. (AA No.)	Material, Aluminum Company of America Code No. (ALCOA No.)	Ultimate strength		Yield strength		Flexural endurance limit s_{ef}, psi	Brinell hardness, 500 kg on 10-mm ball	Elongation in 2 in, %	Remarks and suggested uses
		Tension s_{tu}, psi	Shear s_{su}, psi	Tension s_{ty}, psi	Compression s_{cy}, psi				
Sand-casting alloys:									
	43	19,000	14,000	9,000	9,000	6,500	40	6	High ductility and shock-resistance
	112	23,000	20,000	14,000	17,000	9,000	70	1.5	Easy machining, general-purpose castings
	122-T2	25,000	21,000	20,000	20,000	9,500	75	1	Pump cylinders and parts requiring good wear resistance
	122-T61	36,000	29,000	30,000	43,000	.....	100	1	
	142-T2	27,000	21,000	18,000	18,000	6,500	75	1	Air-cooled engine heads and parts requiring good mechanical properties up to 600°F
	142-T61	37,000	32,000	32,000	47,000	8,000	100	0.5	
	195-T4	31,000	24,000	16,000	16,000	6,000	65	8.5	Heat-treated castings
	212	22,000	20,000	14,000	14,000	8,000	65	2	Good casting qualities, general-purpose castings
Permanent mold castings:									
	B-113	29,000	23,000	19,000	19,000	.....	70	1.5	General-purpose castings
	A-132-T4	38,000	29,000	30,000	30,000	.....	100	1.5	Gas-engine pistons; has lowest coefficient of expansion of the aluminum alloys
Wrought alloys:									
1100	2 SO	13,000	9,500	5,000	.....	5,000	23	45	Low-cost, easily fabricated alloys: most common nonstructural alloy; used for cooking utensils and chemical equipment
	2 SH	24,000	13,000	21,000	.....	8,500	44	15	
3003	3 SO	16,000	11,000	6,000	.....	7,000	28	40	
	3 SH	29,000	16,000	25,000	.....	10,000	55	10	
2014	14 ST	65,000	45,000	59,000	.....	16,000	130	10	Best mechanical properties of the forging alloys
2017	17 SO	26,000	18,000	10,000	.....	16,000	45	22	Corrosion-resisting forgings: most popular heat-treatable alloy used for structural shapes
	17 ST	62,000	36,000	40,000	.....	15,000	100	22	
2024	24 SO	26,000	18,000	10,000	.....	15,000	42	22	
	24 SRT	70,000	42,000	55,000	.....	12,000	116	22	Aeronautic structural shapes
2025	25 ST	55,000	35,000	30,000	.....	15,000	100	16	Easy forging alloy
3032	32 ST	52,000	38,000	40,000	.....	14,000	115	5	High strength at elevated temperatures
5052	52 SO	29,000	18,000	14,000	.....	17,000	45	30	Higher-strength, general-purpose alloy
	52 SH	41,000	24,000	36,000	.....	20,500	85	8	
6053	53 ST	36,000	24,000	30,000	.....	11,000	85	16	Forgings for severe corrosion resistance, as in dairies and breweries

Modulus of elasticity in tension, 10,300,000 psi, in shear 3,850,000 psi.
Yield stress taken at 0.2 per cent permanent deformation.
Endurance limit is based on 500,000,000 cycles of reversed stress.
Bearing strength is 1.8 times the tensile strength.

further chemical action and will replace itself if removed. The alloying elements added to increase the strength lower the corrosion resistance. Foods and beverages may be handled in aluminum containers. Sulfur and concentrated nitric and acetic acids do not attack these alloys readily. Hydrochloric acid and many alkalies dissolve the oxide coating, and corrosion becomes rapid. The resistance to corrosion may be increased by anodizing in a sulfuric or oxalic electrolytic bath.

Aluminum-alloy castings made in sand molds, permanent metal molds, and die-casting machines are generally stronger than the poorer grades of cast iron. Better surface finish, closer dimensional tolerance, and better mechanical properties, together with the savings in machining and finishing costs, make the use of permanent molds and die castings desirable when the quantity justifies the extra cost of equipment.

2-27 Heat Treatment of Aluminum Alloys. Aluminum alloys which have been hardened by cold-working may be softened by annealing. Heat-treatable alloys have their mechanical properties improved by the solution and precipitation processes. The solution process consists of holding the part at temperatures and time intervals determined by the alloy analysis and following with a rapid quench. The precipitation or aging process consists of holding the part at a comparatively low temperature for a period of time determined by the alloy and temperature used. At room temperatures, aging time is about 4 days; at 315 to 325°F, about 18 hr; and at 345 to 355°F, about 8 hr.

2-28 Magnesium Alloys. Magnesium alloys are approximately two-thirds as heavy as aluminum alloys. The common alloys contain from 4 to 12 per cent of aluminum and from 0.1 to 0.3 per cent of manganese; those with more than 6 per cent of aluminum can be heat-treated and aged to increase the yield strength. The alloys are resistant to atmospheric corrosion if kept dry, but when humidity is high, corrosion proceeds slowly with a powder forming on the surface. In very moist or salty atmosphere, or where rain is trapped on the part, surface roughening is pronounced. In general, all magnesium-alloy parts should be given a protective coating.

The magnesium alloys are resistant to attack by most alkalies and by some inorganic substances but are attacked by most acids. Most aqueous salt solutions corrode these alloys rapidly, brines and chloride solutions being extremely corrosive. These alloys are resistant to pure chromic acid, pure concentrated hydrofluoric acid, alkali metal fluorides, chromates, and bichromates.

Alloy F* is used for forgings, sheet, plate, bars, extruded shapes, and similar items. Alloys E and A may be specified when maximum strength

* Symbols used here are the designations of the Dow Chemical Company, manufacturers of Dowmetal.

Table 2-13 Properties of typical magnesium alloys (Mg, Al, Mn, Zn)

Material	Ultimate strength			Tensile yield stress s_{ty}, psi	Flexural endurance limit s_{ef}, psi	Brinell hardness BHN	Modulus of elasticity in tension and compression E, psi	Elongation in 2 in, %	Remarks and suggested uses
	Tension s_{tu}, psi	Compression s_{cu}, psi	Shear s_{su}, psi						
Sand cast:									
A, As cast..........	24,000	44,000	16,000	10,000	6,000	47	6,500,000	3	
Ht-Tr 1..........	32,000	44,000	16,000	10,000	6,000	47	6,500,000	8	
Ht-Tr 2..........	31,000	46,000	17,000	12,000	6,000	52	6,500,000	4	Hard castings, pistons
B, As cast..........	19,000	45,000	15,000	15,000	......	69	6,500,000	0	
Ht-Tr 1..........	20,000	46,000	16,000	12,000	......	66	6,500,000	0.5	
Ht-Tr 3..........	27,000	52,000	17,000	20,000	......	82	6,500,000	0	Sand and permanent-mold castings
G, As cast..........	21,000	47,000	16,000	12,000	......	52	6,500,000	1	
Ht-Tr 1..........	30,000	48,000	18,000	11,000	8,000	49	6,500,000	6	
Ht-Tr 3..........	31,000	52,000	20,000	17,000	9,000	65	6,500,000	1	Pressure-tight castings
C, As cast..........	23,000	......	18,000	14,000	10,000	62	6,500,000	1	
Ht-Tr 1..........	39,000	......	20,000	14,000	10,000	61	6,500,000	10	
Ht-Tr 2..........	38,000	......	22,000	20,000	10,000	67	6,500,000	3	Sand and permanent-mold castings
Die cast:									
K, as cast..........	30,000	......	......	22,000	......	68	6,500,000	1	Thin-section die castings
R, as cast..........	33,000	......	......	20,000	......	66	6,500,000	3	General die castings
Extruded shapes:									
J as extruded......	42,000	......	20,000	27,000	16,000	55	6,500,000	16	Good strength and weldability screw-machine stock
M as extruded......	40,000	......	......	31,000	10,000	42	6,500,000	6	Best salt-water resistance
G as extruded......	50,000	......	22,000	38,000	16,000	70	6,500,000	9	Highest strength and hardness
Forgings:									
J, pressed..........	41,000	......	21,000	25,000	17,000	56	....	9	General forgings, good ductility
M, pressed..........	33,000	......	......	19,000	10,000	43	....	6	Weldable forgings
L, hammered........	37,000	......	16,000	22,000	10,500	51	....	6	

Stress values are average. Minimum values are 2,000 to 4,000 psi lower.

Yield stress taken at 0.2 per cent permanent set.

Ht-Tr 1 indicates the solution heat treatment.

Ht-Tr 2 and Ht-Tr 3 indicate the solution treatment followed by the full aging treatment.

Endurance limit is for 500,000,000 reversals.

41

is required and when the forging is not intricate. Alloy M is used for moderately stressed parts where salt-water corrosion resistance is required.

Structural shapes such as channels, beams, angles, bars, rods, and tubes are usually made from alloy F, which has good extrusion properties. Alloys E and A are more difficult to extrude but are sometimes used. Alloy F is also used for sheet and plate in the hard-rolled and annealed condition.

Castings made of alloy A, without heat treatment, are used for moderate stresses and when cost is a deciding factor. Where higher strength, toughness, and shock resistance are required, sand castings of alloy A are used in the heat-treated condition. Alloy G, cast and heat-treated, has higher yield strength and hardness but lower toughness than alloy A. Its properties can be varied by using different heat treatments. Alloy B has low elongation and toughness but, in the heat-treated condition, has the highest yield strength and hardness and is used where maximum hardness is required. Alloy T should be used only where thermal conductivity is most important. Alloy G is the best die-casting alloy, although alloy M is used when corrosion resistance to salt water and salt atmospheres is more important than mechanical properties.

Magnesium alloys can be welded by experienced operators using the oxyacetylene or oxyhydrogen processes. Electric-resistance welding and spot welding are also possible. The other welding processes are not commercially practical. The alloys are readily machined, but since powdered magnesium is highly inflammable, dust must be removed rapidly and dull tools which might overheat the chips must be avoided. The outstanding characteristic of the magnesium alloys is their light weight, and they are therefore used where weight reduction is a major advantage.

2-29   Heat Treatment of Magnesium Alloys.   Two types of heat treatment are used for castings, the solution treatment and the aging or precipitation treatment. Solution treatment is the holding of the part for 16 to 18 hr at a temperature of about 770°F. This raises the tensile strength, ductility, and toughness. It does not alter the yield strength or hardness. Aging, or precipitation, consists of following the solution treatment by either 4 to 5 hr or 16 to 18 hr at about 35°F, which increases the yield strength and hardness but lowers the ductility and toughness.

2-30   Beryllium, Lead, Tin, Titanium, Zinc, and Other Miscellaneous Alloys.   In addition to the various metals previously mentioned, there are numerous other metals and alloys which may be used to a definite advantage for many applications. *Babbitt* is a tin or lead base material suitable for bearings and for making die castings. The *berylliums* are light materials with Young's modulii 50 per cent greater than those for steel; they have high specific heats and relative high corrosion resist-

Table 2-14 Selected physical properties of beryllium, lead, tin, titanium, zinc, and other special alloys

Material	Tensile ultimate strength s_{tu}, psi	Tensile yield strength s_{ty}, psi	Flexural endurance limit s_{ef}, psi	Brinell hardness BHN	Young's modulus E, psi	Modulus of rigidity G, psi	Elongation of tensile ultimate, %	Remarks
Babbitt, lead base*	10,500	......	4,000	22	4,200,000	....	4	9.3–10.7 % tin
Beryllium†	42,000	32,100	......	120–140	44,000,000	21,500,000	2.3	Vacuum hot-pressed from powder
Beryllium†	83,000	39,500	......	120–140	41,400,000	21,000,000	16	Hot-extruded from hot-pressed powder
Copper beryllium 25 annealed and heat-treated‡	175,000	......	......	352	19,000,000	7,300,000	6	1.9 % Beryllium, 0.25 % Cobalt
Copper beryllium 165 annealed and heat-treated	70,000	......	......	95	19,000,000	7,300,000	45	1.7 % Beryllium, 0.25 % Cobalt
Nickel beryllium Cr-1 cast‡	125,000	70,000	......	285	27,000,000	....	2	2.7 % Beryllium, 0.7 % Chromium, 0.1 % Carbon
Hastelloy C*	125,000	60,000	......	195–225	28,500,000	....	45	Annealed plate
Nimonic 80*	147,000	90,000	......	......	....	....	45	75 % Nickel, 19 % Chromium, 2.3 % Titanium
Soft solder 50-50‡	6,800	......	......	14	....	....	.....	50 % Lead, 50 % Tin
Easy-Flo silver solder†	67,000–89,000	......	......	120–160	....	....	11–29	50 % Silver, 18 % Cadmium, 16.5 % Zinc, 15.5 % Copper
Haynes Stellite, Alloy No. 25‡	140,000	......	......	244	34,200,000	....	60	Wrought, mill-annealed, sheet
Titanium-chromium iron-molybdenum alloy (Ti-140A)‡	155,000	145,000	......	342	16,500,000	6,600,000	10	Heat-treated
Titanium-aluminum-iron-chromium-molybdenum alloy (Ti-155A)‡	160,000	150,000	......	362	16,500,000	6,600,000	10	Heat-treated
Zamak 3*	41,000	31,000	6,875	82	....	....	10	Die-casting alloy

* Samuel Hoyt (ed.), "ASME Handbook—Metals Properties," McGraw-Hill Book Company, New York, 1954.
† *Prod. Eng.—Design Digest Issue*, vol. 31, no. 38, 1960.
‡ "Properties of Some Metals and Alloys," International Nickel Co., Inc., 1959.

ance. The *Hastelloys* are molybdenum-nickel-iron alloys which exhibit very high corrosion resistance and are suitable for use in precision castings. The *Nimonics* are nickel-chromium-titanium-manganese-silicon wrought alloys exhibiting high strength at elevated temperatures. *Solders* are low melting temperature lead-tin alloys in common use for bonding together, wiping, and coating metallic parts. The *stellites* are cobalt-chromium-nickel-manganese-silicon cast alloys having high strength at elevated temperatures; they are difficult to machine and are not easily welded. They are excellent materials for the cutting tools of many machining operations. The *titaniums* are lightweight high-strength materials with good corrosion resistance; they are finding particular use in aircraft and missiles and similar applications where high strength is desired at elevated temperatures. The *Zamaks* and other zinc alloys find particular use in die and sand castings, especially where intricate shapes exist, such as in automotive carburetors and toys. Some of the more important physical properties of these miscellaneous alloys are tabulated in Table 2-14. *Tungsten carbide* is an extremely hard but brittle material. In the granular form it is applied by welding to form an abrasion-resistant surface. In this form it is widely used on the cutting edges of oil-well drag bits, roller-type rock bits, scrapers, excavating shovels, ditch diggers, etc. When ground to a fine powder mixed with powdered cobalt, pressed into form under high pressures, and sintered to form a solid mass, it is used for the tips of high-speed cutting tools. For some types of cutting tools it is combined with tantalum carbide or titanium carbide.

2-31 Molybdenum. Molybdenum has engineering properties of great interest. At 2000°F it has a Young's modulus comparable to that of steel at room temperature. It melts approximately 2000°F higher than iron does. In addition, it has a comparatively low thermal expansion coefficient and a high thermal conductivity. The ultimate tensile strength of stress-relieved unalloyed molybdenum decreases from 100,000 psi at 70°F to 13,000 psi at 2000°F. The ultimate tensile strength of the highest-strength commercially available molybdenum-base alloy decreases from 185,000 psi at 70°F to 53,000 psi at 2000°F.

2-32 Sintered Materials. Certain materials that cannot be alloyed by melting can be formed into useful products by mixing in powdered form, compressing under high pressure, and bonding by *sintering*. After the powders are obtained, they are intimately mixed and compressed in hard steel dies under pressures up to 100 tons/in.^{2}, depending on the materials. The compressed mass usually is weak mechanically and has a density of about 0.8 of that of the solid material. To increase the strength, the compressed powder is sintered by heating in a neutral or reducing atmosphere to a temperature where the grains will grow together to form a strong solid mass. In some cases a bonding powder is used in the mixture.

By varying the treatment, the porosity of the product can be varied from 3 to 30 per cent by volume. By including powders of materials that may be volatilized by heat, a material full of interconnecting pores is obtained that can absorb and retain oil or other liquid and serve as self-lubricating bearings. Graphite may be incorporated in the powder mixture to act as a lubricating material. Bearings, small gears, cams, electric-motor brushes, sintered cutting tool tips, and many other intricate parts are formed by sintering. Quite often items produced by sintering require no machining processes to obtain dimensions meeting specified limits, hence sintering frequently is employed to produce a part which has very little or no machining operations performed on it.

2-33 Metals for High-temperature Service. Metal parts for nuclear reactors, internal-combustion engines, valves, superheated steam equipment, oil stills, chemical and petroleum processes, and similar service are stressed at temperatures ranging from 200 to 1800°F. Ceramics usually are used for higher operating temperatures. The metals used for such parts must be specially selected from those materials which retain a large percentage of their strength at high temperatures and which do not creep excessively.

High-temperature or heat-resisting alloys are made in three principal groups. Alloys of the first group contain over 50 per cent iron with additions of 10 to 30 per cent chromium and varying amounts of copper, nickel, tungsten, and silicon. Alloys of the second group contain up to 25 per cent iron, 50 to 60 per cent nickel and chromium, and small additions of manganese, tungsten, silicon, and molybdenum. Alloys of the third group contain from 0.5 to 6 per cent iron as an impurity, over 80 per cent nickel, considerable chromium, and small amounts of manganese and silicon. In this group may be placed alloys containing 25 to 30 per cent chromium and 70 to 75 per cent cobalt, which show yield stresses of over 40,000 psi at temperatures from 1200 to 1400°F. The alloys of the third group also have excellent resistance to corrosion at temperatures up to 1800°F.

2-34 Metallic Refractory Materials. *Refractory materials* are materials with melting points above approximately 2550°F. The advent of gas turbines, rockets, glassmaking, electronics, and high-temperature industrial processes has necessitated the development of suitable refractory materials. These materials include metals, oxides, carbides, silicides, nitrides, sulfides, borides, intermetallic compounds, and graphite in addition to ceramics. The three most important physical properties usually desired in these materials are high strength at elevated temperatures, resistance to oxidation, and resistance to thermal shock. Molybdenum and its alloys have good refractory properties, but they oxidize readily. Diffusion and cladding in single and multiple coatings of aluminum, chromium, nickel, and silicon improve the high-temperature performance of molybdenum alloys, but perfect coatings are almost impossible to achieve. Metallic oxides retain their strength at higher temperatures than refractory metals, but their resistance to thermal shock is lower. Ceramics bonded with refractory metal provide both high-temperature strength and rather good resistance to thermal shock. In general, the borides, carbides, nitrides, silicides, and sulfides can withstand thermal shock better than metal oxides; metal-bonded titanium carbide appears to be one of the more promising materials for this. From cermet research have come metals reinforced with 1 to 2 per cent oxide, thereby avoiding the poor thermal-shock resistance of the high-ceramic cermets and also raising the load-bearing temperature of the binding metal. For use in rockets, electrodes, furnace linings, and similar applications, graphite gives good performance at high temperatures. A sulfide coating on the graphite almost completely protects the graphite against oxidation. Typical strengths of various metallic refractory materials are listed in Table 2-15.

Table 2-15 **High-temperature tensile ultimate strengths of various metallic refractory materials***

Material†	Approximate tensile ultimate strength s_{tu}, psi, at various temperatures								
	70°F	500°F	1000°F	1500°F	2000°F	2500°F	3000°F	3500°F	4000°F
Al₂O₃	37,000	35,000	33,000	32,500	2,000				
BeO	13,000	12,000	8,500	4,000					
Graphite	3,000	3,500	4,000	4,200	4,350	4,500	5,500	7,000	4,500
Mo	......	......	......	42,000	13,000				
MoSi₂	23,000	31,500	38,000	42,000	42,000				
Mo–0.5 Ti–0.07 Zr	......	......	......	......	32,000				
Re	......	......	......	......	43,000	22,000	11,500		
Ta–10 W	......	......	......	......	19,500	9,000	4,500	2,300	1,000
W	......	......	......	......	30,000	14,500	12,000	3,500	
W–30 Re	......	......	......	......	......	20,000			
W–1 ThO₂	......	......	......	......	47,000	32,500	17,500	8,000	
ZrC	15,000	14,800	14,400	14,000					
ZrO₂	22,500	19,500	17,200	9,000	3,500				

* "Refractory Materials Today," *Prod. Eng.—Design Digest Issue*, vol. 31, no. 38, pp. 106–107, 1960.

† Al, aluminum; Be, beryllium; C, carbon; Mo, molybdenum; O, oxygen; Re, rhenium; Si, silicon; Ta, tantalum; Ti, titanium; Th, thorium; W, tungsten; Zr, zirconium.

2-35 Ceramics. Materials having a high percentage of alumina (Al_2O_3) or steatite ($MgO\cdot SiO_2$) are known as *ceramics*, particularly if they have been fired. They have high thermal and electric resistivity, good chemical resistance, and relatively high hardness and strength, and can be used at temperatures much higher than the average metals. They find applications in electrical insulators, bearings, valve seats, pistons, chemical containers, and internal-combustion turbines.

2-36 Natural and Synthetic Rubber. *Raw,* or *crude, rubber* is obtained by the coagulation of the latex or milky fluid obtained by tapping certain tropical trees and shrubs. Raw rubber is used in the pure state for rubber cements, surgical tape, and similar purposes. Most rubber is vulcanized by adding sulfur and heating. Hard rubber is used for electric insulation, switch handles, bearings, etc. Semihard rubbers are used for rubber belting, flexible tubing, hose, springs, vibration controls, and many other devices.

Synthetic Rubber. *Neoprene* and the *buna* rubbers are the synthetics most like natural rubber. The cost of the synthetics has limited their use, except when compounded for particular properties such as resistance to oils, heat, cold, and oxidation. The properties of a few of the hundreds of compounded synthetic rubbers are given in Table 2-16. Some synthetics, like Thiokol, are thermoplastic, whereas others, like Neoprene and the buna rubbers, are thermosetting. Both types can be compounded for use in molded products.

Table 2-16 **Physical properties of rubber and rubberlike materials**

Material	Form	Specific gravity	Compressive ultimate strength s_{cu}, psi	Tensile ultimate strength s_{tu}, psi	Transverse strength s_b, psi	Hardness, Shore durometer	Maximum temperature for use, °F	Effect of heat	Coefficient linear expansion $\times 10^{-6}$ 32–140 F
Duprene...	......	1.27–3.00	......	200–4,000	....	15–95	300	Stiffens slightly	
Koroseal...	Hard	1.3–1.4	......	2,000–9,000	....	80–100	212	Softens	
Koroseal...	Soft	1.2–1.3	......	500–2,500	....	30–80	190	Softens	
Plioform...	Plastic	1.06	8,500–11,000	4,000–5,000	7,000–9,000	......	160–248	Softens	
Rubber....	Hard	1.12–2	2,000–15,000	1,000–10,000	9,000–15,000	50*–80*	130–160	Softens	35
Rubber....	Soft	0.97–1.25	......	525–600	....	......	150–200	Softens	36
Rubber....	Linings	0.98–1.35	......	....	....	......	190	Softens	

* Scleroscope.

Table 2-17 Physical properties of common plastics

Material	Specific gravity	Tensile ultimate strength s_u, psi	Compressive ultimate strength s_{cu}, psi	Shearing ultimate strength s_{su}, psi	Flexural endurance limit s_{ef}, psi	Elongation in 2 in., %	Impact value, ft-lb Izod	Brinell hardness, 500 kg on 10-mm ball	Young's modulus E, psi	Thermal coefficient of linear expansion in./(in.)(°F)	Maximum operating temperature, °F
Phenolics:											
General purpose	1.40	6,500–8,500	. . . . . .	. . . . . .	8,800–13,000	. . . .	0.13–0.20	20–30	880,000–980,000	. . . .	300
Shock-resistant	1.37	6,300–7,100	. . . . . .	. . . . . .	8,000–11,000	. . . .	1.90–2.70	. . . . .	1,000,000	. . . .	240
Heat-resistant	1.65	4,600–5,500	. . . . . .	. . . . . .	8,000	. . . .	0.13–0.16	. . . . .	. . . .	. . . .	400
Transparent	1.27	8,000	30,000	. . . . . .	16,000	. . . .	0.20	35–40	. . . .	0.00001	225
Laminated, paper base	1.36	12,000	30,000	10,000	21,000	. . . .	0.6–7.6	40	1,500,000	0.00001	255
fabric base	1.38	10,000	40,000	9,000	20,000	. . . .	1.4–15	38	1,000,000	0.00001–	230
Urea-formaldehyde	1.50	5,500–13,000	20,000–30,000	. . . . . .	10,000–15,000	. . . .	0.28–0.32	48–50	1,550,000–1,650,000	0.00002	175
Cellulose nitrate	1.50	6,000–9,000	7,500–30,000	. . . . . .	7,500–	4–20	. . . . . .	10–20	200,000–400,000	0.00065–0.000090	140
Cellulose acetate	1.30	2,800–9,900	7,500–30,000	. . . . . .	3,700–18,000	5–40	0.7–4.2	28–105	200,000–350,000	0.000061–0.000091	140–250
Vinyl acetate	1.35	8,000–10,000	. . . . . .	. . . . . .	10,000–13,000	1–5	0.30–0.60	12–15	350,000–400,000	0.00004	125
Polystyrene	1.07	5,500–6,500	11,500–13,500	. . . . . .	6,500–7,500	. . . .	0.20–0.35	. . . . .	400,000–600,000	0.000035–0.000045	150
Vulcanized fiber	1.30	5,000	20,000	4,500	11,000	. . . .	. . . . . .	8	980,000	0.0000045	110
Borosilicate glass	2.37	9,800	17,900	. . . . . .	. . . . . .	. . . .	. . . . . .	. . . . .	10,400,000	. . . .	

ASTM has established a system of employing letters to designate the more commonly used rubbers. SBR is a styrene-butadiene rubber, formerly known as GR-S or buna S. NBR refers to a nitrile-butadiene rubber often called GR-N or buna N. Chlorophene rubber, commonly called neoprene or GR-A, is designated by CR. IIR implies isobutylene–isoprene rubber called butyl or GR-1. Natural rubber is represented by NR, and synthetic "natural" rubber by IR.

2-37 Plastics. Synthetic organic materials processed by pressure usually accompanied by heat are called *plastics*. Plastic materials usually are divided into *thermoplastic plastics*, which soften under heat and reharden when the heat is removed, and *thermosetting plastics*, which harden or cure under heat, after which more heat has no appreciable softening effect. Often plastic parts are formed by subjecting the material to both heat and pressure. However, pressure is not always employed. Some of the more commonly used plastics and their properties are given in Table 2-17.

3

Stresses and Strains

3-1 Stresses and Strains. When any solid body is subjected to external forces, resisting forces are set up within the body. These internal forces per unit area are called *unit stresses*. The stress is *tensile* where the force tends to elongate fibers in the member collinear with the stress, *compressive* where the force tends to shorten fibers in the member collinear with the stress, and *shear* where the forces in the member tend to make adjacent planes in the member slide relative to each other. It is evident that elements of area in the member may be subjected to a force which has components both parallel and perpendicular to the area such that the stresses at the area may be both shear and tension or compression. In this text, the types of stress are indicated by a subscript, thus, s_c is compressive stress, s_s is shear stress, and s_t is tensile stress, all in pounds per square inch. Additional subscripts will be used at times to further differentiate between stresses.

The forces acting on a body cause changes in the geometry of the body; these changes increase as the forces increase. These changes in the geometry or shape of the body are called deformations or *strains;* if measured on a body of unit dimension or per unit length of the body, they are called *unit strains*.

3-2   Direct and Combined Stresses.   For materials obeying Hooke's law, the tensile, compressive, and shear stresses inside a body as determined by Eqs. (3-1) and (3-2) are examples of simple or *direct stresses;* they are directly obtainable by use of a simple equation considering a single force or moment as the external loading on a body. Stresses within a body obtained by use of equations which combine the effects of two or more external loadings on a body are called *combined stresses.*

$$E = \frac{FL}{Ae} = \frac{s}{\epsilon} \tag{3-1}$$

$$s = E\epsilon = \frac{F}{A} \tag{3-2}$$

where A = area normal to the applied force, in.^{2}
  E = modulus of elasticity, psi
  e = total tensile or compressive strain, in., assuming the unit deformation to be constant throughout the length L
  F = total applied tension or compressive load, lb
  L = total unloaded length collinear with F, in.
  s = tensile or compressive unit stress, psi
  ϵ = unit strain, in./in.

Often, but not always, combined stress equations employ direct stresses as components. For numerous design purposes direct stresses are very useful, but in order to obtain a more nearly correct value of the actual maximum stresses existing within a body, the designer usually finds combined stresses more useful than direct stresses.

3-3   Saint-Venant's Theorem and Stress Concentration.   *Saint-Venant's theorem* essentially states that the stresses indicated by direct stress equations and by combined stress equations are applicable only to those positions of a member of a homogeneous material which are not near the location of any stress raisers. This means that these equations are not directly applicable close to the application of a concentrated force or moment; near a discontinuous change in cross section; near any threads, holes, keyways, or splines; or near the location of any other cause of stress concentration not taken into account in the stress calculations. In the case of beams, this means that the simple equations not taking stress concentration into account are not applicable in that portion of the length of the beam within a distance of approximately the depth of the beam from the point of application of any concentrated load or moment. In the case of a round rod in tension or compression, this distance is approximately equal to the diameter of the bar. Most designers neglect Saint-Venant's theorem and calculate stresses and strains wherever they please in the member; after this is done, some designers make corrections so as to take into account the effects of the stress raisers. This approach will be followed in this book by applying experimentally determined or theoretical *stress-concentration factors* K to obtain more nearly correct values of stresses. These factors, always greater than 1, are the dimensionless ratios of the actual stress to the stress indicated by use of equations that do not take into account stress concentration. See Arts. 3-34 through 3-39 for a discussion of stress concentration.

3-4 Simple Tensile Stresses. If a homogeneous member of uniform cross-sectional area A is loaded by a tensile load F_t acting collinearly with an axis through the centroid of area A and perpendicular to the cross section, the *simple tensile stress* s_t (assumed uniform across the cross-sectional area) is given by

$$s_t = \frac{F_t}{A} \tag{3-3}$$

If F_t does not act collinearly with the centroidal axis of area A, then eccentric loading occurs and simple tension does not exist. See Art. 3-25 for a discussion of the effects of eccentric axial loading.

3-5 Simple Compressive Stresses. If a block of homogeneous material of uniform cross-sectional area A is loaded by a compressive load F_c acting collinearly with the centroidal axis of area A, perpendicular to the cross section, the *simple compressive stress* s_c (assumed uniform throughout the cross section) is

$$s_c = \frac{F_c}{A} \tag{3-4}$$

This is valid provided the length-to-area ratio of the loaded member is sufficiently small that it acts as a compression member without tending to buckle. Simple compression also does not occur when F_c acts eccentrically with the centroidal axis of area A. Article 3-25 discusses the effects of eccentric compressive loading.

3-6 Simple Shear Stresses. As indicated in Fig. 3-1, if two metallic plates lapped over each other are held together by a single bolt or rivet of

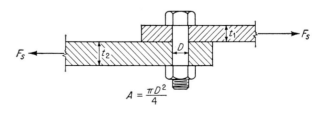

Fig. 3-1 Bolt in simple shear holding together two plates.

cross-sectional area A extending through each of the plates, and if F_s is the total shear force on the bolt resisting slippage of the plates relative to each other, the *simple shear stress* s_s in the bolt (assumed uniform throughout the cross section of the bolt at the plane of shear) is

$$s_s = \frac{F_s}{A} \tag{3-5}$$

3-7 Bearing Stresses. If the thicknesses of the two lapped plates of Fig. 3-1 are t_1 and t_2 in. and the bolt diameter is D in., the compressive load between the bolt and each plate causes *bearing stress* s_{b1} (on plate 1 and the bolt) and s_{b2} (on plate 2 and the bolt) of

$$s_{b1} = \frac{F_s}{t_1 D} \quad \text{and} \quad s_{b2} = \frac{F_s}{t_2 D} \tag{3-6}$$

is obtained. These bearing stresses are assumed constant throughout the projected area of the bolt contacting each plate. Bearing stresses (sometimes called crushing stresses) always are compressive.

3-8 Simple Flexural Stresses in Beams. If an initially straight beam of homogeneous material and prismatic cross section is loaded transversely with coplanar loads and/or moments, flexure or bending of the beam will occur. The resulting *flexural stresses* s_f are given by

$$s_f = \frac{Ec}{R} = \frac{Ec(d^2y/dx^2)}{[1 + (dy/dx)^2]^{1.5}} \tag{3-7}$$

where c = distance from the neutral axis of the beam to the point at which the flexural stress is to be determined, in. (c is measured perpendicular to the centroidal axis of the cross section of the beam normal to the plane of loading; it is plus if measured toward the center of curvature of the deflected neutral axis, minus if measured away from this center of curvature)

E = Young's modulus of the material of the beam, psi

R = radius of curvature of the deflected neutral axis (at point x along the length of the beam), in.

x = a longitudinal coordinate indicating the longitudinal position in the beam at which the flexural stress is to be determined, in.

y = deflection of the loaded neutral axis from its unloaded position (at point x along the length of the beam), in.

For the comparatively small deflections usually encountered in machine members, the $(dy/dx)^2$ term of Eq. (3-7) is negligible in relation to 1, and the equation simplifies into

$$s_f = Ec \frac{d^2y}{dx^2} = \frac{M_x c}{I_x} \tag{3-8}$$

where I_x is the area moment of inertia of the cross section of the beam about the centroidal axis that is normal to the plane of loading (at point x along the length of the beam) in inches fourth, and M_x is the internal bending moment in the beam at point x along the length of the beam, in inch-pounds.

3-9 Longitudinal Shear Stresses in Beams. As the result of flexure of a beam, *longitudinal shear stresses* (often called *horizontal shear stresses*) occur (Fig. 3-2). These stresses s_s may be calculated by

$$s_s = \frac{V_x A' \bar{y}}{I_x b} \tag{3-9}$$

where A' = that portion of the cross-sectional area of the beam (shaded in Fig. 3-2) between the outer fiber of the beam and a line parallel to the neutral axis at which the shear stress is desired, in.^{2}

b = width of the cross section of the beam (parallel to the neutral axis) at which the shear stress is desired, in.

I_x = area moment of inertia of the entire cross section of the beam about the neutral axis at point x along the length of the beam, in.^{4}

V_x = total transverse (vertical) shear on the beam at point x along the length of the beam, lb

$\bar{y}$ = perpendicular distance from the neutral axis to the centroid of area A', in.

The longitudinal shear stress is zero at the outer fibers of the beam. It is uniform across the width b at any given position in the beam. For beams of circular or rectangular cross section, the longitudinal shear stress is maximum at the neutral axis; for other shaped cross sections, the maximum shear stress often does not occur at the neutral axis. For these cases, in order to determine the value of y giving the maximum longitudinal shear stress, Eq. (3-9) may be differentiated with respect to y, set equal to zero, and solved for the desired value of y corresponding to $s_{s\,max}$.

Fig. 3-2 Definition of symbols in Eq. (3-9) for an initially straight prismatic beam of homogeneous material.

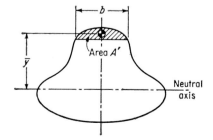

3-10 Deflection of Beams.   The *flexural deflection of a beam* may be obtained by rearranging Eq. (3-8) into the following:

$$\frac{d^2y}{dx^2} = \frac{M_x}{EI_x} \tag{3-10}$$

If Eq. (3-10) is integrated twice between limits, the following equation is obtained for the flexural deflection y at any point x along the length of the beam:

$$y = \int\limits_0^x\!\!\int \frac{M_x \, dx \, dx}{EI_x} \tag{3-11}$$

The maximum deflection of uniform cross-sectioned beams of length L carrying a single concentrated load F is expressible as

$$y_{max} = \frac{K_b F L^3}{EI} \tag{3-12}$$

For a beam carrying a uniformly distributed load w,

$$y_{max} = \frac{K_b w L^4}{EI} \tag{3-13}$$

Typical values of K_b for several commonly encountered beams are:

Simply supported beam with central concentrated load F, $K_b = \frac{1}{48}$
Simply supported beam with uniformly distributed load w, $K_b = \frac{5}{384}$
Cantilever beam with concentrated load F on end, $K_b = \frac{1}{3}$
Cantilever beam with uniformly distributed load w, $K_b = \frac{1}{8}$
Beam with two fixed ends and central concentrated load F, $K_b = \frac{1}{192}$
Beam with two fixed ends and uniformly distributed load w, $K_b = \frac{1}{384}$

3-11 Shear Stresses and Twist in Circular Shafts.   The torsional shear stress s_s in a circular solid or hollow shaft loaded in pure torsion is given by

$$s_s = \frac{Tc}{J} \tag{3-14}$$

where c = radial distance from the polar axis of rotation of the shaft to the point at which the shear stress is to be determined, in.

J = polar area moment of inertia of the cross-sectional area of the shaft about the axis of rotation of the shaft, in.^{4}

T = torque applied to the shaft, in.-lb.

From Eq. (3-14) it is obvious that the maximum torsional shear stress is on the outer circumference of the shaft, that it is zero at the center of the shaft, and that it is equal in magnitude for equal radial distances out from the axis of rotation. The direction of this shear stress is tangent to the circle drawn through the point concerned and concentric with the axis of the shaft.

Equation (3-14) is applicable only to torsional members which are circular in cross section (either solid or hollow with the inner surface concentric with the outer surface). For noncircular cross sections, more complicated equations are necessary.

3-12 Columns. If an initially straight machine member is loaded with a slowly increasing compressive load with no eccentricity, the column acts as an elastic member until, essentially without warning, it collapses or buckles. The load on the column at which buckling occurs is defined as the *critical load* F_{cr}. This load is not a direct function of stress, hence designers normally base their column calculations upon critical load rather than upon direct stress. No one equation is adequate for the design and analysis of all columns, hence several formulas must be employed. Fortunately for the designer, the following three equations cover most situations adequately: the Euler long-column equation for all materials, the J. B. Johnson equation for short columns of ductile materials, and the straight-line equation for short columns for some brittle materials, such as cast iron.

3-13 Euler's Long-column Equation. In 1757, Leonhardt Euler developed the following equation for the critical load on a long column of any homogeneous material:

$$F_{cr} = \frac{n\pi^2 EI}{L^2} = \frac{n\pi^2 EA}{(L/k)^2} \tag{3-15}$$

in which A = constant cross-sectional area of the column, in.^{2}
E = Young's modulus of the material of the column, psi
I = area moment of inertia of the cross section of the column calculated about the axis through the cross section about which buckling occurs, in.^{4}
k = $\sqrt{I/A}$, radius of gyration of the cross section, in.
L = effective length of the column, in.
n = end-fixity factor, the square of the number of 180 deg of approximate sine-curve deflection exhibited by the deflected neutral axis of the column.

Some designers state that if the *slenderness ratio* L/k is approximately 30 or less, the effect of the lateral column deflection is negligible and the member acts essentially as a body in pure compression; for L/k values larger than approximately 30, the effect of column action is not negligible and column equations must be employed. Slenderness ratios between 45 and 130 are in common use by the machine designer.

The load-carrying capacity of a column is dependent upon the condition of restraint of the two ends of the column; this usually is expressed in terms of a dimensionless factor called the *end-fixity factor n*. Determination of the end-fixity factor for any particular design requires the exercising of judgment by the designer. For example, it is almost impossible to achieve complete fixity for the ends since some deflection and rotation will occur in the "rigid" members to which the ends are fixed; values of *n* of 3.0 to 3.5 are common for this case. On the other hand where both ends of the column are pinned, because of friction in the pin joints and other causes, the end-fixity factor never gets as low as 1. Values of 1.1 are common in these instances. End-fixity factors for 5 common types of column deflections are given in Fig. 3-3.

Some designers prefer to base their column calculations upon a significant stress basis. If F_{cr} is divided by A, then Eq. (3-15) may be revised into:

$$\frac{F_{cr}}{A} = \frac{n\pi^2 EI}{AL^2} = \frac{n\pi^2 E}{(L/k)^2} \tag{3-16}$$

As pointed out in Art. 3-14, since the critical load is not a direct function of stress, the significant stress indicated by Eq. (3-16) does not correspond to an actual stress in the column at failure; to some designers it is merely a convenience for calculation, to others it is misleading.

Values of F_{cr} for various slenderness ratios are indicated in Fig. 3-4. According to Eqs. (3-15) and (3-16), for very small slenderness ratios, the

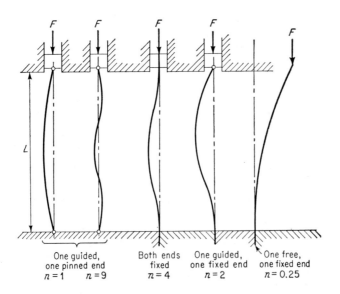

Fig. 3-3 End-fixity factors for various Euler long columns.

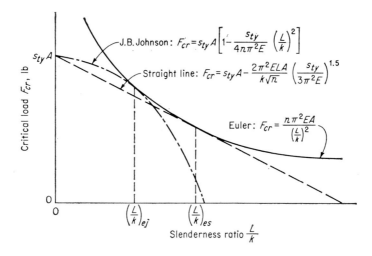

Fig. 3-4 Comparison of the J. B. Johnson and straight-line column equations with the Euler long-column equation.

$$\left(\frac{L}{k}\right)_{ej} = \sqrt{\frac{2n\pi^2 E}{s_{ty}}} \qquad\qquad \left(\frac{L}{k}\right)_{es} = \sqrt{\frac{3n\pi^2 E}{s_{ty}}}$$

critical load approaches infinity. Obviously this does not appear very practical since the critical stress never should be permitted to exceed the yield stress of the material. For this reason, Euler's column equation is not considered applicable to short columns. Euler's equation is applicable to long columns of all materials with relatively high slenderness ratios. Unfortunately for most columns encountered in machine design, the slenderness ratios involved usually are too small for successful use of Euler's equation. For these cases, many semiempirical formulas have been developed, the two most commonly used being the J. B. Johnson and the straight-line equations.

3-14 J. B. Johnson Parabolic Equation for Short Columns. J. B. Johnson noted that the maximum critical stress should not exceed the tensile yield stress s_{ty} of the material and that Euler's equation was not good for columns with low slenderness ratios. As indicated in Fig. 3-4, he suggested that a parabola, symmetrical about the vertical axis going through a zero value of L/k, could be drawn through an ordinate of $s_{ty}A$ for $L/k = 0$ and tangent to the Euler curve at a slenderness ratio $(L/k)_{ej}$ of $\sqrt{2n\pi^2 E/s_y}$, corresponding to F_{cr} of $s_{ty}A/2$. The equation of this parabola is

$$F_{cr} = s_{ty}A\left[1 - \frac{s_{ty}}{4n\pi^2 E}\left(\frac{L}{k}\right)^2\right] \tag{3-17}$$

For L/k values of $\sqrt{2n\pi^2 E/s_{ty}}$, and less, the J. B. Johnson equation is applicable to ductile materials; for higher L/k values of any material, the Euler equation is applicable. The Johnson equation is extensively used by designers to determine the critical load for ductile steel columns.

3-15 Straight-line Equation for Short Columns. For short columns of some of the brittle materials such as cast iron, the J. B. Johnson equation usually gives somewhat too high allowable values of F_{cr}. For these cases, a straight line drawn from $s_{cr}A = s_{ty}A$ at $L/k = 0$ tangent to the Euler curve is more nearly applicable. This line and the Euler curves are tangent at

$$(L/k)_{es} = \sqrt{3n\pi^2 E/s_{ty}} \qquad \text{and} \qquad F_{cr} = s_{ty}A/3$$

The equation of this line is

$$F_{cr} = s_{ty}A - \frac{2\pi^2 ELA}{k\sqrt{n}}\left(\frac{s_{ty}}{3\pi^2 E}\right)^{1.5} \tag{3-18}$$

3-16 Contact Stresses. In most cases in which compressive forces are transmitted from one body to another, the effective area of contact on which these forces act is reasonably well known. For many calculations, the area in contact before the application of the compressive load can be used as a sufficiently accurate approximation of the actual contact area under load. In several instances, however, this procedure yields results which are far in error. For example, when two spheres initially contact each other at a given point, or two parallel cylinders touch each other in a line, the unloaded contact area is a point or line of zero area, yielding according to the above procedure an infinite contact stress under any amount of compressive loading. Infinite contact stresses are obviously impossible; however, contact stresses of several hundred thousand pounds per square inch are common and often can be tolerated.

The original work leading to the actual contact area and contact stresses between two bodies was done by H. Herz.* He stated that the surface of contact between two elastic objects with a total compressive force F_c between them is bounded by an ellipse with a maximum compressive stress s_{max} at the center of the ellipse. Equations for the general case are complicated and tedious to use; for the cases of spheres, cylinders, and planes in contact, simplified equations result.†

Contact between Elastic Sphere and Plane. For an elastic sphere of initially unloaded diameter D contacting a plane, the contact

---

* H. Herz, "Gesammelte Werke," vol. 1, p. 155, 1895.

† S. Timoshenko, "Strength of Materials," part II, 2d ed., pp. 355–361, D. Van Nostrand Co., Inc., Princeton, N.J., 1941.

area is a circle of diameter d. The maximum contact stress at the center of this circular area and the diameter of the contact area are given by

$$s_{max} = 0.98 \sqrt[3]{\frac{F_c}{D^2(1/E_1 + 1/E_2)^2}} \tag{3-19}$$

$$d = 1.396 \sqrt[3]{F_c D \left(\frac{1}{E_1} + \frac{1}{E_2}\right)} \tag{3-20}$$

where E_1 and E_2 are the Young's moduli of the material of the sphere and plane, respectively.

Contact between Two Elastic Spheres. The contact area between two elastic spheres is a circle of diameter d. The maximum contact stress occurs at the center of the circle. This stress and diameter are given by

$$s_{max} = \frac{6F_a}{\pi d^2} = 0.98 \sqrt[3]{\frac{F_a(1/D_1 + 1/D_2)^2}{(1/E_1 + 1/E_2)^2}} \tag{3-21}$$

$$d = 1.396 \sqrt[3]{\frac{F_c(1/E_1 + 1/E_2)}{1/D_1 + 1/D_2}} \tag{3-22}$$

where D_1 and D_2 are the initially unloaded diameters of spheres 1 and 2 respectively, and E_1 and E_2 are Young's moduli of the material of spheres 1 and 2 respectively.

Equations (3-21) and (3-22) are applicable to a sphere of diameter D_1 on the inside of another sphere of diameter D_2 provided in each of the equations a negative value is substituted for D_2.

Contact between Elastic Cylinder and Plane. For an elastic cylinder of initially unloaded diameter D and length L, the contact area is an oblong of length L and width b. The maximum contact stress at the center of the oblong is given by

$$s_{max} = 0.833 \sqrt{\frac{F_c}{LD(1/E_1 + 1/E_2)}} \tag{3-23}$$

$$b = 1.52 \sqrt{\frac{F_c D}{L} \left(\frac{1}{E_1} + \frac{1}{E_2}\right)} \tag{3-24}$$

where E_1 and E_2 are the Young's moduli of the material of the cylinder and the plane respectively.

Contact between Two Elastic Cylinders with Parallel Axes. The contact area between two parallel axis elastic cylinders of contact length L is an oblong of length L and width b. The maximum contact stress at the center of this area is given by

$$s_{\max} = 0.833 \sqrt{\frac{F_c(1/D_1 + 1/D_2)}{L(1/E_1 + 1/E_2)}} \tag{3-25}$$

$$b = 1.52 \sqrt{\frac{F_c(1/E_1 + 1/E_2)}{L(1/D_1 + 1/D_2)}} \tag{3-26}$$

where D_1 and D_2 are the initially unloaded diameters of cylinders 1 and 2 respectively, and E_1 and E_2 are their respective Young's moduli. For internal contact between a cylinder 1 inside a larger diameter cylinder 2, Eqs. (3-25) and (3-26) are applicable provided a negative value is substituted for D_2.

Subsurface Shear Stresses Resulting from Contact Stresses. Although the stress throughout the entire area of the contacting surfaces is compression on the surface of the materials involved, shear stresses exist inside the materials. For the case of two elastic cylinders with parallel axes, this maximum shear stress $s_{s\,\max}$ is given by

$$s_{s\,\max} = 0.304 s_{\max} = 0.253 \sqrt{\frac{F_c(1/D_1 + 1/D_2)}{L(1/E_1 + 1/E_2)}} \tag{3-27}$$

For materials with a Poisson's ratio of 0.3, this maximum shear stress occurs at the center of the contact area at a distance y in. beneath the contacting surface as given by

$$y = 0.39b = 0.583 \sqrt{\frac{F_c(1/E_1 + 1/E_2)}{L(1/D_1 + 1/D_2)}} \tag{3-28}$$

3-17 Induced Stresses from Direct Tensile Loading.   Induced stresses are those tensile, compressive, and shear stresses induced within a body by application of external forces and/or torques onto the body. Most of the stresses previously mentioned in this chapter are induced stresses even though they were calculated by comparatively simple direct stress equations.

Consider the body in Fig. 3-5 to have unit thickness perpendicular to the paper and to be subjected to external forces F_t producing a direct tension stress s_t on the cross-sectional area. Imagine the body to be cut by a plane BC at any angle θ with the line of action F_t. If the left part of the body is removed, forces acting normal and parallel to the plane BC

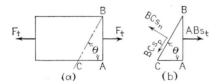

(a)

(b)

Fig. 3-5 (a) Pure tension loading of an elastic member of unit thickness. (b) Forces acting on element ABC of unit thickness.

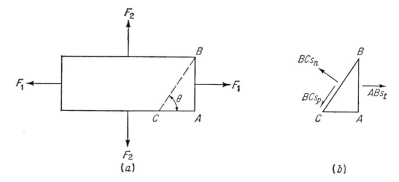

Fig. 3-6 (*a*) Biaxial tensile loading of an elastic member of unit thickness. (*b*) Forces acting on element ABC of unit thickness.

must be supplied to hold the right portion in equilibrium. Let the stress normal to plane BC be designated by s_n, and that parallel to the plane by s_p. Resolving all forces acting on the right portion into components acting parallel to the plane BC,

$$ABs_t \cos \theta - BCs_p = 0 \qquad (3\text{-}29)$$

from which

$$s_p = (s_t \cos \theta) \frac{AB}{BC} = s_t \cos \theta \sin \theta = s_t \frac{\sin 2\theta}{2} \qquad (3\text{-}30)$$

The stress s_p is an induced shear stress along the plane BC and will be maximum when $\sin 2\theta$ is maximum, i.e., when θ is 45 deg. The maximum shear stress $s_{s \, \text{max}}$ induced in a body by direct tensile loading is

$$s_{s \, \text{max}} = \frac{s_t}{2} \qquad (3\text{-}31)$$

Summing forces perpendicular to plane BC yields

$$-s_n BC = s_t \sin \theta \, AB \qquad (3\text{-}32)$$

from which

$$s_n = (s_t \sin \theta) \frac{AB}{BC} = s_t \sin^2 \theta \qquad (3\text{-}33)$$

The stress s_n is an induced tensile stress perpendicular to the plane BC. It will be maximum when θ is 90 deg. The maximum tension induced in a body by direct tensile loading is

$$s_{n \, \text{max}} = s_t \qquad (3\text{-}34)$$

3-18 Stresses Induced from Biaxial Loading. The body in Fig. 3-6 is of unit thickness perpendicular to the paper and is loaded by two

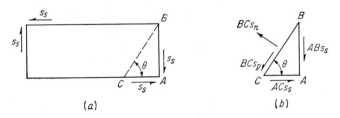

Fig. 3-7   (*a*) Pure shear loading of an elastic member of unit thickness. (*b*) Forces acting on element ABC of unit thickness.

coplanar tensile forces F_1 and F_2 causing direct tensile stresses s_1 and s_2 respectively. Stress s_1 is larger than s_2. By the method of Art. 3-17, it can be shown that

$$s_p = (s_1 - s_2) \sin \theta \cos \theta = (s_1 - s_2) \frac{\sin (2\theta)}{2} \qquad (3\text{-}35)$$

For $\theta = 45$ deg, s_p has a maximum value of

$$s_{s\,max} = \frac{|s_1 - s_2|}{2} \qquad (3\text{-}36)$$

$$s_n = s_1 \sin^2 \theta + s_2 \cos^2 \theta \qquad (3\text{-}37)$$

The maximum tensile induced stress occurs for $\theta = 90$ deg. This gives

$$s_{n\,max} = s_1 \qquad (3\text{-}38)$$

3-19   Stresses Induced from Triaxial Loading.   Consider a body loaded by three mutually perpendicular forces with resulting direct stresses s_1, s_2, and s_3 listed in order of descending algebraic magnitude.

$$s_{s\,max} = \frac{|s_1 - s_2|}{2} \quad \text{or} \quad \frac{|s_1 - s_3|}{2} \quad \text{or} \quad \frac{|s_2 - s_3|}{2} \quad \begin{array}{l} \text{whichever} \\ \text{is largest} \end{array} \quad (3\text{-}39)$$

Determining the maximum normal stress is an elementary matter since s_1, s_2, and s_3 already are the principal stresses. This gives

$$s_{n\,max} = s_1 \qquad (3\text{-}40)$$

3-20   Stresses Induced from Pure Shear.   Consider the body of Fig. 3-7 of unit thickness perpendicular to the paper and loaded by direct shear stresses s_s as shown:

$$s_p = s_s(\cos^2 \theta - \sin^2 \theta) \qquad (3\text{-}41)$$

For $\theta = 45$ deg, the maximum induced shear is

$$s_{s\,max} = s_s \qquad (3\text{-}42)$$
$$s_n = s_s \sin (2\theta) \qquad (3\text{-}43)$$

For $\theta = 45$ deg, the maximum induced tension is

$$s_{n\ max} = s_s \tag{3-44}$$

3-21 Stresses Induced from Combined Shear and Tension.
Consider the body of unit thickness shown in Fig. 3-8 with the forces F_t
and F_s acting as shown to produce direct tension and shear stresses s_t and
s_s respectively. Since the body is in equilibrium, shear forces must also
act on the surfaces BC and DA. Taking moments about any point in the
body,

$$s_s \times AB \times BC = s_2 \times BC \times AB \tag{3-45}$$

and

$$s_s = s_2 \tag{3-46}$$

showing that the unit shear stresses on the surfaces AB and BC are equal
in magnitude.

If an imaginary plane BE cuts away the left part of the body, forces
$s_n \times BE$ and $s_p \times BE$ must be supplied to maintain equilibrium. Resolving all forces acting on the right portion into components parallel to the
plane BE,

$$ABs_t \cos\theta + ABs_s \sin\theta - AEs_s \cos\theta - BEs_p = 0 \tag{3-47}$$

Dividing through by BE and transposing,

$$s_p = s_t \sin\theta \cos\theta + s_s \sin^2\theta - s_s \cos^2\theta \tag{3-48}$$
$$= \tfrac{1}{2}s_t \sin 2\theta - s_s \cos 2\theta$$

Differentiating with respect to θ and equating to zero, s_p is maximum
when

$$\tan 2\theta = -\frac{s_t}{2s_s} \tag{3-49}$$

$$\sin 2\theta = \frac{s_t}{\sqrt{s_t^2 + 4s_s^2}} \tag{3-50}$$

and

$$\cos 2\theta = \frac{-2s_s}{\sqrt{s_t^2 + 4s_s^2}} \tag{3-51}$$

Substituting these values in Eq. (3-48) and noting that s_p is a shear-

Fig. 3-8   (*a*) Combined shear and tension on an elastic member of unit thickness. (*b*) Forces acting on element ABE
of unit thickness.

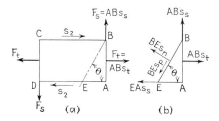

ing stress on the plane BE, the maximum shear induced by the external tension and shearing loads becomes

$$s_{s\,max} = \tfrac{1}{2}\sqrt{s_t^2 + 4s_s^2} \qquad (3\text{-}52)$$

In a similar manner

$$s_n = \frac{s_t[1 - \cos(2\theta)]}{2} + s_s \sin(2\theta) \qquad (3\text{-}53)$$

The maximum value of s_n occurs when

$$\sin(2\theta) = \frac{2s_s}{\sqrt{s_t^2 + 4s_s^2}} \qquad (3\text{-}54)$$

$$\cos(2\theta) = -\frac{s_t}{\sqrt{s_t^2 + 4s_s^2}} \qquad (3\text{-}55)$$

This yields a maximum normal stress of

$$s_{n\,max} = \frac{s_t}{2} + \frac{1}{2}\sqrt{s_t^2 + 4s_s^2} \qquad (3\text{-}56)$$

3-22 Stresses Induced from Combined Biaxial Loading and Shear.   The body of unit thickness shown in Fig. 3-9 is loaded biaxially so as to cause direct stresses s_1 and s_2. In addition, a direct shear stress s_s is induced in the body. Following the method employed in the previous articles,

$$s_p = (s_1 - s_2)\frac{\sin(2\theta)}{2} - s_s \cos(2\theta) \qquad (3\text{-}57)$$

$$s_n = \frac{s_1 + s_2}{2}\frac{(s_2 - s_1)\cos(2\theta)}{2} + s_s \sin(2\theta) \qquad (3\text{-}58)$$

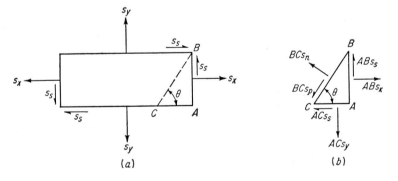

Fig. 3-9   (a) Combined biaxial loading and shear of an elastic member of unit thickness. (b) Forces acting on element ABC of unit thickness.

The maximum value of s_p occurs where

$$\sin (2\theta) = \frac{s_1 - s_2}{\sqrt{(s_1 - s_2)^2 + 4s_s^2}} \tag{3-59}$$

$$\cos (2\theta) = -\frac{2s_s}{\sqrt{(s_1 - s_2)^2 + 4s_s^2}} \tag{3-60}$$

This yields a maximum induced shear stress of

$$s_{s\,\text{max}} = \tfrac{1}{2} \sqrt{(s_1 - s_2)^2 + 4s_s^2} \tag{3-61}$$

The maximum value of s_n occurs where

$$\sin (2\theta) = \frac{2s_s}{\sqrt{(s_1 - s_2)^2 + 4s_s^2}} \tag{3-62}$$

$$\cos (2\theta) = \frac{s_2 - s_1}{\sqrt{(s_1 - s_2)^2 + 4s_s^2}} \tag{3-63}$$

This results in a maximum normal stress of

$$s_{n\,\text{max}} = \frac{s_1 + s_2}{2} + \frac{1}{2} \sqrt{(s_1 - s_2)^2 + 4s_s^2} \tag{3-64}$$

3-23   Principal and Maximum Shear Stresses.   Consider it hypothetically possible at any desired point within a stressed body to locate a very small differential cube and to rotate this cube into any desired position without altering the stresses existing at the point. It always is possible to orient the cube in such a manner that no shear stresses exist on two parallel surfaces of the cube, only normal stresses. *Principal stresses* are those combined stresses existing perpendicular to a plane on which there is no shear stress. There is only one possible orientation position of the cube meeting this condition at any particular point in the body, and the three sets of stresses normal to the three pairs of surfaces on which there are no shear stresses are the three principal stresses. The magnitudes of these three principal stresses usually are not equal. The algebraic largest of these is called the *maximum normal stress;* the smallest is the *minimum normal stress.* These maximum and minimum normal stresses may both be tension or compression, or the maximum normal stress may be tension while the minimum normal stress is compression. The stresses indicated by Eqs. (3-34), (3-38), (3-40), (3-44), (3-56), and (3-64) are maximum normal stresses. As a rule, the principal stress between the maximum normal and the minimum normal stresses is of little use to the designer, hence this third principal stress seldom is determined since it is not critical.

By further rotation of the cube, it is possible to determine a position in which only shear stresses exist on the surfaces of the cube, not normal stresses. The *maximum shear stress* is that shear stress existing on a plane perpendicular to which no normal stress exists. The stresses of Eqs. (3-31), (3-36), (3-39), (3-42), (3-52), and (3-61) are maximum shear stresses.

The stresses which cause failure in a machine member are the principal stresses or the maximum shear stresses, not the direct stresses, therefore these combined stresses are of extreme importance to the designer.

3-24 Combined Axial and Flexural Loading. In Fig. 3-10, a beam is subjected to an axial load F_t, a bending load F_b, and the reactions R_1 and R_2. Perpendicular to section BC there is a tension stress s_t, uniformly distributed over the section, equal to F_t/A and represented in magnitude by the length CD.

The load F_b causes a bending stress on the section BC, which varies from zero at the neutral surface to a maximum in compression at the upper surface. The magnitude of the stress is Mc/I and is represented by the horizontal distance between the lines BC and EG.

On the lower surface, both stresses are tension and in their effect are added together. On the upper surface the bending stress is compression. If no buckling occurs, the combined stresses at the outer surfaces are given by the equation

$$s = \frac{F}{A} \pm \frac{Mc}{I} \tag{3-65}$$

Equation (3-65) is not strictly correct, although for beams of fairly large cross section compared with the length, the results obtained have engineering accuracy. The side loads of an axially loaded beam produce a deflection which then allows the axial load to develop bending moments at the cross sections of the beam. The additional bending moments due to the axial loading will reduce the beam deflection and consequently the bending stress if the axial loading is tension and will increase the deflection and bending stress if the axial loading is compression.

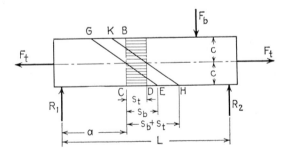

Fig. 3-10 Combined tension and flexural loading of an elastic member.

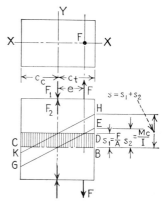

Fig. 3-11 Eccentric axial loading of an elastic member.

3-25 Eccentric Axial Loading. A short member in tension with the external forces acting at a distance e from the center line is shown in Fig. 3-11. Without changing the conditions of equilibrium, the collinear forces F_1 and F_2, equal and opposite to each other and acting along the center line or neutral axis, can be added. If F_1 and F_2 each are made equal to F, then the original force has been replaced by an equal axial force and a couple of magnitude Fe. The couple produces a bending moment in the member, and the same condition exists as in the beam discussed in Art. 3-24. The stress in the member is

$$s = \frac{F}{A} \pm \frac{Mc}{I} = \frac{F}{A} \pm \frac{Fec}{Ak^2} = \frac{F}{A}\left(1 \pm \frac{ec}{k^2}\right) \tag{3-66}$$

where k is the radius of gyration of the cross-sectional area A about the Y axis; c is c_t when the plus sign is used and c_c when the minus sign is used.

If the point of application of the load is not on one of the principal axes of the section, eccentricity in two directions results. This case is illustrated in Fig. 3-12. The stress at any corner of this section is given by the equation

$$\begin{aligned} s &= \frac{F}{A} \pm \frac{Fah}{I_y} \pm \frac{Fbf}{I_x} \\ &= \frac{F}{A}\left(1 \pm \frac{ah}{k_y^2} \pm \frac{bf}{k_x^2}\right) \end{aligned} \tag{3-67}$$

where I_x and I_y are area moments of inertia of the cross section about the

Fig. 3-12 Cross section of eccentrically loaded member of Fig. 3-11.

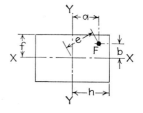

X and Y axes, in inches fourth, and k_x and k_y are radii of gyration of the section about the X and Y axes, in inches.

Equations (3-66) and (3-67) are not strictly correct for the same reasons listed in Art. 3-24.

3-26   Curved Beams.   A *curved beam* is a beam in which the unloaded neutral axis is curved instead of straight. So long as the ratio of the radius of curvature of the deflected neutral axis to the depth of the beam is approximately 10 or less, the conventionally used equations for straight beams usually are of sufficient accuracy. For larger ratios, as is common with curved beams, the resulting stress in the loaded beam is appreciably greater than Mc/I, hence special formulas for these cases are necessary.

A common procedure is to use straight beam formulas and to attempt to correct them for use with curved beams by use of dimensionless correction factors K_c as follows:

$$s = \frac{F}{A} + K_c \left(\frac{Mc}{I} \right) \tag{3-68}$$

Photoelastic data by Dolan and Levine* indicate the following equation to give results in close agreement with actual stresses:

$$s = K_c \left(\frac{F}{A} + \frac{Mc}{I} \right) \tag{3-69}$$

where K_c is determined for the inner fiber of the beam.

Values of K_c for the inner and the outer fibers of a few of the more commonly used cross sections of cuved beams are given in Table 3-1.

3-27   Theories of Failure.   Mechanical properties of materials usually have been determined with testing machines that apply simple direct stresses to the test specimen. For ductile materials most of the properties of the materials are determined by a simple tension test, while for many brittle materials the simple compression test is often used. In most machine and structural parts, the stresses which actually exist are not simple stresses but combined stresses that act in more than one direction. The strength of materials subjected to combined stresses has been investigated experimentally for a comparatively few stress combinations. To interpret these test results and to have a basis for the design of machine parts subjected to combined stresses, various *theories of failure* have been devised, probably none of which is ever exactly correct for any given case of design analysis. For some cases of failure of machine parts, apparently no acceptable theory of failure has been devised as yet. Theories of failure are based upon both static and fatigue strengths of the materials under

---

* T. J. Dolan and R. E. Levine, A Study of the Stresses in Curved Beams, *Proceedings of the Thirteenth Semi-annual Eastern Photoelastic Conference*, June, 1941.

Table 3-1 **Correction factors K_c for use in curved beam Eqs. (3-68) and (3-69)***

Cross section of beam	R/c_i†	Correction factor K_c	
		Inside fiber	Outside fiber
	1.2	3.41	0.54
	1.4	2.40	0.60
	1.6	1.96	0.65
	1.8	1.75	0.68
	2.0	1.62	0.71
	3.0	1.33	0.79
	4.0	1.23	0.84
	6.0	1.14	0.89
	8.0	1.10	0.91
	10.0	1.08	0.93
	1.2	3.28	0.58
	1.4	2.31	0.64
	1.6	1.89	0.68
	1.8	1.70	0.71
	2.0	1.57	0.73
	3.0	1.31	0.81
	4.0	1.21	0.85
	6.0	1.13	0.90
	8.0	1.10	0.92
	10.0	1.07	0.93
	1.2	2.89	0.57
	1.4	2.13	0.63
	1.6	1.79	0.67
	1.8	1.63	0.70
	2.0	1.52	0.73
	3.0	1.30	0.81
	4.0	1.20	0.85
	6.0	1.12	0.90
	8.0	1.09	0.92
	10.0	1.07	0.94
	1.2	3.09	0.56
	1.4	2.25	0.62
	1.6	1.91	0.66
	1.8	1.73	0.70
	2.0	1.61	0.73
	3.0	1.37	0.81
	4.0	1.26	0.86
	6.0	1.17	0.91
	8.0	1.13	0.93
	10.0	1.11	0.95

* F. B. Seely and J. O. Smith, "Advanced Mechanics of Materials," 2d ed., pp. 149–151, John Wiley & Sons, Inc., New York, 1952.

† This ratio is based upon c_i measured to the extreme fiber of the beam closest to the center of curvature regardless of where it is desired to calculate the actual stress.

observation. Unfortunately, at this time there is insufficient information available to formulate theories of failure for combined stresses accompanied by high or low temperatures, impact, and most of the other unusual design conditions. Creep is one exception, since reasonably useful stress theories have been devised to take creep into account.

Most machine elements are presumed to have failed when the maximum stresses within the member exceed the elastic limit of the material involved. In some cases, however, failure is not considered to have occurred until rupture of the machine part has occurred. For ductile materials in simple tension, elastic failure is represented by the tensile yield strength. For both brittle and ductile materials in simple tension, failure often is defined in terms of the ultimate strength. Values of the yield and ultimate strengths of various materials as determined by experimentation are available in handbooks and to a limited extent in Chap. 2. For many of the newer materials, these physical properties are not so easily found in printed form. In many machine members, the stresses do not act in only one direction but in several directions simultaneously; hence a combined stress condition exists within the material rather than a simple stress condition. The general state of stress at any particular point in a body may be designated by the three principal stresses at the point; these stresses may be tension and/or compression. The influence of this combined state of stress upon the strength of materials rather than a simple stress system has been investigated experimentally for a number of materials and various stress combinations, Various theories have been proposed to interpret these test results. Usually the theory which agrees with one set of experimental data will not necessarily agree with another set of experimental data for a different material under apparently the same conditions of loading. These theories define the yield point or the ultimate strengths of materials when they are subjected to combined stresses in terms of the yield or ultimate strengths in a simple tension test.

The more commonly used theories of failure are the maximum normal stress theory, the maximum shear stress theory, the maximum strain energy theory, and the maximum distortion energy theory. Additional theories of failure are based upon plasticity relations, ductility, fatigue stresses, creep, and other considerations. In the four theories of failure which follow, the equations which are given are written in terms of a maximum allowable stress permitted by the theory of failure before inelastic action begins.

3-28 Maximum Normal Stress Theory. This theory of failure, also called the *Rankine theory*, states that inelastic action occurs when the maximum normal stress equals the tensile yield stress of the material as determined in a simple tension test.

For cases of pure tension or compression, or for biaxial or triaxial loading without any direct shear stress, the maximum allowable stress is the tensile yield stress.

For the combination of one direct tensile or compressive stress and one direct shear stress, Eq. (3-56) applies. For combined biaxial loading and one direct shear stress, Eq. (3-64) applies.

The maximum normal stress theory is particularly applicable to brittle materials such as cast iron, hard steels, and glass. The predictions of this theory do not agree very well with experimental data, particularly for biaxial loading of opposite signs.

3-29 Maximum Shear Stress Theory. This theory of failure, also called *Coulomb's law* and the *Guest-Hancock theory*, states that inelastic action occurs when the maximum shear stress equals one-half the tensile yield stress of the material as determined in a simple tension test.

For direct tension or compression without any direct shear, the allowable shear stress is one-half the tensile yield stress.

For biaxial loading without any direct shear

$$s_{s\ max} = \frac{s_{ty}}{2} = \frac{|s_1 - s_2|}{2} \qquad (3\text{-}70)$$

in which either of the principal stresses s_1 and s_2 are known and it is desired to determine the maximum allowable value of the other direct stress.

For triaxial loading with no direct shear stress

$$s_{s\ max} = \frac{s_{ty}}{2} = \frac{|s_1 - s_2|}{2} \quad \text{or} \quad \frac{|s_1 - s_3|}{2} \quad \text{or} \quad \frac{|s_2 - s_3|}{2} \qquad (3\text{-}71)$$

whichever is largest. If any two of the principal stresses s_1, s_2, and s_3 are known, the maximum allowable value of the third stress may be determined by Eq. (3-71).

For combined direct tension or compression and direct shear acting perpendicularly to each other, Eq. (3-52) applies.

For biaxial loading combined with direct shear, Eq. (3-61) applies.

The maximum shear stress theory of failure is applicable to ductile materials such as mild steels, brasses, and most aluminum alloys.

3-30 Maximum Strain Theory. This theory of failure, also called *Saint-Venant's theory*, states that inelastic action occurs when the maximum unit strain in any direction exceeds the unit strain at the tensile yield point for the material as determined in a pure tensile test.

For biaxial loading with no direct shear, the following two equations are applicable:

$$s_{1\ max} = s_{ty} + \mu s_2 \qquad (3\text{-}72)$$
$$s_{2\ max} = s_{ty} + \mu s_1 \qquad (3\text{-}73)$$

For triaxial loading with no direct shear, the following three are applicable:

$$s_{1 \text{ max}} = s_{ty} + \mu(s_2 + s_3) \tag{3-74}$$
$$s_{2 \text{ max}} = s_{ty} + \mu(s_1 + s_3) \tag{3-75}$$
$$s_{3 \text{ max}} = s_{ty} + \mu(s_1 + s_2) \tag{3-76}$$

For cases in which direct shear is involved, the three principal stresses must be determined, and then Eqs. (3-74), (3-75), and (3-76), in which s_1, s_2, and s_3 are the principal stresses, must be employed.

The maximum strain theory of failure is particularly applicable to thick cylinders and to brittle materials.

3-31 Maximum Distortion Energy Theory. This theory is also called the *James Clerk Maxwell theory, Hencky-von Mises theory, Hüber theory*, and the *Beltrami and Haigh theory*. It states that inelastic action occurs when the distortion energy per unit volume of the material equals the distortion energy per unit volume for the material at the tensile yield point as determined by a simple tension test. The distortion energy is the total strain energy minus the energy required to produce a change in volume.

For biaxial loading with no direct shear

$$s_{ty}^2 = s_1^2 - s_1 s_2 + s_2^2 \tag{3-77}$$

in which either of the two direct stresses s_1 or s_2 is known and it is desired to determine the maximum allowable value of the other direct stress.

For triaxial loading

$$s_{ty}^2 = \tfrac{1}{2}[(s_1 - s_2)^2 + (s_1 - s_3)^2 + (s_2 - s_3)^2] \tag{3-78}$$

in which any two of the principal stresses s_1, s_2, and s_3 are known and the maximum allowable value of the third principal stress is desired.

Where a direct shear stress is involved, the simplest procedure is to determine the three principal stresses existing at the point under consideration and then to use Eq. (3-78).

This is one of the best available theories of failure since it agrees rather well with many experimental data. It applies to most ductile materials.

3-32 Temperature Stresses. Most substances expand or increase in volume as their temperature is raised. If the body is free to expand without hindrance, no stress will be set up; but if the tendency to expand is resisted by adjoining bodies or machine members, stresses will be set up. These temperature stresses are of great importance in the design of large internal-combustion engines, high-temperature fluid systems, furnaces, and similar structures.

The linear expansion due to a temperature change is

$$e = c(t_2 - t_1)L \tag{3-79}$$

where c = coefficient of linear expansion, in./(in.)(°F)
$\quad\quad e$ = total increase in length, in.
$\quad\quad L$ = original length at temperature t_1, in.
$t_2 - t_1$ = change in temperature, °F

If the member is held so that it cannot expand, the effect is the same as though a compressive force is applied of sufficient magnitude to produce a compression of e in. Then the stress set up by the temperature change is

$$s = E\frac{e}{L} = cE(t_2 - t_1) \tag{3-80}$$

If two materials having different coefficients of expansion are rigidly fastened together throughout their length, and then subjected to a temperature increase, they will tend to expand different amounts; however, since they must expand equally, the material having the higher coefficient of expansion will be subjected to compressive stresses and the other material will be in tension. The composite part will assume a curvature, the amount of which will depend upon the temperature change. This bimetallic expansion principle is used in the manufacture of metallic thermometers and in the design of temperature-control apparatus.

3-33 Residual Stresses. *Residual stresses* are those internal, inherent, trapped, locked-up body stresses that exist within a material as the result of things other than external loading. These stresses may be microstresses (varying from grain to grain or within a grain) or macrostresses (distributed more or less uniformly over an appreciable area).

Residual stresses usually are caused by cold-working (shot-peening, rolling, cold-drawing, machining), heating or cooling (heat-treating, surface-hardening, and welding), etching, repeated stressing, and electrodeposition (electroplating and similar operations). These stresses exist in many machine members, and though they may be either helpful or detrimental, such stresses are usually not considered desirable because they are often accompanied by objectionable phenomena: warping and distortion after machining or heat-treating, cracking from large reductions in size of the member as a result of cold-working, early failure of machine members, cracks from grinding or quenching, stress corrosion and season cracking, and similar occurrences. Upon occasion, residual stresses can be employed to an advantage, such as the use of shot-peening to increase the endurance limit of machine parts subjected to fatigue loading, sometimes by a factor of 5 or more.

No simple procedure can be described for determining the amount of residual stress which is likely to occur under certain conditions. Typical values of residual stresses for specified conditions usually are determined by experimental procedures employing electrical strain gages, X rays, or special magnetic or electrical devices.

3-34 Stress Concentration and Stress-concentration Factors. As mentioned in Art. 3-3, the magnitudes of stresses obtained by use of the direct stress and the combined stress equations previously given in this chapter usually are lower than the actual stresses that would exist in a real machine member of the same material and dimensions and under the same conditions of loading. This is mainly because the elementary equations do not take into account any stress raisers, such as threads, holes, keyways, splines, notches, grooves, and abrupt changes in cross sections. *Stress concentration* is defined as that condition which causes actual stresses in machine members to be higher than nominal values predicted by the elementary direct and combined stress equations. Some causes of stress concentration are shown in Fig. 3-13. The *stress-concen-*

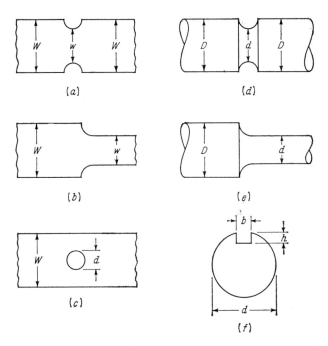

Fig. 3-13 A few of the many causes of stress concentration. (*a*) Notches in a plate of uniform thickness *t*. (*b*) Decrease in width of a plate of uniform thickness *t*. (*c*) Circular hole in a plate of uniform thickness *t*. (*d*) Groove in a shaft of diameter *D*. (*e*) Shaft with filleted decrease in diameter. (*f*) Shaft with rectangular keyway.

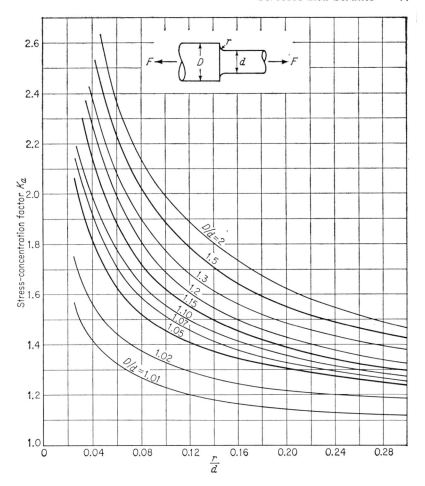

Fig. 3-14 Theoretically determined stress concentration factor K_a for simple tensile static loading of shaft with filleted decrease in diameter. (*From R. E. Peterson, "Stress Concentration Design Factors," 3d printing, John Wiley & Sons, Inc., New York, 1962.*)

tration factor K is defined as the ratio of the stress existing within a machine member if stress raisers are taken into account to the corresponding *nominal stress* resulting from direct or combined stress equations which do not take stress concentration into account.

The stress-concentration factor may be determined from theoretical considerations (usually by use of the Hans Neuber theory*) or from experimental investigations. Stress factors may take into account a single cause of stress concentration or may attempt to combine the effects of

* H. Neuber, "Kerbspannungslehre," Springer, Berlin, 1937.

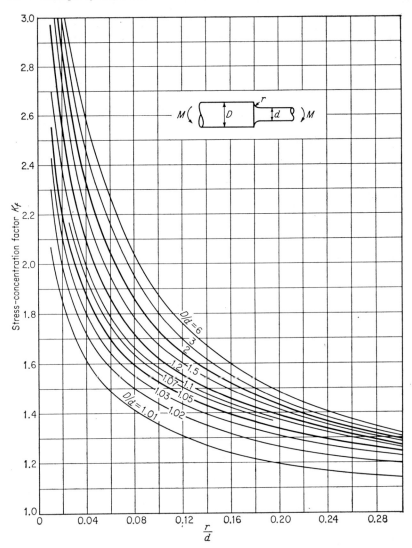

Fig. 3-15 Theoretical stress concentration factor K_f for simple flexural static loading of a shaft with filleted decrease in diameter. (*From R. E. Peterson, "Stress Concentration Design Factors," 3d printing, John Wiley & Sons, Inc., New York, 1962.*)

each of the individual causes of stress concentration which exist at any given point in a body.

3-35 When Stress-concentration Factors Should Be Applied. Experimental data indicate that stress-concentration factors should be applied to all brittle materials whether loaded statically or under fatigue conditions, and to ductile materials under fatigue loading conditions. For static loading of ductile materials, the stress-concentration factor usually may be taken as essentially 1 without making any appreciable error; however, to be on the safe side, some designers prefer to use stress-concentration factors greater than 1 even for static loading of ductile materials, thereby estimating higher stresses than those usually in the actual machine part. This policy may be due in part to the difficulty of differentiation between ductile and brittle materials and the resulting lack

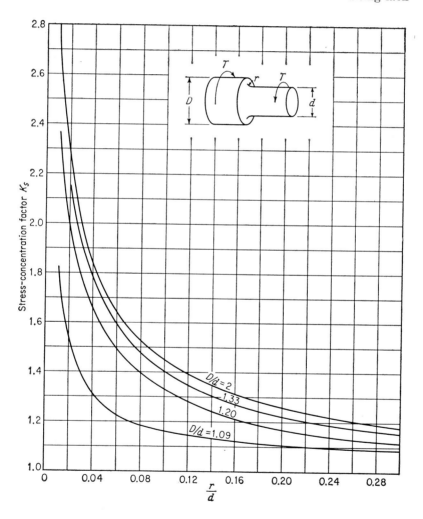

Fig. 3-16 Theoretically determined stress concentration factor K_s for simple torsional static loading of a shaft with filleted decrease in diameter. (*From R. E Peterson, "Stress Concentration Design Factors," 3d printing, John Wiley & Sons, Inc., New York, 1962.*)

of any sharp line of demarcation when a stress-concentration factor greater than 1 must be used.

3-36   Areas Used in Calculating Nominal Stresses.   The nominal stresses upon which stress-concentration factors are based are themselves based upon *nominal areas.* Usually, but not always, these nominal areas

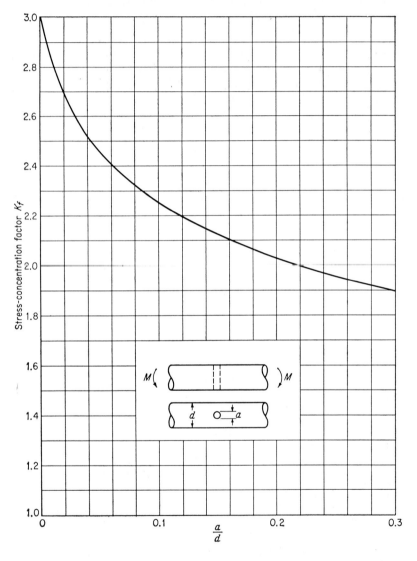

Fig. 3-17   Theoretically determined stress concentration factor K_f for bending of a shaft with a circular transverse hole. (*From R. E. Peterson, "Stress Concentration Design Factors,"* 3d printing, John Wiley & Sons, Inc., New York, 1962.)

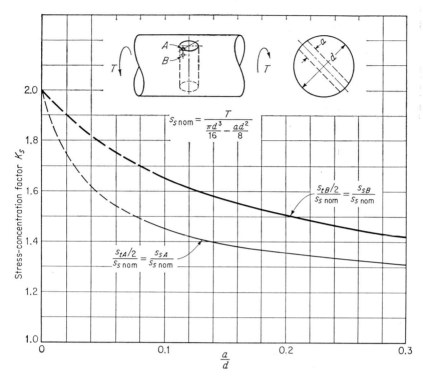

Fig. 3-18 Experimentally determined stress concentration factor K_s for simple torsional static loading of a circular shaft with circular transverse hole. (*From R. E. Peterson, "Stress Concentration Design Factors," 3d printing, John Wiley & Sons, Inc., New York, 1962.*)

are the net cross-sectional areas remaining after the removal of material for a groove or hole, or the smaller of two cross-sectional areas in the case of a machine member with a reduced section. One major exception to this rule is that stress-concentration factors for keyways usually are calculated with a nominal area equal to the gross cross-sectional area of the shaft without the keyway.

3-37 Theoretical Stress-concentration Factors and the von Mises Criterion. Theoretical stress-concentration factors K_a (for axial loading), K_f (for flexural loading), and K_s (for shear loading) usually are based upon the Neuber theory.* Much experimental data indicate that these theoretical factors usually are slightly on the high side. Article 3-31 indicates that the maximum distortion energy theory is in rather good agreement with experimental findings. R. von Mises† was one of the major

* *Ibid.*

† R. von Mises, Mechanik der fester Körper im plastisch deformablen Zustand, *Nachr. Ges. Wiss. Göttinger Jahresber, Geschaftsjahr. Math-phys. Kl.*, p. 582, 1913.

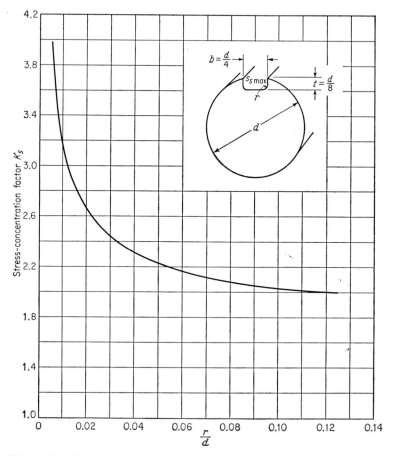

Fig. 3-19 Theoretically determined stress concentration factor K_s for simple torsional static loading of a circular shaft with filleted rectangular keyway. (*From R. E. Peterson, "Stress Concentration Design Factors," 3d printing, John Wiley & Sons, Inc., New York, 1962.*)

investigators suggesting the validity of this theory. By use of this theory, the following stress-concentration factors based upon both the Neuber theory and the von Mises maximum distortion energy criterion are obtained:

$$K_{am} = mK_a \tag{3-81}$$
$$K_{fm} = mK_f \tag{3-82}$$
$$K_{sm} = mK_s \tag{3-83}$$

in which m is a factor taking into account the von Mises criterion of maximum distortion energy failure. Values of m vary between 0.85 minimum and 1. Experimentally determined stress-concentration factors for static or dynamic loading are more nearly in agreement with the K_{am}, K_{fm}, and K_{sm} values given by Eqs. (3-81) through (3-83) than the theoretical K_a,

K_f, and K_s factors from Neuber. Since the difference between these two sets of factors never exceeds 15 per cent and usually is much less, many designers assume a value of 1 or slightly less for m rather than attempt to determine it more correctly. Only theoretical factors will be used in this book.

Typical values of the more commonly used theoretical stress-concentration factors are indicated in Figs. 3-14 through 3-20 from work

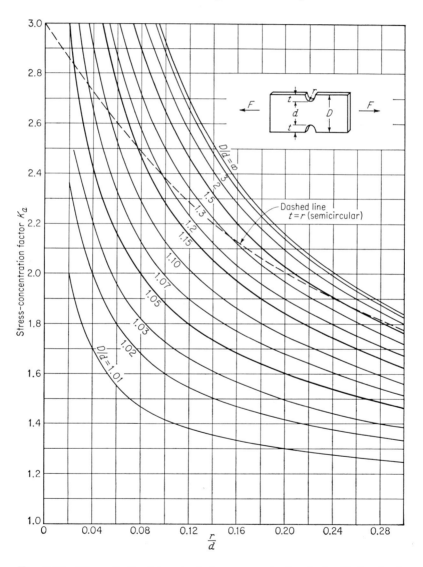

Fig. 3-20 Theoretically determined stress concentration factor K_a for simple tensile static loading of a notched flat bar. (*From R. E. Peterson, "Stress Concentration Design Factors," 3d printing, John Wiley & Sons, Inc., New York, 1962.*)

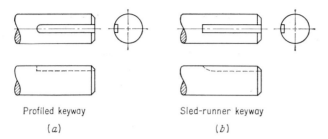

Profiled keyway   Sled-runner keyway

(a)   (b)

Fig. 3-21 Types of standard rectangular keyways. (a) Profiled keyway. (b) Sled-runner keyway.

done by R. E. Peterson* or from information compiled by him. Each of these factors is based upon simple static tensile, bending, or shear loading of the members involved.

In addition the following information concerning gears, keyways, press and shrink fits, and threaded connections should be of use to the designer. K_f values for $14\frac{1}{2}$-deg involute gears range from approximately 1.5 to 2.6, with many designers using 1.6 as an average value for the more common-sized gears. For 20-deg stub-tooth involute gears, the common range is from 1.4 to 2.5, with perhaps 1.5 being the most commonly used value.

Two major types of keyways in shafts are depicted in Fig. 3-21. Actual stress-concentration factors for the sled-runner (Woodruff) keyway shown range from 1.3 to 1.6; 1.5 may be considered as a typically used value. For profiled keyways actual stress-concentration factors vary from approximately 1.6 to 2.0, with 1.8 as the average value commonly employed. Additional information on stress concentration at keyways is given in Table 3-2.

For press and shrink fits, the factors vary from approximately 1.4 to 4.0. It is difficult to give any average values since they vary widely with the degree of the fit and whether or not a shoulder is used. Possible average values may be 1.5 for press fits without any shoulder and 2.3 for press fits in conjunction with a shoulder.

For the threads of threaded connections with the gross area of the threaded shank taken as the nominal area, the K values vary from approximately 1.7 to 4.0. Possible average values for unified and American standard threads are 2.9 for medium-carbon steels and 3.9 for heat-treated nickel steels. Average K values for Whitworth threads may be 1.8 for medium-carbon steels and 3.3 for heat-treated nickel steels.

3-38 Notch Sensitivity. Under conditions of fatigue loading (non-static loading), the stress-concentration effects of keyways, fillets, notches, holes, and similar stress raisers usually are less than those applicable for

* R. E. Peterson, "Stress Concentration Design Factors," 3d printing, John Wiley & Sons, Inc., New York, 1962.

Table 3-2 **Typical actual stress-concentration factors at keyways in shafts***

Condition of material of the shaft	Actual stress-concentration factor in the shaft			
	Profiled keyway		Sled runner (Woodruff) keyway	
	Torsion K_s	Bending K_b	Torsion K_s	Bending K_b
Annealed..............	1.3	1.6	1.3	1.3
Hardened............	1.6	2.0	1.6	1.6

* C. Lipson, G. C. Noll, and L. S. Clock, "Stress and Strength of Manufactured Parts," pp. 117 and 164, McGraw-Hill Book Company, New York, 1950.

static loading conditions. This effect, known as *notch sensitivity*, varies widely with the geometry of the stress raiser and with the material involved. A *notch sensitivity factor q* is defined by Eq. (3-84); this dimensionless factor, between zero and 1, is an indication of the degree of theoretical stress concentration obtained under fatigue conditions. When q is zero, no notch sensitivity effect exists; when q is 1, the full effect of the theoretical stress-concentration factor (as determined from static loading considerations) is applicable.

$$q = \frac{(s_e/s_{en}) - 1}{K_t - 1} \tag{3-84}$$

where K_t is the theoretical stress-concentration factor without taking notch sensitivity into account, s_e is the endurance limit of the material without any stress-concentration notch in the specimen, and s_{en} is the

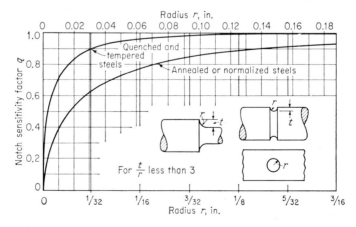

Fig. 3-22 Typical notch sensitivity curves for steel. (*From R. E. Peterson, "Stress Concentration Design Factors," 3d printing, John Wiley & Sons, Inc., New York, 1962.*)

endurance limit of the material with a stress-concentration notch in the specimen.

Unfortunately, at this time there is a marked absence of available data concerning notch sensitivity factors. The data which are available indicate that regardless of whether the loading is axial, flexural, or torsional the notch sensitivity factors appear to be sufficiently close together to be considered as being the same for any specified material and for any given geometric configuration descriptive of the stress raiser. Typically average values of q for both annealed steels and heat-treated steels appear in Fig. 3-22. The following results:

$$K_{actual} = q(K - 1) + 1 \tag{3-85}$$

When the designer does not have access to q data, or when he does not know the conditions of possible loading to a reasonable degree of certainty, it is safe to assume a value of 1 for q for all conditions of loading, both static and fatigue. For the fatigue cases, the resulting stress-concentration factors will be higher than those correctly determined values based upon realistic notch sensitivity factors less than 1.

3-39 Overall Combined Stress-concentration Factor K.   All of the various stress-concentration factors previously mentioned are individual stress-concentration factors, since they result from the consideration of only one cause of stress concentration and are due to only one type of external loading. Frequently in machine parts, two or more individual causes of stress concentration occur at the same point. In addition, two or more different types of external loading may be superposed at the same location in a member. As yet there is no theoretical procedure which adequately predicts the overall combined effect of several coincident individual stress raisers and/or external loadings. Unfortunately, pertinent experimental data are almost nonexistent. These data do indicate, however, that in general the overall combined stress-concentration factor K taking into account the combined effect of the individual stress-concentration factors and/or multiple external loadings is not so large as either the sum or the product of the individual factors. Some experimental findings indicate that as a rough rule of thumb the overall K is approximated by the addition of the highest magnitude individual factor and from one-half to three-fourths of the magnitude of the second highest individual factor, with the other individual factors of smaller magnitude apparently having very little effect upon K. This process gives results that frequently are close to experimentally determined values; in other cases, it may result in a 50 per cent or higher error. Other experimental findings indicate that where the effects of two stress raisers are combined, the overall factor is approximated by 0.8 times the product of the two individual factors.

4

Design Stresses

4-1 The Machine and Its Members. Previous chapters have dealt with the broad aspects of design, some of the materials used and their properties, and the laws and theories of the stresses and strains resulting from various types of loads imposed upon machine members. This chapter will deal with the overall problem of using this information and other considerations in designing the machine member to be used. Consideration must be given to the following:

1. The specific function of the machine.
2. The type of machine indicated, whether it has to be a new device or a rearrangement of known elements to perform the desired result. Particular attention must be given to the selection and arrangement of the "hardware"—the shafts, bearings, gears, nuts, bolts, pulleys, keys, belts, chains, clutches, brakes, retainers, packings, gaskets, cylinders, plates, linkages, frames, etc.
3. The loads, if present, known or unknown, imposed upon the various machine members.
4. The selection of the materials to be used in the various members.
5. The stresses, strains, and effects resulting in each chosen part.

6. The determination of the size of the machine members. This size may be based upon stress analysis, judgment, appearance, the prior decision that the member "must be that size" in order to perform its function, the space it is to occupy, weight, cost, repair, maintenance, assembly, fabrication, operating environment, and operation.

7. Redesign may result from faulty operation, production difficulties, deficient sales appeal, cost, lack of profit, insufficient reliability, and the need for safety features.

The determination of many factors confronting a designer will require ingenuity or just plain "horse sense." Mathematical, graphical, statistical, and testing techniques are available tools of the designer, but thinking, judgment, and experience often are more useful. *The designer should have an excellent mathematical and basic physical science background, as well as creative ability.*

Automation is becoming more important in all phases of life; it should be remembered that *simplicity* has not lost its appeal but is often difficult to obtain. The more controls, gadgets, devices, and gimmicks that are used, the greater is the likelihood of failure, increased maintenance, and lower reliability.

The advent of missile and space technology has emphasized the importance of *reliability*. Electronics was the first field to give real consideration to reliability, but the design of mechanical components and systems must also take cognizance of it. Reliability may be defined as the probability that a device will perform, without failure, a specified function under given conditions for a specified period of time. The mechanical designer must concern himself with this concept. Military specifications and bid requests frequently include reliability requirements. In the future, industry as a whole will require that reliability be designed into the product.

Safety is another factor which the designer always keeps uppermost in his mind. Will the machine and each of its members perform its function without failure? What will be the effect of failure? Will it fail safe? Does the failure result from faulty design, lack of maintenance, or tampering? The designer must consider the consequence of tampering, lack of adjustment, improper lubrication, and injurious environment. Self-contained sealed units or assemblies are being used more and more in all machines and particularly in consumer products. The size and location of access openings for repairs and maintenance must be considered. If a machine member fails, the complete failure of the entire machine may result, or only one or more members may lose their usefulness. The use of devices to cause a machine to fail safe, such as a throw-out pin operated by centrifugal force to stop a turbine whose governor fails to function properly, has been standard practice for some time. What happens when the power fails with the use of power brakes? The hazardous effect of a failure should be considered by a designer, and if at all possible, means should be provided to prevent a total failure which might result in loss of life or property.

The determination of the stresses in a machine member is not an exact science. The safety built in the member is often questionable. The remaining discussion in this chapter will be a consideration of the design stresses.

4-2 Stress-strain Considerations. A designer in determining the stress-strain effects in a machine member considers the following factors:

1. The load
2. The material
3. The relationship of the factors affecting stress-strain
4. The deformation or strain
5. The design stress or working stress
6. The shape or size

The load may be static (dead load) or dynamic (live load). The load may be suddenly applied (impact or shock) or gradually applied. The magnitude of the load may fluctuate from a minimum (which may be zero) to a maximum value, in one direction or in more than one direction. This fluctuation may be in equal or varying cycles. These cycles may or may not follow a definite pattern. A dynamic load may be superimposed upon a static load. These loads will produce strains and stresses in the machine member. Residual stresses caused by fabrication or assembly may be present to supplement the stresses produced by the loads. In some cases it is difficult to determine the exact static load. Dynamic load determinations are more difficult to evaluate. The point of application, the direction of action, and the magnitude of the forces resulting from the loading cause the designer a great deal of concern. In some cases only an estimation is possible or necessary.

The material to be used in a particular case will not only depend upon cost, availability, weight, fabrication possibilities, appearance, wear, resistance to environmental effects with possible protection, and lubrication qualities but upon load-carrying capacity with the accompanying deformation and stress. The load-carrying capacity of a material is indicated by the ultimate strength, the yield stress, the elongation, reduction in area, hardness, modulus of elasticity, and endurance limit. The effect of temperature upon these properties must be considered. Shock and impact loading may have an effect. Ductility and brittleness are important considerations. Is the material loaded within the elastic limit? Is the usefulness of the member impaired by stresses beyond the elastic limit but not exceeding the ultimate strength? What effect does manufacture have upon the properties? Does size and/or shape cause the carrying capacity of the material to be different from the value obtained by standard tests? Are the properties in compression and shear as well known as in tension? How definite are the bearing properties? Is the material homogeneous? From the above discussion and questions raised it should be apparent that the selection of the material to be used usually is not an easy matter. The designer should be alert to the increasing supply of special-purpose materials and the improvement in old materials and fabrication methods.

The relationships of the factors affecting stress-strain are not definite and are very often inaccurate. These relationships commonly designated as equations or formulas are obtained from theories of failure, experimentation, practice, and codes or standards. Many are empirical. In the derivation of these formulas, assumptions are made, factors are assumed to be constant, and conditions are idealized. The conditions in the designer's actual problem may only approximate those assumed in obtaining the formula. It behooves the designer to understand, and take into account, the parameters and limits of accuracy of the formulas. The final proof of the formula is whether or not the machine element designed by the formula performs the function for which it was designed. Experimental techniques are useful in determining the accuracy of the relationship of the factors used in obtaining the size of machine members.

4-3   Design and Working Stresses.   The selection of the design stress to be used in computing the size of a machine member is one of the most important problems to be met by the designer. At the same time, it is one of the most vague and difficult problems. The experienced designer arrives at the design stress by an analysis of the service conditions that must be met, by an evaluation of the strength of the material under these conditions, and by deciding on the margin of safety between service conditions and actual failure.

Service conditions include such factors as the magnitude of the loads applied; the method of load application, whether steady, variable, or subject to impact; the type of stress, whether tensile, compressive, shear, or combined; the environment; and the temperature at which the member is to operate. Other factors such as appearance, the use of standard parts, ease of assembly, ease of repair, and the wearing qualities must also be considered; and in many cases these may be so important that the problem of strength may be given only minor consideration.

Tensile and fatigue tests are the ones from which failure can be most readily predicted. Other tests provide information that can be used to supplement these. Failure may mean actual rupture, a sudden increase in the strain without an increase in stress, a stress producing a permanent strain, or a stress exceeding the proportional limit.

For ductile materials, the yield stress is usually considered to be the criterion of failure. When there is no well-defined yield point, the stress corresponding to a permanent elongation of 0.2 per cent is accepted as the yield stress. For brittle materials, the breaking stress (ultimate strength) is the criterion of failure. Ductile materials subjected to fluctuating stresses behave like brittle materials so far as failure is concerned, and the endurance limit is the criterion of failure.

Lack of stiffness or rigidity may be considered to constitute failure, in which case stress may be relatively unimportant. For example, if the deformation of a steel member is to be limited, it may not be advisable to use the high-strength alloy steels, since the modulus of elasticity, which is the measure of rigidity, is practically the same for all steels. In order to use advantageously the high strength of the alloy steels, higher unit stresses must be used, and the deformation will increase in direct proportion to the stress. If, when the member is proportioned to give the permissible deformation, the stress is within the limits permitted with the lower-strength steels, then the plain carbon steels should be used unless alloy steels are required by other considerations.

The *design stress* is the stress value which is used in the mathematical determination of the required size of the machine member. It may be considered to be the stress that the designer hopes will not be exceeded under operating conditions. When the properties of the material are definitely known and when the stress can be accurately determined, this stress may be as high as 80 per cent of the yield stress, but 50 per cent is the usual value for nonshock and nonfatigue conditions. The *working stress*, as distinguished from the design stress, is the stress actually occurring under operating conditions.

The stress equations developed in texts on mechanics and strengths of materials are directly applicable only when the forces on the machine member can be accurately determined and analyzed. In service, however, many machine members are subjected to load combinations that do not permit accurate stress determinations, and more or less satisfactory approximations must be used. The designer bases his computations on the principal or predominating stress and makes allowances for minor stresses by an intelligent modification of the permissible or design stress.

When the stress varies across the section, the maximum value should be the one controlling the design, except in cases of high local stress in ductile material under static load conditions. The high local stress, if reaching the elastic limit, will cause local plastic flow of the material at the section with a consequent readjustment of the stresses without yielding or failure of the entire member. Because of this readjustment, high local stresses often are neglected in the design of members of ductile materials. Note that this statement does not apply to shock- or fatigue-loading conditions. Under these conditions, the existence of high local stresses requires serious consideration and careful design.

When machines were operated at low speeds, their design required only a knowledge of simple statics and kinematics. Designers developed a highly perfected judgment of static stresses, and the use of an apparent factor of safety to cover the ignorance of the real stresses was not serious. The increasing use of high speeds and the increasing size of individual machines have introduced dynamic problems which cannot be handled in such a simple manner, and the determination of the design stress is becoming more dependent on critical analysis of the type of loading and the stress distribution.

The usual method of determining the design stress to be used in computations is to divide the ultimate strength of the material by a factor of safety:

$$s_d = \frac{s_u}{FS} \quad \text{or} \quad F_d = \frac{F_u}{FS} \tag{4-1}$$

where s_d = design stress, psi
$\quad s_u$ = ultimate strength, psi
$\quad FS$ = factor of safety
$\quad F_u$ = ultimate load, lb
$\quad F_d$ = working load, lb

In chain and wire rope the factor of safety is based upon the ultimate load. For columns the factor of safety is based upon the critical load which is often assumed to be the load on the column causing a stress equivalent to the yield stress.

Specific procedures for obtaining the design stress are used for certain machine members such as gears, rolling contact bearings, chains, tubes, cylinders, and other parts. These will be presented when the particular element is discussed.

A factor of safety ordinarily is not used in determining the allowable deformation. The deformation or strain is held within certain limits.

The term factor of safety is misleading and causes much confusion. The legal and most code definitions are as given in Eq. (4-1). Some design and materials texts base the factor of safety on the yield stress. Others use such terms as design factor, apparent factor of safety, margin of safety, and real factor of safety. The term factor of ignorance has been suggested. In the succeeding pages a procedure will be presented for obtaining the design stress by the use of the factors which a designer must consider in order to obtain a more rational design stress. The *design factor df,* which is not the factor of safety as defined in Eq. (4-1), is used to take into account unknowns, contingencies, and emergencies. It is a factor which the designer uses to allow for unreliable material, such as dirty or seamed steel, castings subject to blowholes, and shrinkage stresses; the probable effects of unknown, unforeseen, or accidental overloads; temperature effects; the probable accuracy of the mathematical analysis; stresses

encountered during machining, assembly, shipping, or intentional overloading during acceptance tests; and the chances of failure leading to
injury, loss of life, costly shutdowns, and expensive repairs. Average
values are 1.2 to 2 for ductile materials of uniform structure, and 2 to 3 for
brittle materials.

The working stress in a machine member might be increased by
suddenly applying the load, with resulting shock. Account must also be
taken of such operating conditions as whether it will be in intermittent or
continuous service, lubrication, and protection it will have from injurious
agents. In determining the design stress, the *service factor a* will be used to
make allowance for the above condition. As an example of an impact
stress, the maximum stress produced by a falling body on a machine
member is

$$s_i = \frac{F}{A}\left[1 + \left(1 + \frac{2h}{y_{st}}\right)^{1/2}\right] = s_{st}\left[1 + \left(1 + \frac{2h}{y_{st}}\right)^{1/2}\right] = s_{st}a \qquad (4\text{-}2)$$

where F = applied load, lb
A = cross-sectional area, in.^{2}
a = service factor
y_{st} = deformation produced by load F under static conditions, in.
h = height of free fall to produce the velocity of impact, in.
s_i = maximum impact stress produced, psi
s_{st} = static stress F/A, psi

Since s_{st} is the static stress

$$a = 1 + \left(1 + \frac{2h}{y_{st}}\right)^{1/2} \qquad (4\text{-}3)$$

For torsional loads h and y_{st} must be expressed as the impact travel and
angular deformation in radians.

When the load is steady or gradually applied, the factor a will be
unity. When the load is applied suddenly without impact or initial
velocity, h becomes zero and the factor a will be 2, indicating that suddenly applied loads are twice as severe as the same loads under static
conditions. Equation (4-2), however, is based on the assumption that the
supporting members are absolutely rigid, which is never true in an operating machine. Since the supporting members deform and absorb some
of the shock energy, the stresses are less than indicated by the equation.
The actual magnitude of impact stresses can be accurately determined only in very simple cases such as longitudinal impact on rounded-
end bars, impact of elastic spheres, and similar arrangements. It is therefore necessary to rely on experimental and service data in modifying the
value of the factor a as computed. When impact causes a series of oscillations about a mean value that represents the mean stress, the factor a
may be estimated by comparing the maximum oscillation with the mean

value or static strain. Wheel loads moving on tracks are examples of this type of loading. Average values of the factor a are 1.2 to 1.5 for light shock; 1.5 to 2 for moderate shock; 2 to 3 for very heavy shock; and as high as 5 in a few extreme cases.

Since the load-carrying capacity of a material may be the yield stress or the endurance limit as well as the ultimate stress, the procedure for determining the design stress will be presented separately for the static and dynamic loads.

4-4 Static Loading. In static loading the ultimate stress or the yield stress is considered as the criterion of the load-carrying capacity of a machine member. If the member's usefulness is impaired by causing a permanent deformation exceeding that at the proportional limit, the yield stress should be used. When the yielding of the member beyond the proportional limit causes a permanent deformation which has no effect upon the position of the member, the ultimate stress may be used. The tightening of a nut on a bolt may produce an acceptable stress beyond the yield stress without failure.

$$s_d = \frac{s}{aK(df)} \tag{4-4}$$

where s_d = design stress, psi

s = stress at which the material is considered to fail, psi

= $s_{tu}, s_{cu}, s_{su}, s_{ty}, s_{cy}, s_{sy}$, or s_b (as defined in Chaps. 2 and 3), psi depending upon whether the failing stress is in tension, compression, shear, or bearing and upon whether a failing condition exists when the material is stressed to the yield point or ultimate strength

a = service factor as discussed in Art. 4-3

K = 1.0 for ductile materials, or the overall combined stress concentration factor for brittle materials

df = design factor as defined in Art. 4-3

4-5 Fatigue Loading. The criterion of the load-carrying capacity of a machine element under fatigue loading is the endurance limit:

$$s_d = \frac{s_e}{aK(df)} \tag{4-5}$$

where s_d = design stress, psi

s_e = actual endurance limit of the material for the desired number of load cycles under condition of flexure, shear, or axial loading considering the effects of surface finish and environment, psi (see Art. 2-11)

a = service factor as discussed in Art. 4-3

K = overall combined stress concentration factor used for both brittle and ductile materials (see Arts. 3-34 to 3-39)

df = design factor as defined in Art. 4-3

Normally the value of s_e used in this equation should not be higher than the yield stress of the material under static load conditions. There may be other factors to be considered in determining the design stress for fatigue loading.

4-6 Allowance for Corrosion and Wear. * Some materials deteriorate with age, others are subject to corrosion by gases and liquids, and others are subject to wear and abrasion under operating conditions. In the case of castings, the shifting of cores or poor molding procedure may result in variations in the thickness of vital parts of the casting. The designer must recognize these conditions and allow for them in size computations. The ordinary factor of safety or the selected design stress does not provide for those items, and an additional amount of metal must be provided, over and above the computed size of the machine member. The allowance varies from $\frac{1}{8}$ in. on small castings, to $\frac{3}{8}$ in. on large castings. Practical experience is a very useful guide to the designer in correctly taking these items into account.

* E. Rabinowicz, Predict Wear of Metal Parts, *Prod. Eng.*, vol. 29, no. 25, 1958. D. J. Myatt, Wear Life of Rolling Surfaces, *Prod. Eng.*, vol. 31, no. 19, 1960.

5

Keys, Cotters, Retainers, and Fasteners

5-1 Keys. Pulleys, gears, sprockets, levers, couplings, and similar devices are employed to transmit torque to or from shafts and usually are rigidly attached to the shaft by shrink fits, setscrews, keys, splines, or cotters. Shrink fits are suitable for permanent assemblies, setscrews for light service, and cotters for axial loads. When the parts must be disassembled, keys or splines generally are used. A *key* is a machine member employed at the interface of a pair of mating male and female circular cross-sectioned members to prevent relative angular motion between these mating members. The key fits into mating grooves in the shaft and mating member called the *keyway* and transmits torque by shear across the key. The cutting of the keyway into the shaft reduces its strength and rigidity by an amount depending upon the shape and size of the keyway. This is discussed in Art. 8-3.

5-2 Types of Keys. Keys may be classified according to whether they are constant or variable in cross section. The *constant cross-sectioned keys* are square, flat, round, or Barth keys as illustrated in Fig. 5-1. The *square key*, with the key sunk half in the shaft and half in the hub, is the

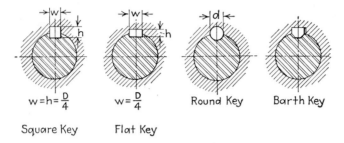

Fig. 5-1 Constant cross-sectioned keys.

type most commonly used. *Flat keys* are used where the weakening of the shaft by the keyway is serious. The key is sunk half in each of the two mating members. Although there is no universal standard, the width of square and flat keys usually is approximately one-fourth the shaft diameter. Keyways for *round keys* may be drilled and reamed after assembly of the mating parts. Small round keys are used for fastening cranks, handwheels, and other parts that do not transmit heavy torques. A few manufacturers employ round keys for heavy-duty shafts over 6 in. in diameter because the absence of the sharp corners reduces the stress concentration below that which would exist had a square or flat key been used. The *Barth key* is a square key with bottom two corners beveled. This double beveling ensures that the key will fit tightly against the top of the keyway when the drive is in either direction, and lessens the tendency to twist. This key does not require a tight fit, and the small clearance permits easy assembly and removal.

Variable cross-sectioned keys include the Woodruff, round taper, gib-head taper, and the saddle keys. The *Woodruff key* consists of one-half of a circular disk fitting into a rectangular keyway in the female member and a semicircular keyway in the male member. It has the advantage that it weakens the shaft less than does a square or flat key of equal torque capacity. In addition, Woodruff keys can act as blind keys; that is, the female member can completely hide the key and the keyway in the male member. *Round tapered keys* usually have a taper in diameter of $\frac{1}{8}$ in. per

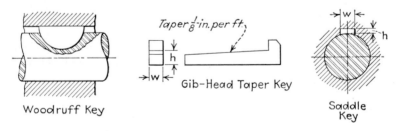

Fig. 5-2 Variable cross-sectioned keys.

foot, and a nominal diameter of one-fourth the shaft diameter for shafts under 6 in. in diameter, or a nominal diameter of one-fifth the shaft diameter for larger shafts. Often they are used with constant-diameter round keyways. Because of their taper, they can easily be driven into the keyway until they are very tight. *Gib-head taper keys* are flat keys with varying height and with a special gib-head to facilitate easy driving and removal of the key. Usually they are used with constant dimension flat keyways and have a taper in height of $\frac{1}{8}$ in. per foot. *Saddle keys* are flat keys used without a keyway in the shaft. Since the torque is transmitted by friction on the shaft rather than by shear of the key, the top or outside of the key usually is slightly tapered in height to ensure a large radial pressure onto the shaft. They are employed for light work or in cases where relative motion between the shaft and its mating hub is required for adjustment and a keyway cannot be cut into both. They also are employed to hold parts together during assembly until the permanent keyway can be located and machined. Figure 5-2 depicts Woodruff, gib-head tapered, and saddle keys.

A *feather key* is one which has a tight fit into one member and a loose sliding fit in the other mating member. It is used when there must be relative axial motion between the shaft and the mating hub. The bearing loading on feather keys should not exceed 1,000 psi; if the members are to slide under load, the bearing loading should be under 1,000 psi.

Dimensions of ASME standard keys and keyways are given in Table 5-1.

5-3 Number of Keys Used. Usually only one key is employed between the two mating members. For heavy duty shafts or those whose direction of motion is subject to reversal, two keys angularly spaced 90 deg apart or three keys 120 deg apart are used. Experience has shown that when two keys are used, the 90-deg spacing is preferable to a 180-deg spacing. When multiple keys are employed, the keys are usually square or flat and often tapered. Three keys are employed for maximum torque capacity or where concentricity of the mating members is of great importance. Splines are employed when it is desired to obtain a torque capacity greater than that available by the use of three keys.

5-4 Splines. Splines are permanent keys made integral with the shaft and fitting into keyways broached into the mating hub. They are used where maximum torque capacity is desired or where axial motion of the mating parts under load is expected. The contour of the sides of the splines may be either straight or involute, both of which are standardized.

5-5 Stresses in Keys and Splines. When a single square or flat keyway is cut in both the shaft and the hub, force is transmitted by com-

Table 5-1   **Dimensions of ASME standard keys and keyways**

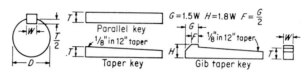

$G = 1.5W$ $H = 1.8W$ $F = \dfrac{G}{2}$

Diam. of shaft D, in.	Width of key W, in.	Thickness of key T, in. Square	Flat	Depth of keyway $T/2$, in. Square	Flat	Height of gib H, in.	Length of gib G, in.	Length of flat F, in.	Tolerance on stock keys (—), in.	Size of set-screw to be used with parallel key, in.
$\frac{1}{2}$ to $\frac{9}{16}$	$\frac{1}{8}$	$\frac{1}{8}$	$\frac{3}{32}$	$\frac{1}{16}$	$\frac{3}{64}$	$\frac{1}{4}$	$\frac{1}{4}$	$\frac{1}{8}$	0.0020	$\frac{1}{4}$
$\frac{5}{8}$ to $\frac{7}{8}$	$\frac{3}{16}$	$\frac{3}{16}$	$\frac{1}{8}$	$\frac{3}{32}$	$\frac{1}{16}$	$\frac{3}{8}$	$\frac{5}{16}$	$\frac{3}{16}$	0.0020	$\frac{1}{4}$
$\frac{15}{16}$ to $1\frac{1}{4}$	$\frac{1}{4}$	$\frac{1}{4}$	$\frac{3}{16}$	$\frac{1}{8}$	$\frac{3}{32}$	$\frac{7}{16}$	$\frac{3}{8}$	$\frac{3}{16}$	0.0020	$\frac{3}{8}$
$1\frac{5}{16}$ to $1\frac{3}{4}$	$\frac{3}{8}$	$\frac{3}{8}$	$\frac{1}{4}$	$\frac{3}{16}$	$\frac{1}{8}$	$\frac{11}{16}$	$\frac{11}{16}$	$\frac{3}{8}$	0.0020	$\frac{3}{8}$
$1\frac{13}{16}$ to $2\frac{1}{4}$	$\frac{1}{2}$	$\frac{1}{2}$	$\frac{3}{8}$	$\frac{1}{4}$	$\frac{3}{16}$	$\frac{7}{8}$	$\frac{3}{4}$	$\frac{3}{8}$	0.0025	$\frac{1}{2}$
$2\frac{5}{16}$ to $2\frac{3}{4}$	$\frac{5}{8}$	$\frac{5}{8}$	$\frac{7}{16}$	$\frac{5}{16}$	$\frac{7}{32}$	$1\frac{1}{8}$	$\frac{15}{16}$	$\frac{1}{2}$	0.0025	$\frac{5}{8}$
$2\frac{7}{8}$ to $3\frac{1}{4}$	$\frac{3}{4}$	$\frac{3}{4}$	$\frac{1}{2}$	$\frac{3}{8}$	$\frac{1}{4}$	$1\frac{3}{8}$	$1\frac{1}{8}$	$\frac{9}{16}$	0.0025	$\frac{5}{8}$
$3\frac{3}{8}$ to $3\frac{3}{4}$	$\frac{7}{8}$	$\frac{7}{8}$	$\frac{5}{8}$	$\frac{7}{16}$	$\frac{5}{16}$	$1\frac{1}{2}$	$1\frac{1}{4}$	$\frac{5}{8}$	0.0030	$\frac{3}{4}$
$3\frac{7}{8}$ to $4\frac{1}{2}$	1	1	$\frac{3}{4}$	$\frac{1}{2}$	$\frac{3}{8}$	$1\frac{7}{8}$	$1\frac{1}{2}$	$\frac{3}{4}$	0.0030	$\frac{3}{4}$
$4\frac{3}{4}$ to $5\frac{1}{4}$	$1\frac{1}{4}$	$1\frac{1}{4}$	$\frac{7}{8}$	$\frac{5}{8}$	$\frac{7}{16}$	$2\frac{1}{4}$	$1\frac{7}{8}$	$\frac{15}{16}$	0.0030	$\frac{7}{8}$
$5\frac{3}{4}$ to $7\frac{3}{4}$	$1\frac{1}{2}$	$1\frac{1}{2}$	1	$\frac{3}{4}$	$\frac{1}{2}$	$2\frac{5}{8}$	$2\frac{1}{4}$	$1\frac{1}{8}$	0.0030	1
$7\frac{1}{2}$ to $9\frac{7}{8}$	$1\frac{3}{4}$	$1\frac{3}{4}$	$\cdots$	$\frac{7}{8}$	$\cdots$	$3\frac{1}{8}$	$2\frac{5}{8}$	$1\frac{5}{16}$	0.0030	$1\frac{1}{4}$
10 to $12\frac{1}{2}$	2	2	$\cdots$	1	$\cdots$	$3\frac{5}{8}$	3	$1\frac{1}{2}$	0.0030	$1\frac{1}{4}$

pression on the bearing surfaces ab and de as shown in Fig. 5-3. These compression forces act as a couple tending to roll the key, and, if the key is fitted on all four sides, induce a resisting couple acting on the surfaces cd and af as indicated by the forces marked F'. The crushing or bearing force is found approximately by considering the force F to act at the circumference of the shaft.

Let T = torque transmitted, in.-lb
  D = shaft diameter, in.
  L = length of the key, in.
  w = width of the key, in.
  h = depth of the key, in.

Then

$$F = \frac{2T}{D}$$

(5-1)

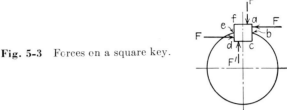

Fig. 5-3   Forces on a square key.

The crushing stress on the surfaces *ab* and *de* is

$$s_c = \frac{2F}{Lh} = \frac{4T}{DL\bar{h}} \tag{5-2}$$

and the shearing stress on the area *eb* is

$$s_s = \frac{F}{Lw} = \frac{2T}{DLw} \tag{5-3}$$

The ideal key should be equally strong in crushing and in shear; this condition is satisfied if Eqs. (5-2) and (5-3) are solved simultaneously for the torsional moments and equated. Hence

$$\frac{DLhs_c}{4} = \frac{DLws_s}{2} \tag{5-4}$$

from which

$$\frac{h}{w} = \frac{2s_s}{s_c} \tag{5-5}$$

When the key is fitted on all four sides, the permissible crushing stress s_c for the usual key materials is at least twice the permissible stress in shear. Assuming s_c equal to $2s_s$, the equation indicates that a square key is the proper shape. In this case, it is necessary to check the key for shear strength only. When the key is not fitted on all four sides, experience has shown that the permissible crushing stress is about 1.7 times the permissible shear stress, and the key must be checked for crushing. When the key is made of the same material as the shaft, the length of key required to transmit the full power capacity of the shaft is determined by equating the shear strength of the key to the torsional shear strength of the shaft. Hence

$$\frac{2T}{DLw} = \frac{Tc}{J} \times \frac{1}{0.75} = \frac{16T}{\pi D^3} \times \frac{1}{0.75} \tag{5-6}$$

and if $w = \dfrac{D}{4}$

$$L = 1.18D \qquad \text{or approximately} \qquad L = 1.2D \qquad (5\text{-}7)$$

In this equation, the value 0.75 is the estimated weakening effect of the keyway on the torsional strength of the shaft.

Where multiple keys or splines are used, it is almost impossible to make each key or spline carry its proportional share of the total torque, hence some keys or splines may be more heavily loaded than others. When two keys are employed, some designers size keys on the basis that each key can carry 20 per cent more than its proportional one-half of the total torque; for three keys, a 15 per cent overload often is used. When torque is transmitted by splines, perhaps a 10 per cent overload factor is applicable.

5-6   Setscrews.   *Setscrews* are threaded members used to prevent angular rotation between a gear, pulley, sheave, collar, or similar machine element and the shaft on which it is mounted. They also are used to prevent axial movement between the mating members. Upon occasion, a setscrew may fit tightly into one member and loosely into a groove in the mating member, with the setscrew and the groove effecting a guided relative motion between the mating members.

The standard types of setscrew points are illustrated in Fig. 5-4. The *cup point* is widely used, particularly for gears, pulleys, and sheaves. The *cone point* is used where permanent location of parts is required. This type of point penetrates the shaft material greater than any other point, hence its holding power is greatest. Where frequent adjustment is necessary and it is desired to protect the shaft against excessive deformation, the *oval point* is used. This type has the lowest holding power. Where frequent readjustment is necessary but greater holding power is needed than

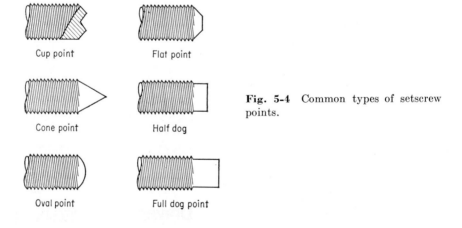

Cup point   Flat point

Cone point   Half dog

Oval point   Full dog point

Fig. 5-4   Common types of setscrew points.

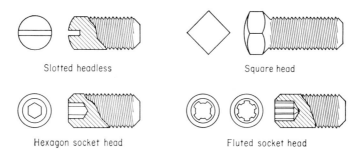

Fig. 5-5 Standard types of setscrew heads.

that provided by an oval point, the *flat point* is used. This type often is used in thin-walled mating members. The *half-dog* and *full-dog points* are employed when the cylindrical point of the setscrew projects into a drilled hole in the male members. Both types are used with hollow tubing. The half-dog is in more common use than the full dog point.

The standard types of setscrew heads are illustrated in Fig. 5-5. The *square-head setscrew* has the largest tightening possibility and hence the highest holding capacity for any given point type; however, it has disadvantages because of the protrusion of the head. All other types of setscrews can be completely screwed into the threaded hole. Choice between the *hexagonal-socket* (Allen), *fluted-socket* (Bristo), or *slotted-headless* (screwdriver) heads depends mainly upon the type of driver desired by the designer.

Since setscrews essentially are compression devices, usually they are rated in terms of the pounds force required to slip the mating members in an axial direction. This force is called the *axial holding force* H_t, the subscript t referring to a particular set of test conditions. Although it usually is considered to be an undesirable practice, setscrews occasionally are called upon to transmit torque. The torque capacity of a single setscrew is obtained by multiplying this axial holding force by the radius of the shaft. For any given setscrew application, the magnitude of the axial holding force depends mainly upon the amount of tightening of the setscrew, the type of point used, the hardness of the shaft against which the point of the setscrew is compressed, and the number of setscrews used. This is illustrated in Eq. (5-8).

$$H_a = H_t K_h K_n K_p K_t \tag{5-8}$$

where H_a = actual holding force of the setscrew application, lb
$\quad H_t$ = axial holding force for a single setscrew at a given set of test conditions, lb, from Table 5-2
$\quad K_h$ = $1.094 - 0.00623 R_c$, a shaft hardness factor

K_n = factor taking into account whether one or two setscrews are employed

= 1.0 if one setscrew is employed

= 2.0 − 0.0030θ if two setscrews are employed

K_p = point factor

= 1.0 for cup points

= 1.07 for cone points

= 0.92 for flat, half-dog, or full-dog points

= 0.90 for oval points

K_t = seating torque factor

= T_a/T_t

R_c = Rockwell C hardness of the shaft

T_a = actual seating torque employed in tightening the setscrew, in.-lb

T_t = seating torque under test conditions listed in Table 5-2, in.-lb

θ = angle between the centerline of the two setscrews, measured in a plane perpendicular to the axis of the shaft, degrees

Typical axial holding forces of various-sized cup point setscrews on shafts with 15 Rockwell C hardness are given in Table 5-2.

Table 5-2 Approximate axial holding forces of cup point setscrews on shafts with 15 Rockwell C hardness*

Nominal size	Seating torque T_t, in.-lb	Axial holding force H_t, lb
No. 0	0.5	50
No. 1	1.5	65
No. 2	1.5	85
No. 3	5	120
No. 4	5	160
No. 5	9	200
No. 6	9	250
No. 8	20	385
No. 10	33	540
$\frac{1}{4}$ in.	87	1,000
$\frac{5}{16}$ in.	165	1,500
$\frac{3}{8}$ in.	290	2,000
$\frac{7}{16}$ in.	430	2,500
$\frac{1}{2}$ in.	620	3,000
$\frac{9}{16}$ in.	620	3,500
$\frac{5}{8}$ in.	1,225	4,000
$\frac{3}{4}$ in.	2,125	5,000
$\frac{7}{8}$ in.	5,000	6,000
1 in.	7,000	7,000

* Francis R. Kull, Set-screws, *Machine Design*, p. 60, Sept. 29, 1960.

The axial holding forces in Table 5-2 are based upon tightening torques which give approximately 0.01 in. radial deflection between the shaft and the collar at the location of the setscrew, with Class 3A threads on the setscrews and Class 2B threads in the tapped holes. Between 5 and 10 per cent increase in axial holding forces can be obtained above those values in Table 5-2 by employing a suitable thread lubricant when tightening the setscrew, or by plating either the setscrew threads or those in the female member with a soft material such as zinc or cadmium.

Setscrew retention often is a problem, particularly under vibrational conditions. As soon as a setscrew loosens the least bit, usually the fastened mating members move relative to each other, unless the setscrew fits into a hole in the male member. The use of a flat spot on the shaft does not appreciably decrease the probability of the setscrew loosening under conditions of use. The cup point has better antiloosening properties than do the other types of common points. Some manufacturers have available setscrews with special self-locking features, such as ratchetlike teeth on the face of the point surface. In general, if a particular size of setscrew is found to loosen repeatedly, the use of a setscrew one or two sizes larger often will prevent loosening, all other factors being held constant.

5-7 Locknuts. A *locknut* can be used to an advantage in a bolted connection if the joint members are subjected to varying loads, if the entire joint is subjected to vibratory conditions, if accurate preloading of the loads in the threaded members is difficult because of resilience of the parts, if joint members are too fragile to withstand preload, as an added safety factor provided by the addition of a locking device because of unknown service conditions, or for accurate positioning of the nut along a threaded element, such as spacer applications where parts must be free to rotate without end play.

The four principal types of locknuts are jam, slotted, free-spinning, and prevailing-torque nuts. As illustrated in Fig. 5-6, a *jam nut* is a thin nut used on a threaded male member under a full nut. Jam nuts sometimes

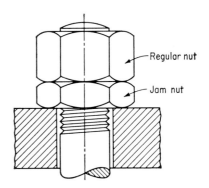

Fig. 5-6 Jam nut used as a locknut.

are used for assemblies where long travel of the nut along the thread under load is required to bring the mating parts properly into position, such as in brake drums, clamps, and certain spring assemblies. Jam nuts are no longer in frequent use.

Slotted nuts have radial slots to receive a cotter pin which passes through a diametral drilled hole in the bolt. As indicated in Fig. 5-7, the three main types of slotted nuts are *full slotted, thick slotted,* and *castle nuts.* Slotted nuts are used primarily as a safety locking device. The presence of a hole in the threaded male member makes their locking action somewhat more positive than that of some of the other locknuts. Sometimes a saftey wire is employed instead of a cotter pin, particularly if several slotted nuts are used in close proximity to each other.

Free-spinning nuts are designed to turn freely onto a male threaded member until they seat against a base surface; further tightening of the nut then produces a locking action. Although nuts of various manufacturers differ in the details by which this is accomplished, the principles of operation are similar. When the locking element contacts the base surface, a spring or wedging action is produced to create a tight friction grip against the bolt threads, thereby in effect locking the nut from angular rotation. Some designs consist of two mating parts which wedge together when tightened against each other to develop an inward pressure grip on the bolt. Still other single-element designs have a recessed or concave bottom face and an upper section that is slotted or grooved. When these nuts are tightened against the base surface, a beam action is produced in the nut, causing the upper threaded section of the nut to bend inward and grip the bolt under pressure. So long as tightening stresses are kept within recommended design limits, these nuts will resume their original shape

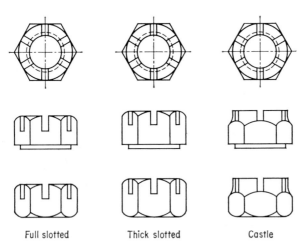

Full slotted   Thick slotted   Castle

Fig. 5-7 Major types of slotted nuts.

Integral lock washer
and nut

Spring nut
(with square or circular flange)

Fig. 5-8 Free-spinning nuts.

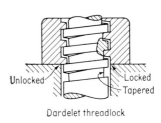

Dardelet threadlock

when disassembled and can be readily removed and reused if desired. The nuts should be seated to manufacturers' recommended torque values. Several types of free-spinning nuts are depicted in Fig. 5-8.

The free-spinning nut is advantageously used where long travel under load is required to seat the nut. So long as the load on the nut is not sufficient to develop its locking action, the nut will rotate freely into position without danger of damage to the locking feature. For all free-spinning nuts, preloading is essential to develop proper locking action. These nuts should not be employed on fragile parts which cannot withstand the preloading necessary to develop locking action.

Prevailing-torque locknuts have built-in features which develop full locking action as soon as these locking features are engaged with the bolt threads. These nuts turn freely onto the bolt until the locking feature is engaged; then the nuts develop locking action and must be wrenched to final seated position under locking conditions. Usually these nuts will retain their locking action upon one or more cycles of disassembly and reuse.

Several methods of preloading are employed. In one procedure of preloading, the threads in the top portion of the nut body are preclosed to a noncircular shape to reduce the pitch diameter without changing the angular relationship between the mating threads. When the nut is turned onto the bolt, the preclosed section grips the bolt threads to produce a tight friction clamp. On other designs, the upper portion of the nut is slotted and pressed radially inward slightly, causing these locking fingers to provide a frictional grip on the bolt threads.

Other types of prevailing-torque locknuts employ nonmetallic or soft metal plugs, strips, collars, or inserts which are retained in the nut. These

Elastic insert
lock nut

Formed locking
segment lock nut

Fig. 5-9   Prevailing-torque locknuts.

Upset threads
lock nut

Expansion nut

Table 5-3   **Locking ability of steel prevailing-torque locknuts***

Nominal size	Maximum prevailing torque, in.-lb		Minimum breakaway torque, in.-lb			
			1st removal		5th removal	
	Grades A and B	Grades A and B	Grades A and B	Grades A and B	Grades A and B	Grades A and B
$\frac{1}{4}$	30	40	5.0	6.0	3.5	4.5
$\frac{5}{16}$	60	80	8.0	10.5	5.5	7.5
$\frac{3}{8}$	80	107	12	16	8.5	11.5
$\frac{7}{16}$	100	133	17	23	12	16
$\frac{1}{2}$	150	200	22	30	15	20
$\frac{9}{16}$	200	266	30	40	21	28
$\frac{5}{8}$	300	400	39	52	27	36
$\frac{3}{4}$	400	533	58	78	41	54
$\frac{7}{8}$	600	800	88	117	62	82
1	800	1,070	120	160	84	112
$1\frac{1}{8}$	900	1,200	150	200	105	140
$1\frac{1}{4}$	1,000	1,333	188	250	132	176
$1\frac{3}{8}$	1,200	1,600	220	293	154	205
$1\frac{1}{2}$	1,350	1,800	260	346	182	242

* Specifications for Hexagon Locknuts Prevailing-Torque Type Steel, Locknut Section, Industrial Fasteners Institute, May 28, 1958.

As defined by this specification, "a prevailing torque type locknut is one in which the locking feature is self-contained, resists screwing on and off, and does not depend on bolt load for locking." Steel locknuts of this type are classified under four grades, A, B, C, and D, according to strength and performance characteristics defined by the specification. In addition, it is recommended that the different locknut grades be used with the following bolt-material grades: Locknut Grade A with SAE bolt grades 0, 1, and 2; Grade B with SAE 3, 4, and 5; Grades C and D with SAE 6, 7, and 8.

soft locking materials are plastically deformed by the bolt threads, producing an interference fit action to resist rotation of the nut on the bolt threads. Another type has a projecting wire or pin which travels between the bolt threads to provide a rachetlike locking action. Still another type employs a locking action that results from deflection of a few threads; these threads on the nut are deformed when the nut is screwed onto the bolt.

Prevailing-torque locknuts are in wide usage. They should be used with the minimum thread engagement necessary to develop the holding strength required of the nut. Torquing these nuts over long thread travel under load should be avoided to prevent damage to the locking feature. Unpointed bolts should not be employed with this type of locknut since the starting thread on such bolts may damage the prevailing-torque locking feature. Prevailing-torque locknuts are also employed as stop nuts and as spacers. Typical locking abilities of steel prevailing-torque locknuts are presented in Table 5-3. Examples of these locknuts are depicted in Fig. 5-9.

5-8   Locknuts for Ball and Roller Bearings.   Frequently ball and roller bearings are held axially against a shoulder on a shaft by use of a special *bearing locknut* with threads mating with those on the shaft. As indicated in Fig. 5-10, in addition to the shaft being threaded, a shallow axial keyway is cut into the shaft. During assembly, the bearing is pressed onto the shaft until it stops against the shoulder on the shaft, the special *bearing lock washer* is placed on the shaft with its inside tab projecting into the keyway in the shaft, the bearing locknut is placed on the shaft and tightened against the lock washer, and one or more external tabs on the lock washer are bent into one or more slots in the locknut. Since the locknut cannot loosen with the lock washer so installed, the bearing is held tightly in position until it is desired to disassemble the bearing from the shaft. Bearing locknuts and lock washers have been standardized in size and shape; typical examples of these are pictured in Fig. 5-11. Bearing locknuts are available for all standard ball or roller bearings with either four or eight equally spaced slots in the periphery of the locknuts.

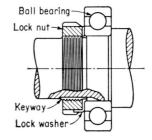

Fig. 5-10   Ball bearing held against shoulder on shaft by use of locknut, lock washer, and keyway.

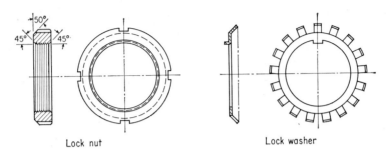

Lock nut   Lock washer

Fig. 5-11 Typical ball bearing locknut and lock washer.

5-9 Washers. *Washers* are commonly employed in fastener applications to give a better bearing for nuts or screw faces, to distribute loads
over greater areas, to maintain spring-resistance loading, to provide a
seal, to span an oversize clearance hole, to guard surfaces against marring,
and to perform other duties as desired by the designer. In general, a
washer is sized by the diameter of the male member which fits into its
center hole.

Plain washers, also called *flat washers*, usually are circular in shape,
the center hole concentric with the outer perimeter, and are made with
flat annular surfaces. They are employed to cover large clearance holes, to
distribute fastener loads over a larger area than would result if no washer
were used, and to provide bearing surfaces for the heads of bolts or
nuts. The reader is referred to American Standard Plain Washer, ASA
B27.2-1958, for dimensions of plain washers.

Conical washers are similar to plain washers except the annular areas
are surfaces of a cone rather than flat surfaces. They are used as spring
take-up devices to compensate partially for any stretch or other relaxation
of load in a threaded member. They generally are designed so that they
are loaded to above their elastic limit, making their free height after
removal of load less than its original free height. Cone washers do not have
any appreciable locking features other than friction. In their flattened
position, they are equivalent to plain washers. Cone washers with teeth on
the lower edge do not add appreciably to the locking of the bolt; however,
the penetration of the teeth into a bearing surface tends to resist lateral
shifting of position.

Helical spring lock washers are so commonly used that often they are
referred to simply as lock washers. They are made of wire slightly trapezoidal in cross section and formed into a helix of approximately one coil so
that the free height is approximately twice the thickness of the washer
section. They are available with plain ends or as a barbed type with
turned-up ends. These turned-up ends tend to produce a gouging or
milling action on the fastener bearing face, an action which is desirable in

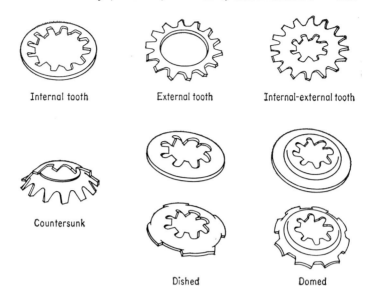

Internal tooth   External tooth   Internal-external tooth

Countersunk   Dished   Domed

Fig. 5-12   Tooth lock washers.

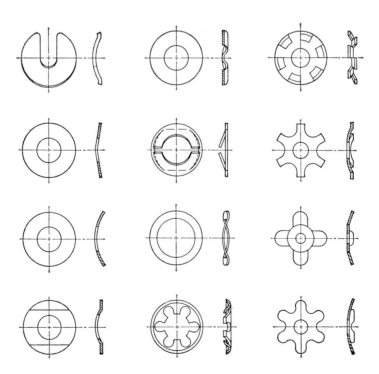

Fig. 5-13   Several of the many types of spring washers.

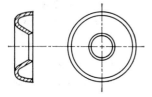

Finishing washer

Fig. 5-14   Special-purpose washers.

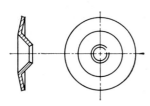

Fairing washer

certain applications. As indicated in ASA B27.1-1958, helical spring lock washers come in light, medium, heavy, and extra-heavy types for screw sizes from No. 2 through 3 in. They are used as spring take-up devices and as hardened thrust bearings to facilitate assembly and disassembly of threaded connections.

Tooth lockwashers are available in a variety of different types as indicated in Fig. 5-12. They are made of hardened steel with the teeth twisted so that sharp cutting edges are presented to both the workpiece and the bearing face of the bolt head or nut. Usually these teeth produce a slight milling action on the bearing surfaces they contact. Their locking action is usually greater than that of any other type of washer of comparable size.

Spring washers are available in so many different varieties that no definite standards for them have yet been established. A few of their types are depicted in Fig. 5-13. They are employed as take-up devices or to produce a preloading on the fastened members.

Special-purpose washers are produced by many manufacturers, the finishing and fairing washers in Fig. 5-14 being two such examples. *Finishing washers* are designed for use with oval heads and eliminate the necessity of countersinking holes. They are used extensively for fastening fabric and wood. The *fairing washer* is employed by the aircraft industry to fasten aluminum skins with flat-head screws, thereby spreading the load over a rather large area on the aluminum skin. Washers with seals and washers made of electrical insulating material are other special-purpose types.

5-10 Retaining Rings. *Retaining rings,* often referred to as *snap rings,* provide a removable shoulder to locate, lock, or retain components on shafts and in bored housings. In addition, some retaining rings have spring action whereby end play resulting from accumulated tolerances or wear of parts may be adequately compensated. In each case, the retaining ring snaps into a retaining groove in the member on which it is located. There are only two major types of retaining rings, wire-formed and stamped rings.

Wire-formed retaining rings are split rings formed and cut from spring wire of uniform cross section and size. Typical cross sections of wire-formed retaining ring wire are given in Fig. 5-15. Of the cross sections shown, by far the most commonly used shapes are the circular and the rectangular cross sections.

The more common shapes of wire-formed retaining rings are given in Fig. 5-16. The general-purpose rectangular section external rings are used widely to retain standard rolling element bearings onto shafts. They usually are available in sizes from $\frac{1}{8}$ in. to 5 in. in diameter. The general-purpose rectangular section internal rings are employed to position rolling-element bearings inside bored housings and come in $\frac{7}{8}$ in. to 5 in. diameters. The rectangular-section external clip rings, the round-section external clip rings, and the general-purpose round-section external rings are general-purpose rings for axial location of elements onto shafts. The round-section rings have less thrust capacity than the general-purpose rectangular-section rings. Where wider bearing surfaces are desired, external clip rings with extra-wide rectangular sections can be used; they are available in $\frac{1}{4}$ in. to $\frac{13}{16}$ in. sizes. For light loads the annealed rectangular-section external clip ring can be used. When the ring is installed it is deformed into the more nearly closed shape.

Typical thrust capacities of wire-formed thrust rings are given in Table 5-4.

Under conditions of dynamic loading, the thrust capacity of a wire-formed ring is 50 to 60 per cent of its static capacity; hence the thrust capacities of Table 5-4 should be reduced accordingly for dynamic loading.

The diameter of the groove into which the retaining ring fits may be

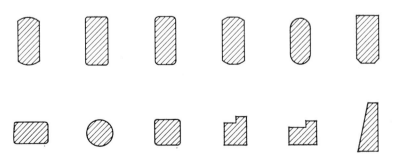

Fig. 5-15 Typical cross sections for wire-formed retaining rings.

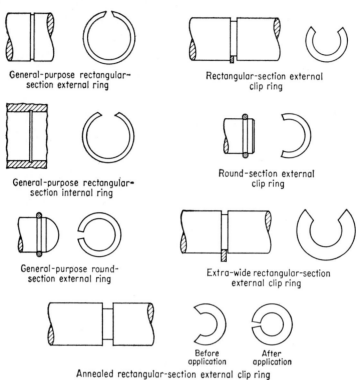

General-purpose rectangular-
section external ring

Rectangular-section external
clip ring

General-purpose rectangular-
section internal ring

Round-section external
clip ring

General-purpose round-
section external ring

Extra-wide rectangular-section
external clip ring

Before
application

After
application

Annealed rectangular-section external clip ring

Fig. 5-16   Common types of wire-formed retaining rings.

obtained from the following two equations. For rings on male members

$$d_g = d - 0.6w \tag{5-9}$$

For rings inside bored housings

$$d_g = d + 0.6w \tag{5-10}$$

In the above equations,

d = diameter of male member or bore of housing, in.
d_g = diameter of retaining ring groove, in.
w = width of retaining ring, in.

Stamped retaining rings have a tapered radial width which decreases symmetrically from the center section to the ends in contrast to the uniform cross section of wire-formed retaining rings. As implied by their name, they almost always are produced by stamping from sheet material. Their tapered construction permits them to remain circular when they are expanded for assembly over a shaft or contracted for insertion into a

Table 5-4 **Static thrust capacities of wire-formed retaining rings***

Shaft or bore diameter d, in.	Section height h, in.	Section width w, in.	Static thrust capacity F_t, lb
		External rings	
1	0.047	0.094	1,530
2	0.062	0.125	3,520
3	0.094	0.188	9,180
4	0.109	0.219	13,100
5	0.156	0.312	24,480
		Internal rings	
1	0.042	0.078	1,450
2	0.062	0.125	3,520
3	0.094	0.188	9,180
4	0.109	0.219	13,100
5	0.141	0.281	21,092

* R. J. Munsey, Wire-formed Retaining Rings, *Machine Design*, p. 87, Sept. 29, 1960.

bore or housing. This constant circularity assures maximum contact surface with the bottom of the groove. Stamped retaining rings are illustrated in Fig. 5-17. *Axially assembled retaining rings* usually have holes in the lugs at the free ends. Special pliers fit into these holes and expand or contract the rings for installation or removal. *Radially assembled retaining rings* usually are installed with a screwdriver or other simple hand tool rather than a special tool made for the purpose. *Self-locking retaining rings* are produced in several varieties for locking over male members or inside female bores or housings. They require no groove whereas both the axially assembled and the radially assembled rings require grooves.

The maximum static thrust load capacity of the rings F_{sr} and for the groove F_{sg} may be obtained from the following equations:

$$F_{sr} = \frac{\pi dt s_{su} K_r}{df} \tag{5-11}$$

$$F_{sg} = \frac{\pi dh s_{ty} K_g}{df} \tag{5-12}$$

where d = diameter of the shaft or housing, in.
df = design factor
F_{sg} = static thrust load capacity of the groove, lb
F_{sr} = static thrust capacity of the ring, lb

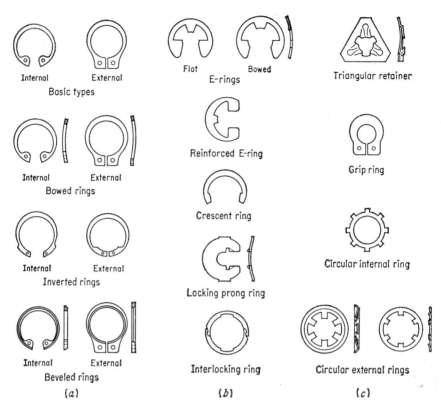

Fig. 5-17 Common types of stamped retaining rings. (*a*) Radially assembled into grooves. (*b*) Axially assembled into grooves. (*c*) Rings not requiring grooves (self-locking).

h = height (radial depth) of groove, in.
K_g = factor depending upon the type of ring, from Table 5-5
K_r = factor depending upon the type of ring, from Table 5-5
s_{su} = shearing ultimate strength of the material of the ring, psi
s_{ty} = tensile yield strength of the material of the groove, psi
t = thickness of the ring, in.

Under dynamic conditions, the maximum allowable impact loads on retaining rings and their grooves are one-half to two-thirds the values given by Eqs. (5-11) and (5-12). In some assemblies where relative rotation can occur between the ring and its groove, it is possible that the ring can become unseated from its groove. The maximum allowable load on the ring under conditions of rotation is given by

$$F_{rr} = \frac{s_w t D^2}{18 f d} \tag{5-13}$$

Table 5-5   **Static loading ring factors K_g and K_r***

Type of ring	K_g	K_r
Basic internal, external............................	1.00	1.00
Inverted internal, external.......................	0.67	0.33
E rings (including reinforced and bowed types)....	0.33	0.67
Crescent..	0.50	0.50
Interlocking....................................	0.75	0.75
Beveled internal, external.......................	1.00	0.50

* Retaining Rings, *Design News*, p. 30, Oct. 12, 1959.

where D = largest diameter or section of the ring, in.

F_{rr} = maximum allowable thrust load on the ring under conditions of rotation, lb

f = coefficient of friction between the ring and its groove (0.15 to 0.20 for dry, unlubricated parts)

s_w = allowable maximum working stress of the ring under conditions of maximum contraction, psi (less than the tensile yield strength of the material of the ring)

and all other symbols are as previously defined.

Typical applications of retaining rings are indicated in Fig. 5-18.

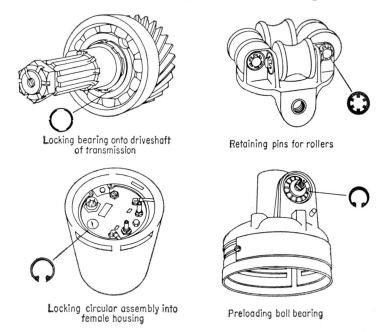

Locking bearing onto driveshaft of transmission

Retaining pins for rollers

Locking circular assembly into female housing

Preloading ball bearing

Fig. 5-18   Typical applications of retaining rings.

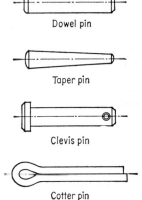

Dowel pin

Taper pin

Clevis pin

Cotter pin

Fig. 5-19   Machine pins.

5-11   Pin Fasteners.   For loading that is primarily shear, either radial or axial, various sorts of pin devices known as *pin fasteners* often can be used to an advantage over other types of fasteners. Pin fasteners can be designated as either *semipermanent pin fasteners* or *quick-release pin fasteners*. The majority of pin fasteners in use are of the semipermanent type and are either *machine pins* or *radial-locking pins*.

Machine pins include the *hardened and ground dowel pins*, the *taper pins*, the *clevis pins*, and the *cotter pins* (often called *cotter keys*); they are shown in Fig. 5-19.

The hardened and ground dowel pins are cylindrical in shape. They should be installed with a tap or press fit into reamed holes to assure tightness of the assembly. They find wide usage in fastening of machine parts where accuracy of alignment is important, in locking components on shafts, in holding laminated sections together, as a hinge or wristpin, or as a stub shaft in moving parts. They come in standard diameters between $\frac{1}{8}$ and $\frac{7}{8}$ in. inclusive.

Taper pins have a standard taper of $\frac{1}{4}$ in. of diameter change per foot of length. They come with large end diameters between 0.0625 in. and 0.7060 in. Taper pins usually are driven into reamed holes with a diameter equal to that of the small end of the pin. They are used for the attachment of levers, wheels, and similar devices onto shafts for light-duty service. Usually they are installed so that they are in double shear.

Clevis pins are cylindrical in shape with a head on one end and a radial through hole in the other end. They come in standard cylindrical diameters of $\frac{3}{16}$ in. to 1 in. They find usage in connecting mating fork or yoke to eye members in knuckle-joint assemblies. They are held into position by use of a small cotter pin, lockwire, or similar device through the radial hole.

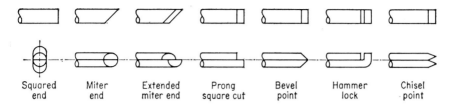

| Squared | Miter | Extended | Prong | Bevel | Hammer | Chisel |
| end | end | miter end | square cut | point | lock | point |

Fig. 5-20 Standard point types for cotter pins.

Cotter pins have standard sizes varying from $\frac{1}{32}$ in. to $\frac{3}{4}$ in. in diameter and come with a variety of point types as indicated in Fig. 5-20. They are used almost exclusively as locking devices for other fasteners.

Many of the machine pins have been standarized by the American Standards Association as given in ASA B5.20-1958.

Radial locking pins have high resistance to vibration and impact loads, are easy to install, and are relatively low in cost. They are divided into *grooved straight pins* and *spring pins*.

The locking action of grooved straight pins is provided by parallel, longitudinal grooves uniformly spaced around the surface of the pin as indicated in Fig. 5-21. The grooves effectively increase the diameter of the pins. When these pins are driven into a drilled hole corresponding to the nominal diameter of the pin, the deformation of the raised groove edges produces a force fit with the walls of the hole, thereby securing the pin into the hole. ASA has designated six different types of grooved straight pins. Type A with full-length grooves, is a general-purpose pin. Type B has a groove only half its total length and is used as a hinge or linkage pin where a locking action over only half the length of the pin is desired. Type C has a full-length groove with a pilot section at one end; it is frequently used under conditions of severe shock or vibration and when the maximum locking effect is desired. Type D has a reverse taper groove one-half the length of the pin and is used in blind holes in much

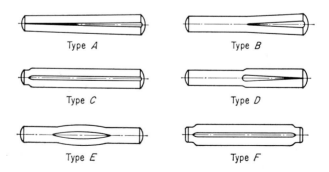

Type *A*   Type *B*

Type *C*   Type *D*

Type *E*   Type *F*

Fig. 5-21 Types of grooved straight pins.

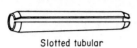

Slotted tubular

Fig. 5-22   Spring pins.

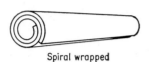

Spiral wrapped

the same applications as a type B pin; that is, where locking action is desired over half the length of the pin. Type E has a half-length groove centered axially on the pin; it is used when an artificial shoulder or a locking fit over the center portion of the pin is desired. Type F is identical to type C in construction and usage except that it has pilot sections on both ends. Standard sizes are $\frac{3}{64}$ to $\frac{1}{2}$ in. in diameter.

   Spring pins employ the resilience of hollow cylinders under radial compression to hold the pin tightly in place. *Slotted tubular spring pins* and *spiral-wrapped spring pins* are the two major types of these fasteners and are depicted in Fig. 5-22. They are available in outside diameters of $\frac{1}{32}$ to $\frac{1}{2}$ in. Slotted tubular spring pins are easily driven into holes punched or drilled without reaming to accurate size. Under most conditions of impact or vibration, these pins remain in position. There is a double shear strength for standard heat-treated carbon steel pins of this design ranging from 425 lb for $\frac{1}{16}$-in.-diameter pins to 25,800 lb for $\frac{1}{2}$-in.-diameter pins. The pins should be installed in such a manner that the direction of vibration does not coincide with the axis of the pin. The shear plane of the pin should be kept a minimum distance of one pin diameter from the edge of the pin. Upon occasion, the use of two or more pins in different holes may be advisable.

   Spiral-wrapped spring pins are available in three different grades:

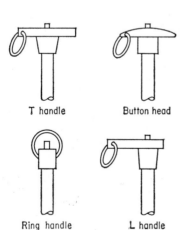

T handle   Button head

Ring handle   L handle

Fig. 5-23   Quick-release pins with various handles.

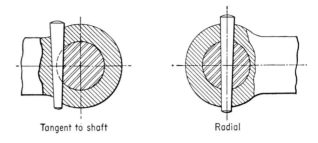

Tangent to shaft   Radial

Fig. 5-24 Taper pins used with solid shafts.

light-duty pins for low shear loading, usually static conditions; medium-duty pins for medium-intensity shear loads and conditions of shock or vibration above static conditions; and heavy-duty pins for extreme service conditions where shock and vibration loads are severe.

Quick-release pins are designed for a clearance fit in holes formed to nominal diameter and are commercially available in a wide variety of head styles, lengths, and types of locking and release devices. A few of the more common types are illustrated in Fig. 5-23. In general, quick-release pins may be divided into positive-locking and push-pull types. They find application where it is desired to have an easily removed pin fastener and yet one which will stay in place under conditions of usage.

Standard taper pins may be used as fasteners for light work by placing them tangent to the shaft or on a diameter as shown in Fig. 5-24. Hollow shafts are frequently connected by means of a sleeve and taper pins, as in Fig. 5-25. The pin holes should be drilled and reamed with a taper reamer after the parts have been assembled. The diameter of the large end of the pin should be one-fourth the shaft diameter, and the taper is $\frac{1}{4}$ in./ft.

For light work, the hub may be taper-bored and fitted with a *taper bushing*. The bushing is usually split axially into two or three parts so that when it is pressed into the hub it can exert a large radial pressure on the shaft. Since the power is transmitted by friction, these bushings are not suitable when accurate alignment must be maintained.

In a *cotter joint*, the key or cotter transmits power by shear on an area perpendicular to the length of the key instead of by shear on an area parallel to the length. The cotter is usually a flat bar tapered on one side to ensure a tight fit. Several types of cotter joints are shown in Figs. 5-24 through 5-26.

Fig. 5-25 Taper pins used with hollow shafts.

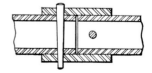

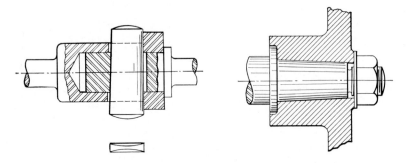

Fig. 5-26   Cotter joint.   **Fig. 5-27**   Taper joint and nut.

5-12   Taper Joint and Nut.   The piston rod of reciprocating pumps often is joined to the piston by a tapered rod end provided with a nut as shown in Fig. 5-27. In a joint of this type, the nut takes the tension load, and the taper and collar take the compression load.

5-13   Pin or Knuckle Joints.   These joints are used to connect two rods or bars when a small amount of flexibility or angular movement is necessary. Common uses are with valve and eccentric rods, diagonal stays, tension links in bridge structures, and lever and rod connections of many kinds. A typical rod end and forked knuckle joint are shown in Fig. 5-28.

5-14   Practical Considerations of Keys, Shafts, and Cotters. Shaft diameters are usually determined by deflection limits and are in general stronger than necessary; hence keys designed to transmit the full power capacity of the shaft may be excessive in size. Practical considerations require that the hub length should be at least $1.5D$ to obtain a good grip and to prevent rocking on the shaft. In general, this is the minimum length of key that should be used.

To facilitate removal, keys are often made with gib heads and tapered. In some cases, a tapped hole is provided in the end to accommodate a drawbolt. When it is necessary to drive the key out, the point may be hardened to resist the battering action. Gib-head keys should be provided with a cover or guard to prevent the possibility of injury when hands or clothing come into contact with the rotating shaft.

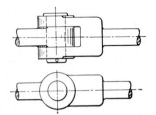

Fig. 5-28   Rod end and forked knuckle joint.

When a tapered key is used, a high pressure that tends to rupture the hub is exerted. This also forces the shaft against the opposite side of the bore, and, if a single key is used and the bore is slightly oversize, the hub will be forced into a position eccentric to the shaft. Also, since the shaft and hub are held only at two contact lines, there is a tendency for the hub to rock on the shaft, a condition which may eventually loosen the key. To overcome this tendency, the hub is often bored eccentric and fitted with two keys at 90 deg, thus obtaining a three-line contact. Rocking and eccentricity may also be prevented by using a light press fit between the shaft and the hub.

Cotters are usually driven in to make a tight joint and the initial stresses are indeterminate, and, as in the case of bolts, the applied loads may or may not increase these stresses.

5-15   Quick-operating   Fasteners. Upon   occasion   the   designer desires to employ a fastener which can be operated quickly; this is particularly true when a number of fasteners are employed on one application and it is necessary for each of the fasteners to be operated in order to effect an assembly or disassembly of the fastened items, such as access doors and panels. There are numerous occasions when it is desired to employ a fastener which must be operated many times over its useful life. In each of these cases the designer could employ a *quick-operating fastener*.

Quick-operating fasteners may be divided into lever-actuated, turn-operated, slide-action, and magnetic fasteners. These are illustrated in Fig. 5-29. The lever-actuated fasteners consist of the *draw-pull catch* (useful in edge-to-edge applications and providing good leverage for pulling parts together) and the cam-action fastener (a good quick-disconnect device with good resistance to vibration and shock). The turn-operated fasteners consist of the *pawl fastener* (a quarter-turn device used mainly for doors), the *adjustable pawl fastener* (a spring-loaded quarter-turn device useful on doors, particularly those with gaskets), the *standoff thumb screw* (widely used in electronics application, particularly in attaching panels to equipment), *the screw fastener* (a simple device for holding two members together and requiring two to three turns of the screw for complete assembly or disassembly), and the *quarter-turn* or *aircraft-type fastener* (a very rapidly hand-operated fastener requiring one-quarter of a turn to operate; widely used in aircraft applications, such as on inspection plates and doors). The *sliding latch* is the most common type of slide-action fastener and is widely used on doors and similar applications. The *magnetic fastener* consists of a permanent magnet, a piece of soft iron attracted by the magnet, and suitable means of fastening the magnet and the iron to the two members being fastened. Magnetic fasteners are widely used for refrigerator and cabinet doors. Only a few of the various available quick-operating fasteners have been mentioned.

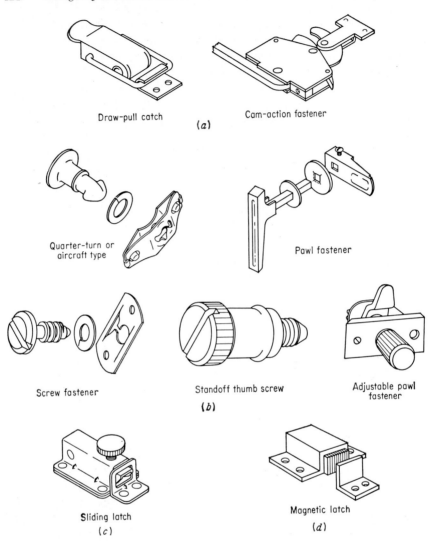

Draw-pull catch

Cam-action fastener

(*a*)

Quarter-turn or aircraft type

Pawl fastener

Screw fastener

Standoff thumb screw

Adjustable pawl fastener

(*b*)

Sliding latch

(*c*)

Magnetic latch

(*d*)

Fig. 5-29 Quick-operating fasteners. (*a*) Lever-actuated. (*b*) Turn-operated. (*c*) Slide-action. (*d*) Magnetic.

5-16 Small Rivets. For permanent fastening of items produced in large volumes, usually at low cost, *small rivets* find wide application. As illustrated in Fig. 5-30, the major types of small rivets are the *tubular*, *semitubular*, *split*, and *compression rivets*. The tubular type consists of a hollow-shank female member, the end of which is bradded over to form the second head. The semitubular rivet is the same as the tubular rivet except the cylindrical hole in the shank extends over only a small portion

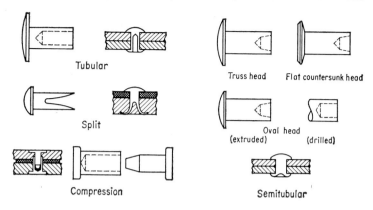

Fig. 5-30 Types of small rivets.

of the length of the shank. Under ideal installation, the portion of the shank of the semitubular rivet inside the members being fastened is solid, thereby giving a greater shear area than that for a similar-sized tubular rivet. The shanks of split rivets are sawed or punched. During installation, the two leaves of the split shank are spread, thereby forming the head which holds the rivet in place. The compression rivet consists of two mating cylindrical members with heads. These cylindrical members fit together with interference as they are installed. They present a pleasing appearance on both sides of the installed rivet.

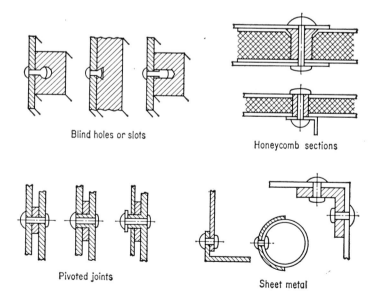

Fig. 5-31 Some uses of blind rivets.

Although small rivets carrying shear can be sized according to their shear resistance, most small rivets are selected "by eye" according to what the designer believes appears to be suitable for adequately holding together the fastened members.

Clinch allowance is the length of shank which projects beyond the members being fastened together before riveting this projecting shank into a rivet head. Clinch allowances vary from 50 to 70 per cent of the shank diameter for semitubular rivets to one times the shank diameter for full-tubular and split rivets. *Edge clearance* between the center of the rivet hole and the edge of the material being fastened should be at least two rivet diameters, preferably more. In general, the minimum *pitch* between the centers of adjacent rivets is three times the rivet diameter.

Blind rivets, unlike regular rivets, are inserted and clinched from the same side. Since no standards have been established for blind rivets, they vary considerably according to core style, blind-end configuration, and method of setting. In general, they are selected and sized in much the same manner as regular small rivets. Fig. 5-31 depicts several applications of blind rivets.

6
Threaded Members

6-1 Uses of Threads. Threaded bolts and screws are used to hold the removable heads of cylinders, machine members that must be readily disassembled, and parts of large machines that must be built in small units for ease in manufacturing, assembling, or shipping. Screws are also used for the transmission of power; examples are the lead screws on machine tools and screws on presses. Screws are sometimes used for adjusting or obtaining accurate movement in measuring instruments such as micrometers.

6-2 Thread Forms. The form and proportions of common screw threads are shown in Figs. 6-1 to 6-8. The sharp tops and bottoms of the V thread tend to weaken the screw, especially when it is subjected to shock and repeated stresses. It is also difficult to maintain a sharp point on the cutting tool used in forming the thread and on the top of a thread cut in cast iron. In the American National thread the tops and bottoms are flattened to increase strength and to make it easier to keep the threading tool sharp.

Table 6-1 gives the basic dimensions for the Unified* (UNC, UNF,

* Unified Screw Threads, ASA B1.1-1960.

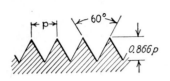

Fig. 6-1   V thread.

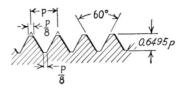

Fig. 6-2 Sellers screw thread. American National Standard; Society of Automotive Engineers Standard.

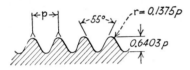

Fig. 6-3 Whitworth thread.

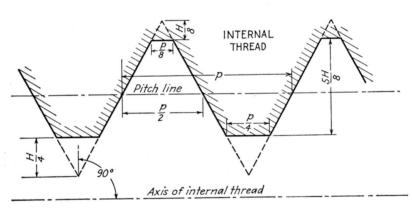

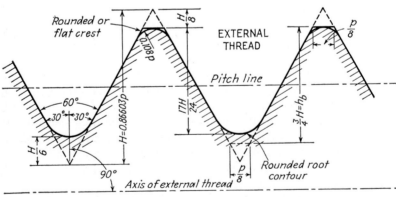

Fig. 6-4   Basic form of internal and external Unified and American screw threads (maximum metal condition).

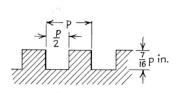

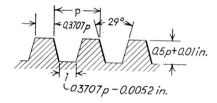

Fig. 6-5 Sellers square thread. **Fig. 6-6** Acme thread.

Fig. 6-7 Buttress thread.

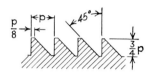

and UNEF) and the American National (NC, NF, and NEF) threads for the numbered sizes from 0 to 12 and for standard sizes from $\frac{1}{4}$ to 4 in.; sizes up to 6 in. are also standardized. In addition to the coarse (UNC), fine (UNF), and extra-fine (UNEF) series found in Table 6-1, the standard lists various constant-pitch series (UN) with 4, 6, 8, 12, 16, 20, 28, and 32 threads per inch. A new thread series designated as Unified Miniature Screw Threads* (UNM) has been introduced. Diameters range from 0.30 to 1.40 mm (0.0118 to 0.0551 in.) and supplement the Unified series which begins at 0.060 in. (number 0 in the fine thread series). Essentially the transition from the American National threads to the Unified standard has been completed. The American National threads are included in Table 6-1 for comparison and for those designs where still used. The

* Unified Miniature Screw Threads, ASA B1.10-1958.

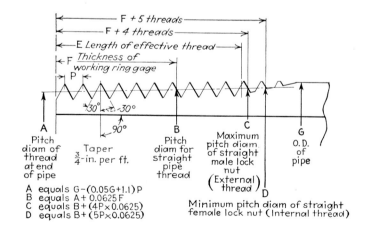

Fig. 6-8 American, or Briggs, standard pipe threads.

Table 6-1   **Unified and American National threads, coarse, fine, and extra-fine series***

Nominal size and diameter	Coarse thread series (UNC and NC)			Fine thread series (UNF and NF)			Extra-fine thread series (UNEF and NEF)		
	Threads per in.	Minor diam., in.	Stress area, in.^{2}	Threads per in.	Minor diam., in.	Stress area, in.^{2}	Threads per in.	Minor diam., in.	Stress area, in.^{2}
0 (0.060)				80	0.0447	0.0018			
1 (0.073)	64	0.0538	0.0026	72	0.0560	0.0027			
2 (0.086)	56	0.0641	0.0037	64	0.0668	0.0039			
3 (0.099)	48	0.0734	0.0048	56	0.0771	0.0052			
4 (0.112)	40	0.0813	0.0060	48	0.0864	0.0066			
5 (0.125)	40	0.0943	0.0079	44	0.0971	0.0083			
6 (0.138)	32	0.0997	0.0091	40	0.1073	0.0101			
8 (0.164)	32	0.1257	0.0140	36	0.1299	0.0147			
10 (0.190)	24	0.1389	0.0175	32	0.1517	0.0200			
12 (0.216)	24	0.1649	0.0242	28	0.1722	0.0258	32	0.1777	0.0270
$\frac{1}{4}$	20	0.1887	0.0318	28	0.2062	0.0364	32	0.2117	0.0379
$\frac{5}{16}$	18	0.2443	0.0524	24	0.2614	0.0580	32	0.2742	0.0625
$\frac{3}{8}$	16	0.2983	0.0775	24	0.3239	0.0878	32	0.3367	0.0932
$\frac{7}{16}$	14	0.3499	0.1063	20	0.3762	0.1187	28	0.3937	0.1274
$\frac{1}{2}$	13	0.4056	0.1419	20	0.4387	0.1599	28	0.4562	0.170
$\frac{9}{16}$	12	0.4603	0.182	18	0.4943	0.203	24	0.5114	0.214
$\frac{5}{8}$	11	0.5135	0.226	18	0.5568	0.256	24	0.5739	0.268
$\frac{11}{16}$							24	0.6364	0.329
$\frac{3}{4}$	10	0.6273	0.334	16	0.6733	0.373	20	0.6887	0.386
$\frac{13}{16}$							20	0.7512	0.458
$\frac{7}{8}$	9	0.7387	0.462	14	0.7874	0.509	20	0.8137	0.536
$\frac{15}{16}$							20	0.8762	0.620
1	8	0.8466	0.606	12	0.8978	0.663	20	0.9387	0.711
$1\frac{1}{16}$							18	0.9943	0.799
$1\frac{1}{8}$	7	0.9497	0.763	12	1.0228	0.856	18	1.0568	0.901
$1\frac{3}{16}$							18	1.1193	1.009
$1\frac{1}{4}$	7	1.0747	0.969	12	1.1478	1.073	18	1.1818	1.123
$1\frac{5}{16}$							18	1.2443	1.244
$1\frac{3}{8}$	6	1.1705	1.155	12	1.2728	1.315	18	1.3068	1.370
$1\frac{7}{16}$							18	1.3693	1.503
$1\frac{1}{2}$	6	1.2955	1.405	12	1.3978	1.581	18	1.4318	1.64
$1\frac{9}{16}$							18	1.4943	1.79
$1\frac{5}{8}$							18	1.5568	1.94
$1\frac{11}{16}$							18	1.6193	2.10
$1\frac{3}{4}$	5	1.5046	1.90						
2	$4\frac{1}{2}$	1.7274	2.50						
$2\frac{1}{4}$	$4\frac{1}{2}$	1.9774	3.25						
$2\frac{1}{2}$	4	2.1933	4.00						
$2\frac{3}{4}$	4	2.4433	4.93						
3	4	2.6933	5.97						
$3\frac{1}{4}$	4	2.9433	7.10						
$3\frac{1}{2}$	4	3.1933	8.33						
$3\frac{3}{4}$	4	3.4433	9.66						
4	4	3.6933	11.08						

* Extracted from American Standard Unified Screw Threads (ASA B1.1-1960) with the permission of the publisher, The American Society of Mechanical Engineers.

Unified standard resulted from the need of interchangeability of threaded parts among the United States, the Dominion of Canada, and the United Kingdom, and the need for modification of the allowances and tolerances of the American thread so as to prevent the mating threads from approaching the same basic dimensions, thus causing assembly difficulties. The Unified standard prevents tight fits during wrenching and seizing at high temperatures and allows clearance for plating. The Unified and American threads are interchangeable. Three classes of fits, 1A and 1B, 2A and 2B, and 3A and 3B, are provided by the new standard. Classes 2 and 3 of the American standard have been retained. Classes 1A, 2A, and 3A apply to external threads; 1B, 2B, and 3B apply to internal threads; and Classes 2 and 3 apply to both internal and external threads. Classes 2A and 2B are recognized standards for the normal production of bolts, screws, and nuts. The holding power of a fastener has not been affected by the new standards.

The fine thread is used where greater strength is required at the root of the threads, but the coarse thread is desirable where the material around a tapped hole is the weaker. The fine thread is desirable for fastenings requiring fine adjustments or where vibration is present. The extra-fine thread is suitable for tough strong materials and is used in thin light sections where fine adjustments are needed or where vibration is present.

The *pitch* p is the axial distance between corresponding points on adjacent threads and is equal to the reciprocal of the number of threads per inch.

The *lead* is the axial distance a thread advances in one revolution. A single thread is one on which the lead equals the pitch; a double thread is one on which the lead equals twice the pitch; and a triple thread is one on which the lead equals three times the pitch.

Major diameter is the outside, or largest, diameter of the threads and is the nominal diameter. *Minor diameter* is the smallest diameter of the threads and is commonly called the root diameter. *Pitch diameter* is the mean of the major and minor diameters. *Stress area* is the area of an imaginary circle whose diameter is the mean of the pitch and minor diameters. This area is used for the purpose of computing the tensile strength.

A 1-in. Unified Coarse Thread Series right-hand thread with eight threads per inch and Class 2A fit is designated: 1 in.-8UNC-2A. If this thread is a left-hand thread it is designated: 1 in.-8UNC-2A-LH.

The square, the Acme, and the buttress threads are used for power screws, being more efficient than the 60-deg Sellers thread. The square thread has the highest efficiency, but is comparatively costly to make, and adjustment for wear is difficult. The Acme thread is not so costly, and adjustment for wear can be accomplished by using nuts split length-

Table 6-2   Sellers standard-square threads

Bolt diam., in.	Threads per in.	Root diam., in.	Root area, sq in.
$\frac{1}{4}$	10	0.1625	0.0207
$\frac{5}{16}$	9	0.2153	0.0375
$\frac{3}{8}$	8	0.2658	0.0555
$\frac{7}{16}$	7	0.3125	0.0767
$\frac{1}{2}$	$6\frac{1}{2}$	0.3656	0.1049
$\frac{9}{16}$	6	0.4167	0.1364
$\frac{5}{8}$	$5\frac{1}{2}$	0.4666	0.1709
$\frac{11}{16}$	5	0.5125	0.2063
$\frac{3}{4}$	5	0.5750	0.2597
$\frac{13}{16}$	$4\frac{1}{2}$	0.6181	0.3000
$\frac{7}{8}$	$4\frac{1}{2}$	0.6806	0.3638
$\frac{15}{16}$	4	0.7188	0.4058
1	4	0.7813	0.4804
$1\frac{1}{8}$	$3\frac{1}{2}$	0.8750	0.6013
$1\frac{1}{4}$	$3\frac{1}{2}$	1.0000	0.7854
$1\frac{3}{8}$	3	1.0834	0.9201
$1\frac{1}{2}$	3	1.2084	1.1462
$1\frac{5}{8}$	$2\frac{3}{4}$	1.307	1.3414
$1\frac{3}{4}$	$2\frac{1}{2}$	1.400	1.5394
$1\frac{7}{8}$	$2\frac{1}{2}$	1.525	1.8265
2	$2\frac{1}{4}$	1.612	2.0422
$2\frac{1}{4}$	$2\frac{1}{4}$	1.862	2.7245
$2\frac{1}{2}$	2	2.063	3.3410
$2\frac{3}{4}$	2	2.313	4.2000
3	$1\frac{3}{4}$	2.500	4.9087
$3\frac{1}{4}$	$1\frac{3}{4}$	2.750	5.9396
$3\frac{1}{2}$	$1\frac{5}{8}$	2.962	6.8930
$3\frac{3}{4}$	$1\frac{1}{2}$	3.168	7.8853
4	$1\frac{1}{2}$	3.418	9.1756

Table 6-3   Acme screw threads

Threads per in.	Depth of thread, in.	Thickness at root of thread, in.
1	0.5100	0.6345
$1\frac{1}{2}$	0.3850	0.4772
2	0.2600	0.3199
3	0.1767	0.2150
4	0.1350	0.1625
5	0.1100	0.1311
6	0.0933	0.1101
7	0.0814	0.0951
8	0.0725	0.0839
9	0.0655	0.0751
10	0.0600	0.0681

wise; hence it is used for power drives where there must be little or no backlash, such as feed screws and lead screws of machine tools. Its efficiency is less than that of the square thread. When the power transmission is in one direction only, the buttress thread is used, the flat driving side retaining the high efficiency of the square thread, and the sloping side permitting adjustment by means of a split nut.

6-3   Threaded Fasteners.   Bolt sizes are designated by the outside diameter of the thread and by the length under the head. Stock sizes vary in length by $\frac{1}{4}$ in. from 1 to 5 in., by $\frac{1}{2}$ in. from $5\frac{1}{2}$ to 12 in., and by inches for all longer bolts, other lengths being obtained by special order. The threaded length is about $1\frac{1}{2}$ times the diameter.

Through bolts are used where both the head and nut can be made accessible by the use of flange connections and are the most satisfactory form of screw fastenings, since they can be easily renewed when broken or when the threads strip. Machine bolts have rough bodies and either rough or finished heads and nuts. Since the body of the bolt is not finished, the hole should be drilled $\frac{1}{16}$ in. larger than the bolt. Coupling bolts are through bolts having finished bodies, and are used with reamed holes. When the bolt is subjected to shear loads, this is the proper construction.

Cap screws have no nuts, the screw passing through a clearance hole in one member and threading into the mating member. They are desirable where lack of space or other considerations prevent the use of through bolts, but should not be used when they must be removed frequently as this may ruin the thread in the tapped hole. Cap screws are threaded for a length of two diameters plus $\frac{1}{4}$ in. and should enter the tapped hole at least one diameter in steel, at least $1\frac{1}{2}$ diameters in cast iron, and three diameters in aluminum.

Machine screws are small screws, in the American Coarse and Fine Thread Series, designated by size numbers up to No. 12 (0.216 in.). The larger sizes, $\frac{1}{4}$, $\frac{5}{16}$, and $\frac{3}{8}$ in. diameter, are seldom used. The Phillips cross-slot head machine screws are obtainable in sizes up to $\frac{3}{8}$ in. and are used with flat, round, oval fillister, binding, and truss heads. The clutch head is available on standard machine, wood, and thread-forming screws. It may be assembled with either an ordinary screw driver or a specially built assembly bit which locks bit and screw together by a slight twist of screw or bit and releases the screw when driving starts. The clutch head is available in sizes up to $\frac{3}{8}$ in. and is used with round, flat, oval, fillister, binding, and truss heads. The most common types of heads for cap and machine screws are shown in Fig. 6-9.

Studs are threaded on both ends and are used where through bolts are undesirable. The threads are either of the Coarse or Fine series, depending

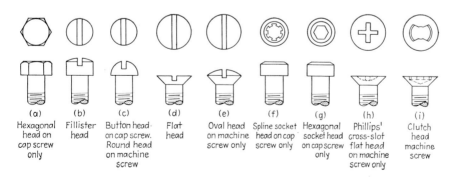

(a)	(b)	(c)	(d)	(e)	(f)	(g)	(h)	(i)
Hexagonal head on cap screw only	Fillister head	Button head on cap screw. Round head on machine screw	Flat head	Oval head on machine screw only	Spline socket head on cap screw only	Hexagonal socket head on cap screw only	Phillips' cross-slot flat head on machine screw only	Clutch head machine screw

Fig. 6-9 Forms of cap- and machine-screw heads. Cap-screw ends are flat and chamfered. Machine-screw ends are flat.

on the material into which the stud is screwed. Studs are preferable to cap screws since they need not be removed when the joints are disassembled. To secure frictional resistance against turning when the nuts are removed, studs should enter the tapped holes at least $1\frac{1}{4}$ diameters in steel, $1\frac{1}{2}$ diameters in brittle materials, and three diameters in soft material such as aluminum. Oversize tap drills should be used in soft metals as the tap will roll the material and form a deeper thread. The holes should be drilled at least one-half diameter deeper than the threaded length of the studs.

Setscrews and nut-locking devices are discussed in Chap. 5. There are many special types of bolts such as plow, carriage, eye, U, J, and hook bolts.

6-4 Effect of Initial Tension. As a result of the turning of the nut, a bolt is subjected to direct tension, compression on the threads, shear across the threads, and torsional shear in the body of the bolt.* Since none of these stresses can be accurately determined, bolts are designed on the basis of the direct tension stress with a comparatively high factor of safety to allow for the undetermined stresses. Experiments conducted at Cornell University† and confirmed by similar experiments in Germany‡ showed that an experienced mechanic tightens a nut so that the initial tension load in a bolt, in pounds, is about 16,000 times the nominal bolt diameter. Applying this relationship to a $\frac{1}{2}$-in.-13UNC bolt, the tensile stress due to tightening is 56,500 psi. This should indicate that bolts of less than $\frac{5}{8}$ in. diameter should rarely be used where serious effects would result from the failure of the bolted connection unless there is specific control of the tightening or the bolts are made of alloy steel.

The use of the torque wrench for assembly of bolted connections allows better control of tightening. The relation between the torque applied to the nut and the axial tension load in a bolt may be determined in terms of the coefficient of friction between the sliding parts and the dimensions of the nut, bolt, and threads. Equation (6-7) is a theoretically correct expression for the tensile force F_a, in pounds, when the applied torque is T, in inch-pounds. Neglecting the effect of the helix angle, assuming $f = f_c = 0.15$ and that D_c = the mean diameter of nut force = approximately $\frac{4}{3}$ (pitch diameter of bolt),

$$T = 0.2F_a \qquad \text{(nominal diameter of bolt)} \qquad (6\text{-}1)$$

The above assumptions are reasonable since the coefficient of friction is

* V. L. Doughtie and W. J. Carter, Bolted Assemblies, *Machine Design*, pp. 127–134, February, 1950.

† D. S. Kimball and J. H. Barr, "Elements of Machine Design," 3d ed., p. 278, John Wiley & Sons, Inc., New York, 1935.

‡ E. Bock, Z. *Ver. deut. Ing.*, vol. 78, p. 780, 1934.

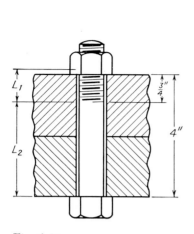

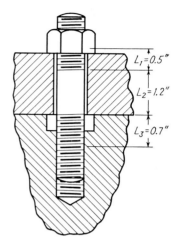

Fig. 6-10   Fig. 6-11

subject to considerable variation, being affected by the condition and fit
of the thread surface, the lubrication, and the magnitude of the unit pressure between sliding surfaces. Also, the coefficient of running friction is
less than the coefficient of starting friction. Thus a nut tightened to the
specified torque by a series of small angular movements of the nut may
produce a much lower tension in the bolt than that resulting from a continuous angular rotation of the nut until the specified torque is reached.

If consistent results are to be achieved when using the torque wrench,
nut and bolt threads must be in good condition and the thread should be
lubricated, preferably with graphite grease. For threads so lubricated the
coefficient of friction will be about 0.10.

When through bolts are used, the bolt tightening load may be determined by measuring the stretch of the bolt during tightening. This
procedure may be more easily explained by the use of an example.

Example 1. Determine the stretch that should be given a $\frac{3}{4}$-in.-16UNF
steel bolt (Fig. 6-10) when tightened to produce a direct tension stress of 40,000
psi in the body of the bolt.

Solution. For a stress of 40,000 psi in the bolt body, the stress at the
stress area of the threads $s_{t1} = 0.442 \times 40{,}000/0.3723 = 47{,}500$ psi, where
0.442 is the area of the $\frac{3}{4}$-in. shank. Assuming that the effective length of the
threaded section being elongated extends from the third engaged thread in the
nut to the end of the threaded section, $L_1 = 0.75 + 3(\frac{1}{16}) = 0.9375$ in. Deformation of the threaded section $= s_{t1}L_1/E = 47{,}500 \times 0.9375/30{,}000{,}000 =
0.001484$ in. Neglecting the small deformation of the bolt head, the deformation
in the unthreaded shank $= s_{t2}L_2/E = 40{,}000 \times 3.25/30{,}000{,}000 = 0.00433$ in.
Total elongation $= 0.001484 + 0.00433 = 0.0058$ in. This value will be the
elongation if the members fastened together are rigid. If the members are not
rigid, the assumed compression of the members should be added to this value.

When studs are used to fasten members, both ends cannot be reached.   In this case the angle of twist of the nut may be measured in order to determine the tightening load.

Example 2.   Determine the rotation of the nut for tightening the $\frac{3}{4}$-in. stud shown in Fig. 6-11 so that the tightening load will be 8,000 lb. The nut thread is UNF and the stud is screwed into cast iron with a UNC thread.

Solution.   Because of the difference in the values of the modulus of elasticity of cast iron and steel, an approximation of the effective length of the bottom threaded section will be taken as the distance from the beginning of the threaded section to the fifth engaged thread in the cast iron.

Stress in top section:

$$s_{t1} = \frac{8,000}{0.373} = 21,500 \text{ psi}$$

and

$$\text{Deformation} = \frac{s_{t1}L_1}{E} = \frac{21,500 \times 0.5}{30,000,000} = 0.00036 \text{ in.}$$

Stress for the unthreaded shank:

$$s_{t2} = \frac{8,000}{0.442} = 18,100 \text{ psi}$$

and

$$\text{Deformation} = \frac{18,100 \times 1.2}{30,000,000} = 0.000724 \text{ in.}$$

Stress for the bottom section:

$$s_{t3} = \frac{8,000}{0.334} = 23,900 \text{ psi}$$

and

$$\text{Deformation} = \frac{23,900 \times 0.7}{30,000,000} = 0.000558 \text{ in.}$$

The total elongation = 0.001642 in. The required nut rotation, measured from the hand-tight position = (0.001642 × 360)/0.0625 = 9.47 deg.

When using this procedure, the members to be fastened together must be clean and accurately fitted and all slack must be taken up by the nut before measuring the angular rotation.

6-5   Effect of Applied Loads on Bolt Stresses.   A bolted joint is shown diagrammatically in Fig. 6-12, the gasket being represented by the coiled springs.

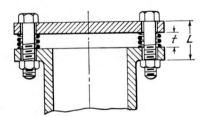

Fig. 6-12

Let F = final load on the bolt, lb
$\quad F_a$ = external or applied load, lb
$\quad F_i$ = initial load due to tightening, lb
$\quad t$ = gasket thickness, in.
$\quad L$ = length of bolt from nut to head, in.
$\quad E_g$ = modulus of elasticity of gasket, psi
$\quad E_b$ = modulus of elasticity of bolt, psi
$\quad A_g$ = loaded area of gasket, in.²
$\quad A_b$ = cross-sectional area of bolt, in.²
$\quad \Delta$ = increase in length of bolt and gasket due to application of load
$\qquad F_a$, in.

The increase in bolt load due to change in length = $\Delta E_b A_b / L$ and the decrease in gasket load = $\Delta E_g A_g / t$.

Since the summation of the forces acting = 0,

$$F_a + \left(F_i - \frac{\Delta E_g A_g}{t} \right) - \left(F_i + \frac{\Delta E_b A_b}{L} \right) = 0$$

From which

$$\Delta = \frac{F_a}{E_b A_b / L + E_g A_g / t}$$

The final load on the bolt is

$$F = \frac{\Delta E_b A_b}{L} + F_i = F_a \left(\frac{E_b A_b / L}{E_b A_b / L + E_g A_g / t} \right) + F_i$$

Or using K equal to the term in the parentheses

$$F = KF_a + F_i \tag{6-2}$$

The quantity K may have values between zero and 1. If the gasket is hard and thin and of large area, the term $E_g A_g / t$ will be large compared to $E_b A_b / L$ and K will approach zero. For gaskets that are soft, the term $E_g A_g / t$ may become so small compared to $E_b A_b / L$ that the value of K approaches 1. If no gasket is used between the members, the value of K is zero. The absence of a gasket between the members is the same as having a gasket of infinite stiffness, that is, $E_g A_g / t$ becomes infinity. It should be noted that Eq. (6-2) is valid only as long as the gasket remains in contact with the bolt-connected members. If the bolt should be stretched beyond this point, $F = F_a$.

Table 6-4 may be used as a guide in selecting the proper value of K when the final load on the bolt is to be determined.

6-6 Stresses in Tension Bolts. The torsional shear, combined with the direct tensile stress in the bolt, produces an equivalent tension somewhat greater than the initial tensile stress. It is evident that the actual

Table 6-4   **Values of K for Eq. (6-2)**

Type of Joint	K
Soft packing with studs.............................	1.00
Soft packing with through bolts......................	0.75
Asbestos..	0.60
Soft-copper gasket with long through bolts............	0.50
Hard-copper gasket with long through bolts............	0.25
Metal-to-metal joints with through bolts..............	0.00

loads and stresses imposed on bolts may be very indefinite, and that the actual design in some cases must be based on empirical formulas that reduce the apparent working stresses as the bolt diameter is reduced. One formula that gives reasonable results when used for bolts made of steel containing from 0.08 to 0.25 per cent carbon and with diameters of $\frac{3}{4}$ in. and over is

$$s_w = C(A_r)^{0.418} \tag{6-3}$$

from which the total load capacity of the bolt is

$$F_a = s_w A_r = C(A_r)^{1.418} \tag{6-4}$$

where F_a is the applied load (not including the initial tightening load), s_w is the permissible working stress, and A_r is the stress area.

The constant C may be taken as 5,000 for carbon-steel bolts of 60,000 psi ultimate tensile strength, and up to 15,000 for alloy-steel bolts, increasing in direct proportion to the ultimate strength of the steel. For bronze bolts, C may be 1,000.

Bolts 2 in. and larger are generally designed for a stress of 7,000 to 8,000 psi with carbon steels, and up to 20,000 psi with alloy steels, the initial tension being disregarded.

When the actual stresses in bolts can be fairly accurately determined, the working stress of the bolt may be slightly under the yield stress.

Example 1.   Determine the size of bolt required for the head of a 12-in. cylinder containing steam at 200 psi. Assume a hard gasket used in making up the joint.

The diameter of the bolt circle will be approximately 14 in., and 12 bolts may be assumed. Assuming that steam may enter under the head to the bolt circle, the load per bolt is

$$F_a = \frac{\pi 14^2}{4} \times \frac{200}{12} = 2,560 \text{ lb}$$

Equation (6-4) gives

$$A_r = \left(\frac{2,560}{5,000}\right)^{1/1.418} = 0.625 \text{ in.}^2$$

Table 3-1 gives a $1\frac{1}{8}$-in. UNC bolt with a stress area of 0.763 in.²

The initial stress in this bolt will be approximately 23,600 psi, and the applied stress will be 3,350 psi. The probable final stress, using Eq. (6-2), is

$$s_t = \frac{KF_a}{A_r} + \frac{F_i}{A_r} = 0.25 \times 3,350 + 23,600 = 24,438$$

This stress is less than the yield stress of this material and is safe. A stress slightly higher than the yield stress might also be considered safe, since a slight tightening of the nut would take up any stretch of the bolt without causing failure. The distance from the cylinder wall to the bolt centers should be at least equal to the bolt diameter; hence the bolt-circle diameter should be increased to $14\frac{1}{4}$ or $14\frac{1}{2}$ in. The bolt spacing should be checked by means of Fig. 6-13.

Example 2.   The cylinder of a stationary engine is $4\frac{3}{4}$ in. in diameter and is held to the crankcase by $\frac{1}{2}$-in. nickel steel bolts with American Coarse threads. The maximum gas pressure in the cylinder is 500 psi. Assume the ultimate strength of this steel to be 110,000 psi and the yield stress to be 85,000 psi. Determine the number of bolts required.

Since these bolts are less than $\frac{3}{4}$ in. in diameter, Eq. (6-4) does not apply directly, and it is necessary to consider the initial tightening stress. For a $\frac{1}{2}$-in. bolt, s_i is 56,500 psi. Subtracting this stress from the yield stress leaves 28,500 psi available for the applied gas load. Assuming a real factor of safety of 2 (based on the applied load) and a gasket factor of 0.75, the number of bolts is determined as follows. The load capacity per bolt is

$$F_a = A_r s_a = 0.1419 \times \frac{28,500}{2} = 2,020 \text{ lb}$$

The total load on all bolts is

$$F_t = \frac{\pi \times 4.75^2}{4} \times 500 = 8,850 \text{ lb}$$

and the number of bolts required is

$$N = \frac{8,850}{2,020} = 4.37 \qquad \text{say 5 bolts}$$

The probable stress in each bolt is

$$56,500 + \frac{0.75 \times 8,850}{0.1416 \times 5} = 56,500 + 9,370 = 65,870 \text{ psi}$$

which is about 77 per cent of the yield stress and therefore satisfactory. In order to maintain an oil-tight joint it may be necessary to use eight or nine bolts.

6-7 Bolt Spacing.

To permit the proper use of wrenches on the bolt heads and nuts, the spacing of the bolts should not be closer than the values indicated in Fig. 6-13. These values may be decreased slightly by the use of special wrenches. To maintain uniform contact pressure in a gasketed joint, the pitch should not be excessive, say five to six bolt diameters.

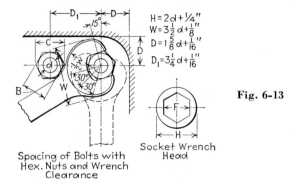

$$H = 2d + \tfrac{1}{4}''$$
$$W = 3\tfrac{1}{2}d + \tfrac{1}{8}''$$
$$D = 1\tfrac{5}{8}d + \tfrac{1}{16}''$$
$$D_1 = 3\tfrac{1}{4}d + \tfrac{1}{16}''$$

Fig. 6-13

Spacing of Bolts with
Hex. Nuts and Wrench
Clearance

Socket Wrench
Head

6-8 Effect of Dynamic Loading. The tightening load of any
threaded fastener should be equal to or greater than the applied load.
When this condition prevails, the load variation will be a minimum and
the stress in the bolt will never be more than the stress due to tightening.
This is true only when the fastened members are absolutely rigid [see
Eq. (6-2)]. Therefore, an idealized joint is one in which assembled parts
are as rigid as possible and the bolt is elastic. The threaded portion of the
bolt is weaker than the shank, not only because of the reduced area, but
also because the groove bound by the thread is a region of highly localized
stress, a condition which is undesirable when shock and repeated loads are
encountered. Tests conducted by Moore and Henwood* give the stress-
concentration factors shown in Table 6-5.

Table 6-5   Stress-concentration factors for threads

Material	Stress-concentration factor	
	Whitworth	American
Medium-carbon steel.........	1.76	2.84
SAE 2320, heated-treated.....	3.32	3.85

Lipson, Noll, and Clock† suggest the fatigue stress-concentration
factors for annealed steel shown in Table 6-6.

It is probable that the stress-concentration factor for the Unified
screw threads, which may have rounded root contours, will approach the
values for the Whitworth thread.

* H. F. Moore and P. E. Henwood, Strength of Screw Threads under Repeated
Tension, *Univ. Illinois Eng. Exp. Sta. Bull.* 264, 1938.
† C. Lipson, G. C. Noll, and L. S. Clock, "Stress and Strength of Manufactured
Parts," p. 118, McGraw-Hill Book Company, New York, 1950.

Table 6-6 **Stress-concentration factors for threads**

Thread type	Bending or tension	
	Rolled	Cut
American................	2.2	2.8
Whitworth..............	1.4	1.8
Dardelet................	1.8	2.3
Aero Stud..............	1.2	1.5

Several suggestions have been made for the reduction of fatigue failure. One is to increase the number of free threads under the nut to at least the diameter of the bolt.* Another is that the runout angle of the thread should not exceed 15 deg.† Fatigue cracks at the junction of shank and head may be prevented by providing a radius under the head of not less than 0.08 times the bolt diameter for small bolts and 0.1 times the bolt diameter for $\frac{3}{4}$-in. and larger bolts. A rounded circumferential groove in the shank immediately below the threads will be beneficial in the resistance to shock. The axial length of this groove should be at least half the nominal diameter, and the diameter of bolt at ground section should be 0.9 of the minor thread diameter.

A reduction of the shank of a bolt to a value equal to or slightly less than the minor diameter of the thread increases the capacity of the bolt to absorb strain energy. This results from the higher stress which may be developed in the bolt body for any given maximum stress to which the threaded area may be subjected. An incidental advantage accruing from a reduction in shank diameter is the greater total stretch in the bolt during tightening. Thus a small loosening rotation of the nut of slight wear under the head or nut will not result in as great a percentage decrease in the tightening load of the bolt as would be the case for a bolt with full body diameter. A bolt with reduced shank diameter is as strong as a regular bolt when subjected to a static load and much stronger when subjected to impact loads. If any incidental bending of the bolt occurs, a reduced shank bolt will develop smaller bending stresses than a bolt of full-diameter shank. Figure 6-14 shows this type of bolt in a connecting-rod end.

Example. Design the through bolts for fastening the head of a pressure vessel to the flange of the vessel. The inside diameter of the vessel is 10 in. A

---

* O. J. Horger and T. V. Buckwalter, Photoelasticity as Applied to Design Problems, *Iron Age*, p. 42, May 23, 1940.

† S. M. Arnold, Effect of Screw Threads on Fatigue, *Mech. Eng.*, p. 500, July, 1943.

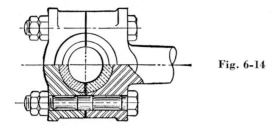

Fig. 6-14

rubberized fabric gasket 1 in. wide and $\frac{1}{8}$ in. thick is used for making the joint. The recommended contact pressure for the gasket is 3,000 psi. Assume the factor $K = 0.5$. The inside of the gasket is flush with the inner wall of the vessel. The pressure in the chamber varies from 200 to 1,000 psi 60 times per minute. The bolts are made of SAE 2320 steel quenched and drawn to 800°F for which $s_{ty} = $ 110,000 psi and $s_{ef} = 68,000$ psi.

Solution.   Assuming the effective sealing of the pressure occurs at the middle of the gasket, the maximum pressure on the cover, $F_a = \dfrac{\pi(11)^2 \times 1,000}{4} = $ 95,000 lb. Initial tightening load, $F_i = F_a +$ load due to 3,000 psi on outer half of gasket $= 95,000 + 0.5(11.5\pi)(3,000) = 149,200$ lb. From Eq. (6-2),

$$F = 0.5 \times 95,000 + 149,200 = 196,700 \text{ lb}$$

Average internal force on cover $= \dfrac{\pi(11)^2}{4}\left(\dfrac{1,000 + 200}{2}\right) = 57,000$ lb. The mean load on bolts, $F_m = 0.5 \times 57,000 + 149,200 = 177,700$ lb. The load variation from the mean, $F_v = 196,700 - 177,700 = 19,000$ lb.

The design stress is determined by applying the method of Art. 2-15 as follows: Neither the mean stress nor the variable stress is known, but the ratio of stresses will be the same as the ratio of loads. Therefore,

$$\frac{s_v}{s_m} = \frac{F_v}{F_m} = \frac{19,000}{177,700} = 0.107$$

In Fig. 6-15, draw the line OB with a slope of 0.107. Assuming a stress-concentration factor of 3.85 and a design factor of 1.75, lay off on the vertical axis.

$$\frac{s_{ef}}{3.85 \times 1.75} = \frac{68,000}{3.85 \times 1.75} = 10,090 \text{ psi}$$

and lay off on the horizontal axis

$$\frac{s_{ty}}{1.75} = \frac{110,000}{1.75} = 62,900 \text{ psi}$$

Join these two points with a straight line. At the intersection of this line and OB, s_v and s_m are located as shown. Then s_{ef} for design purposes $= s_m + s_v = 39,700 + 3,800 = 43,500$ psi. The required stress area for all bolts $= 196,700/43,500 = 4.52$ in.² Assuming 14 bolts, the stress area per bolt $= 4.52/14 = 0.323$ in.² Use $\frac{3}{4}$-in. UNF bolts whose stress area $= 0.373$ in.²

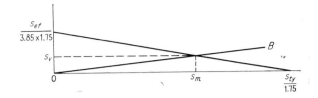

Fig. 6-15

In the above solution it should be noted that the stress-concentration factor is applied to s_{ef} and not to s_{ty}. It might be argued that the stress-concentration factor should be applied to both s_{ef} and s_{ty}, but it is probably more correct to use the stress-concentration factor only with s_{ef} because when the stress reaches the yield point the stresses are redistributed. If the stress-concentration factor had been applied to s_{ty} as well as to s_{ef}, $1\frac{1}{8}$-in. UNF bolts would be required. The student should check this value.

6-9 Bolts Subject to Shear. Bolts subjected to shear loads should be fitted tightly into reamed holes, and the plane of shear should never be across the threaded portion of the shank. In the best designs, the diameter of the shank is made slightly larger than the threaded portion in order to prevent injury to the threads by bearing pressure. When the bolt is subjected to both tension and shear loads, the root diameter can be approximated from the tension load, and the diameter of the shank may be approximated from the shear load. A diameter slightly larger than that required for either tension or shear can then be assumed, and the stresses due to combined tension and shear can be computed as outlined in Art. 3-21.

6-10 Eccentric Loading. In the bracket shown in Fig. 6-16, the load tends to rotate the bracket about its lower edge, and the bolts are not equally stressed, if the load is such as to cause the surfaces to separate. As long as the load is less than the tightening load, the load on each bolt will be the tightening load.

Fig. 6-16

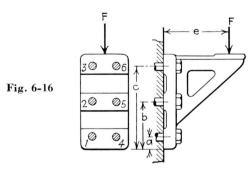

When the flange is heavy, it may be considered to be a rigid body, and the elongation of the bolts will be proportional to their distances from the lower edge. Hence the stresses will also be proportional to the distances from the lower edge. In practice, the bolts would all be made the same size as the most heavily loaded bolt. Then

$$T_1 = T_4 = \frac{a}{c} T_6$$

and

$$T_2 = T_5 = \frac{b}{c} T_6$$

Taking moments about the lower edge,

$$Fe = (T_1 + T_4)a + (T_2 + T_5)b + (T_3 + T_6)c \tag{6-5}$$

Substitution in these equations will determine the maximum tensile load on any bolt in the group. There is also a shear load that may be considered to be equally distributed between the bolts if they are fitted in reamed holes, and distributed between two bolts if rough bolts are used in clearance holes.

The methods discussed in Art. 7-18 in connection with rivets may be applied to eccentrically loaded bolted assemblies.

6-11   Power Screws.   Screws used for the transmission of power develop considerable friction; and efficiency, wear, and heating become prime considerations in their design. It has already been noted that the square thread is used because of its higher efficiency, and that the Acme or buttress threads are used when adjustment of the nut to prevent back-

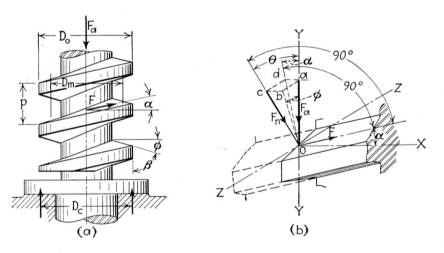

(a)   (b)

Fig. 6-17

lash is necessary. Figure 6-17 shows a power screw arranged so that when a turning moment is applied to the nut, the screw will be advanced against an axial load F_a. The relation between the applied torque and the resisting load is expressed by the equation

$$T = \frac{F_a p}{2\pi e} \tag{6-6}$$

where the symbols have the meanings given below.

In the discussions that follow, let

A_n = total normal area of threads, in.^{2}
D_o = outside diameter of screw, in.
D_m = mean diameter of screw, in.
D_c = mean diameter of thrust collar, in.
e = efficiency, between 0 and 1.0
f = coefficient of friction on threads
f_c = coefficient of friction on thrust collar
F_a = total axial resisting load, lb
F_n = total force normal to thread surface, lb
p = lead of threads, in.
p_n = unit normal force on threads, psi
W = work done, in.-lb; subscripts u, f, and c refer to useful work, thread friction, and collar friction
T = turning moment or torque required to turn the nut (or screw), in.-lb
α = lead angle of threads at mean diameter, deg
ϕ = one-half the included thread angle measured on a plane through axis, deg
θ = angle between the normal to thread surface and a line parallel to axis, deg

The forces acting on the screw threads are shown in Fig. 6-17. The total axial force exerted must equal the algebraic sum of the axial components of the forces normal to the thread surface and the friction forces. Hence

$$F_a = A_n(p_n \cos \theta - f p_n \sin \alpha) = F_n(\cos \theta - f \sin \alpha),$$

from which

$$F_n = \frac{F_a}{\cos \theta - f \sin \alpha}$$

When a torque T is applied to the nut by means of a lever, a gear, or any other turning medium, the screw will move forward p inches for one revolution of the nut. Hence, during one revolution

$W_u = F_a p = F_n(\cos \theta - f \sin \alpha)p = $ useful work

$W_f = \dfrac{fF_n p}{\sin \alpha} = \dfrac{fF_a p}{\sin \alpha(\cos \theta - f \sin \alpha)} = $ work in overcoming thread friction

$W_c = f_c F_a \pi D_c = $ work in overcoming collar friction
$W = 2\pi T = $ work applied to turn the nut

It is evident that

$$W = W_u + W_f + W_c$$

and by substitution

$$T = \frac{F_a p}{2\pi}\left[1 + \frac{f}{\sin \alpha(\cos \theta - f \sin \alpha)} + \frac{f_c \pi D_c}{p}\right] \tag{6-7}$$

The angle θ is dependent upon the lead and the included angle of the threads. From Fig. 6-17 it is found that

$$\cos \theta = \frac{1}{\sqrt{1 + \tan^2 \phi + \tan^2 \alpha}} \tag{6-8}$$

Equation (6-7) applies directly to horizontal screws and to "lifting screws" having a vertical axis, the load lifted being the axial force F_a. When the load is being lowered, the friction forces are reversed; hence for a "lowering screw"

$$T = \frac{F_a p}{2\pi}\left[\frac{f}{\sin \alpha(\cos \theta + f \sin \alpha)} + \frac{f_c \pi D_c}{p} - 1\right] \tag{6-9}$$

When the lead is large or the coefficient of friction small, the axial load may be sufficient to turn the nut, and the screw is said to "overhaul." In the limiting condition, no torque is required to lower the load, the axial load just balances the friction forces, and the right-hand side of Eq. (6-9) becomes a zero. Hence the overhauling condition is reached when

$$1 - \frac{f}{\sin \alpha(\cos \theta + f \sin \alpha)} = \frac{f_c \pi D_c}{p} = \frac{f_c D_c}{D_m \tan \alpha}$$

or

$$\tan \alpha = \frac{f \cos \alpha + \dfrac{f_c D_c}{D_m}(\cos \theta + f \sin \alpha)}{\cos \theta} \tag{6-10}$$

This equation applies to any type of power screw. For the special case of the square-thread screw, ϕ is zero, $\tan \phi$ is zero, and $\cos \theta$ becomes $\cos \alpha$. Substituting these values in the equation and transposing, the overhauling condition for square threads is found to be

$$\tan \alpha \gtreqless \frac{fD_m + f_c D_c}{D_m - ff_c D_c} \tag{6-11}$$

In the usual construction, f_c is equal to f. When a ball or roller thrust bearing is used, f_c becomes practically zero, and the screw will overhaul when tan α is equal to f.

6-12 Efficiency of Screw Threads. The efficiency of a screw thread is the ratio of the useful work to the work input, or

$$
\begin{aligned}
e &= \frac{W_u}{W} = \frac{F_a p}{2\pi T} = \frac{W_u}{W_u + W_f + W_c} \\
&= \cfrac{1}{1 + \cfrac{f}{\sin\alpha(\cos\theta - f\sin\alpha)} + \cfrac{f_c D_c}{D_m \tan\alpha}} \\
&= \cfrac{\tan\alpha(\cos\theta - f\sin\alpha)}{\tan\alpha\cos\theta + f\cos\alpha + \cfrac{f_c D_c}{D_m}(\cos\theta - f\sin\alpha)}
\end{aligned}
\tag{6-12}
$$

For the special case of square threads, the efficiency is

$$
e = \frac{\tan\alpha(1 - f\tan\alpha)}{\tan\alpha + f + \dfrac{f_c D_c}{D_m}(1 - f\tan\alpha)}
\tag{6-13}
$$

To show more clearly the relation between efficiency and helix angle, efficiency of square threads against helix angles has been plotted in Fig. 6-18. The curve reveals that for the ordinary values of f, the efficiency rises rapidly as the angle increases to 15 or 20 deg and more slowly until the maximum is reached at angles between 40 and 45 deg. However, as the helix angle is increased, the screw becomes more difficult to machine, and the mechanical advantage decreases. It should also be noted that any screw having an efficiency of over 0.5 will overhaul, and will not support an axial load without an applied torque.

If a power screw such as a screw jack is used to lift a weight W a distance h, the useful work done would equal Wh. The input necessary, considering the efficiency, would be Wh/e, where e is the efficiency. The

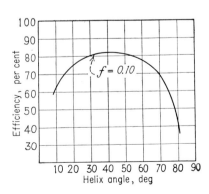

Fig. 6-18 Efficiency of square threads not including thrust collars.

amount of energy required to overcome friction is

$$\frac{Wh}{e} - Wh = Wh \left(\frac{1}{e} - 1\right) = Wh \left(\frac{1 - e}{e}\right)$$

Since the same load would be on the screw in lowering as in raising, it must be assumed that the energy expended in overcoming friction would be the same in lowering the weight a distance h as in raising it the same amount. It is obvious then that overhauling may occur when the potential energy available, Wh, is equal to or greater than the frictional work

$$Wh \geqq Wh \left(\frac{1 - e}{e}\right)$$

or

$$e \geqq (1 - e)$$

The actual helix angle selected must be a compromise based on the particular service requirements, and, in practice, angles as high as 30 deg are seldom used. Since $\cos \theta$ decreases with an increase in the included thread angle, it follows that triangular threads are less efficient than square threads.

6-13 Coefficient of Friction. The coefficient of friction varies with the quality of lubrication, with the materials used, and with the unit pressure on the threads. Most power screws are made of steel with nuts of cast iron or bronze. When the unit pressure is less than 14,000 psi and the rubbing velocity less than 50 fpm, the coefficients shown in Table 6-7 will be obtained with average lubrication.

Table 6-7 **Coefficient of friction for power screws**

Lubricant	Coefficient of friction
Machine oil and graphite.............	0.07
Lard oil............................	0.11
Heavy machine oil...................	0.14

Tests made at the University of Illinois[*] indicate that the average values of the coefficient of friction for plain thrust collars used with power screws are as shown in Table 6-8.

In these tests, the coefficient of friction was found to be independent of the load and speed within the ranges used in common practice. Hardened

[*] Clarence W. Ham and David G. Ryan, *Univ. Illinois Bull.*, vol. 29, no. 81, June 7, 1932.

Table 6-8 **Coefficient of friction on thrust collars**

Material	Coefficient of running friction	Coefficient of starting friction
Soft steel on cast iron............	0.121	0.170
Hardened steel on cast iron........	0.092	0.147
Soft steel on bronze..............	0.084	0.101
Hardened steel on bronze.........	0.063	0.081

steel on soft steel was found to be unsatisfactory, since galling, or seizing occurred at fairly low pressures.

The ball-bearing screw and nut, originally used in the steering mechanisms of automobiles, trucks, and buses, is applied to power screws. Polished steel balls roll in semicircular cross-sectional helical races in the screw and nut, providing rolling friction instead of sliding friction. When torque is applied to either the nut or screw, the balls travel in the grooves, and when a ball runs out, it goes through a return tube to travel again through nut and screw races. Since the coefficient of friction of hardened steel balls on races is from 0.0003 to 0.0015, the efficiency of this screw exceeds 90 per cent.*

6-14   Differential Screws.† For some types of service a very slow advance of the screw is required, whereas in other services a very rapid movement is required. With a single screw, slow movement is obtained by using a small helix angle and hence a small pitch, which gives a weak thread. A rapid movement requires a large helix angle with the attendant mechanical difficulties in machining. To minimize these difficulties, an arrangement similar to Fig. 6-19 may be used, in which the revolving member is threaded on the outside as well as on the inside. When the threads are of the same "hand" and of different pitch the driven screw moves slowly, and the arrangement is called a differential screw. When the threads are of opposite hand, the driven screw moves rapidly and the arrangement is called a compound screw. As previously indicated, a square thread will not be self-locking when tan α is greater than f; hence when rapid movement with self-locking properties is desired, the compound screw with its smaller helix angles may be used to advantage. When self-locking is not required, the choice between a single screw and a compound screw must be made on the basis of space requirements, cost, and efficiency.

* V. L. Maleev, Design of Groove-form Screws with Ball-bearing Nuts, *Prod. Eng.*, pp. 471–473, June, 1945.

† A. R. Kingman, Determining Efficiency of Differential Screws, *Machine Design*, p. 25, April, 1934.

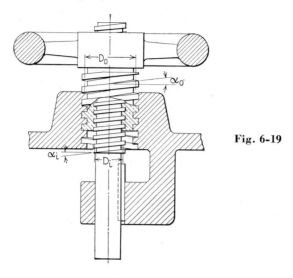

Fig. 6-19

The efficiency of a compound screw, not including the collar friction, is given by the equation

$$e = \frac{D_o \tan \alpha_o + D_i \tan \alpha_i}{D_o \dfrac{\tan \alpha_o + f_o}{1 - f_o \tan \alpha_o} + D_i \dfrac{\tan \alpha_i + f_i}{1 - f_i \tan \alpha_i}} \qquad (6\text{-}14)$$

The efficiency of a differential screw is given by the equation

$$e = \frac{D_o \tan \alpha_o - D_i \tan \alpha_i}{D_o \dfrac{\tan \alpha_o + f_o}{1 - f_o \tan \alpha_o} - D_i \dfrac{\tan \alpha_i - f_i}{1 + f_i \tan \alpha_i}} \qquad (6\text{-}15)$$

6-15 Stresses in Power Screws. A power screw is subjected to an axial load and to a turning moment, which induce in the screw direct tension or compression, torsional shear, shear across the root of the threads, and compression between the thread surfaces. If the axial load is compressive and the unsupported length is more than six or eight times the root diameter, the screw must be treated as a column. If possible, the thrust collar should be placed so that the screw is in tension.

The compressive stress or bearing pressure between the thread surfaces must be limited in order to reduce wear. In general, pressures of 2,000 psi should not be exceeded if the rubbing velocity is greater than 50 fpm, although pressures as high as 10,000 psi have been used with low velocities and adequate lubrication. With feed screws and the like, where accuracy and the maintaining of small backlash are important, pressures should be less than 200 psi.

Example. A screw press is to exert a force of 12,000 lb with an applied torque of 5,000 in.-lb. The unsupported length of the screw is 18 in. and a thrust bearing of hardened steel on cast iron is provided at the power end. The screw is to be made of SAE 1045 steel having an ultimate strength of 76,000 psi and a yield stress of 38,000 psi. The design stresses are to be 12,500 psi in tension and compression, 7,500 psi in shear, and 2,000 psi in thread bearing. The nut is cast iron, and the permissible shear is 3,000 psi. Determine the dimensions of the screw and nut.

Neglecting any column action, the root diameter must be at least

$$D_r = \left(\frac{12,000}{\frac{\pi}{4} \times 12,500} \right)^{1/2} = 1.11 \text{ in.}$$

Allowing for the increase in stress due to column action and torsional shear, a trial mean diameter of $1\frac{1}{2}$ in. is assumed. The efficiency of the screw and thrust collar will be approximately 0.15, and the torque converted into useful axial work will be 0.15 times 5,000 or 750 in.-lb. Equating the useful work per revolution to the work equivalent of this torque

$$12,000p = 2\pi750$$

and

$$p = 0.395 \text{ in.}$$

Assuming square threads and a pitch of three threads per inch, the depth of the thread will be 0.167 in. and the outside diameter will be $1\frac{1}{2}$ plus 0.167 or 1.667 in. The next larger commercial size is $1\frac{3}{4}$ in., giving

$$D_o = 1.75 \text{ in.}$$
$$D_r = 1.75 - 2 \times 0.167 = 1.416 \text{ in.}$$
$$D_m = 1.75 - 0.167 = 1.583 \text{ in.}$$

The mean diameter of the thrust collar varies with the construction and may be assumed in this case to be 2 in. Using these dimensions

$$\tan \alpha = \frac{p}{\pi D_m} = \frac{0.333}{\pi \times 1.583} = 0.067$$

and the efficiency from Eq. (6-13) is

$$e = \frac{0.067(1 - 0.14 \times 0.067)}{0.067 + 0.14 + (0.14 \times 2/1.583)(1 - 0.14 \times 0.067)}$$
$$= 0.175, \text{ or } 17.5 \text{ per cent}$$

which is greater than the assumed value of 0.15.

A portion of the applied torque is absorbed in friction in the thrust collar, and the torque transmitted to the screw is

$$T - \frac{f_c F_a D_c}{2} = 5,000 - 0.14 \times 12,000 \times 1 = 3,320 \text{ in.-lb}$$

This torque produces a torsional shear stress at the root of the threads.

$$s_s = \frac{Tc}{J} = \frac{3,320 \times 16}{\pi \times 1.416^3} = 5,940 \text{ psi}$$

Note that in some cases the torque may be applied in such a manner that the entire torque is transmitted through the screw.

The stress due to column action, assuming one end to be fixed and the other pivoted, is, from Eq. (3-17),

$$s_c = \frac{F_a}{A_r}\left[\frac{1}{1 - \dfrac{s_{ty}}{4n\pi^2 E}\left(\dfrac{L}{k}\right)^2}\right]$$

$$= \frac{12,000}{1.58}\left[\frac{1}{1 - \dfrac{38,000}{4 \times 1 \times \pi^2 \times 30 \times 10^6}\left(\dfrac{18}{0.354}\right)^2}\right] = 8,300 \text{ psi}$$

The combined stresses at the root of the threads are

$$s_{s\,max} = \tfrac{1}{2}\sqrt{8,300^2 + 4 \times 5,940^2} = 7,270 \text{ psi}$$

and

$$s_{t\,max} = \tfrac{1}{2}[8,300 + \sqrt{8,300^2 + 4 \times 5,940^2}] = 11,420 \text{ psi}$$

both of which are less than the permissible stresses.

Since the bearing pressure on the threads is limited, the length of the nut must be determined. The bearing capacity of one turn of the thread is

$$As_b = 2,000\,\frac{\pi}{4}\,(D_o^2 - D_i^2) = 2,000\,\frac{\pi}{4}\,(1.75^2 - 1.416^2) = 1,680 \text{ lb}$$

and the minimum length of nut is

$$\frac{F_a p}{As_b} = \frac{12,000 \times 0.333}{1,680}$$

$$= 2.38, \text{ say } 2\tfrac{3}{8} \text{ in.}$$

The shear stress across the threads of the nut is

$$s_s = \frac{F_a}{\pi D_o}\frac{2}{L} = \frac{12,000 \times 2}{\pi \times 1.75 \times 2.375} = 1,840 \text{ psi}$$

which is less than the stress permitted for the cast iron.

The student should rework this problem using Sellers standard square threads.

7

Welded and Riveted Connections

7-1 Types of Welding. Welding may be divided into two general types, the first requiring both heat and pressure and the second requiring heat only. *Fusion welding* is the name frequently given to processes not requiring pressure; these include thermit welding, gas welding, and arc welding. Processes requiring pressure in addition to heat include forge welding, spot welding, seam welding, projection welding, flash welding, and upset welding.

In *thermit welding* the weld metal is essentially cast steel fused into the parts welded. This process is principally used in repairing heavy machine parts and in building up defective castings.

Gas welding utilizes the heat produced by the combustion of either acetylene or hydrogen in a stream of pure oxygen. Acetylene welding is the most common form of gas welding and is widely used in repair work, in welding thin plates, and in welding gas, steam, and hydraulic pipelines.

In *arc welding* heat is supplied by a continuous arc drawn between two electrodes. The metallic-arc process is the most common electric welding process. The work forms one electrode, and the welding rod forms the other.

Molten steel has an affinity for oxygen and nitrogen, which make up the air; hence the weld metal is likely to contain gas pockets and nitrides, which weaken the weld and reduce the corrosion resistance. To prevent this, a shielded arc may be used. The welding rod is heavily coated with a material which, in the heat of the arc, gives off large quantities of inactive gas, thereby protecting the weld metal from contact with the air. Welds made in this manner are 20 per cent or more stronger than those made with bare welding rods.

The *atomic-hydrogen* and *helium arc-welding* processes are used to prevent oxidation of the metal. A reducing atmosphere is created by forcing a jet of hydrogen or helium through the arc drawn between two tungsten electrodes. The heat of the arc separates the hydrogen into the atoms, which later recombine giving back the heat of disassociation.

In *electrical-resistance welding* the parts are brought into contact, and a heavy current at low voltage is passed through the junction. Because of the high electrical resistance at the junction, the metal is rapidly brought up to the fusion temperature. Pressure, applied mechanically, forces the parts together and forms the weld. *Butt* and *flash welding* are economical where mass production justifies the special equipment required for each individual job. This process is widely used in the assembly of the bodies of automobiles, refrigerators, and other pressed-steel parts. The ordinary *spot welding* requires little special equipment and is used extensively in the manufacture of such parts as gear housings, switch housings, lamp reflectors, and other similar parts built in small lots. The field of resistance welding has been extended by the development of electronic controls, the use of which simplifies the welding of many dissimilar metals such as copper to aluminum, bronze to steel, and copper to steel. *Seam, projection*, and *upset welding* are some of the newer processes requiring both pressure and heat, simultaneously or sequentially.

7-2   Welding Properties of Materials.   Most metals can be welded by some process, but some are more readily welded than others, and the properties of the weld depend upon many factors. At the temperatures reached, structural changes in the metal that change the physical properties and the corrosion resistance may take place. Some elements in the base metal, such as zinc, may vaporize during the welding and cause porous weld metal. Gaseous oxides may cause blowholes, soluble oxides in the molten metal reduce the strength and toughness of the weld, and insoluble oxides cause slag inclusions in the weld. Metals of high thermal expansion and low thermal conductivity are subject to high cooling stresses in the weld.

All the plain carbon steels except spring steel and tool steel (with carbon contents from 0.75 to 1.50 per cent) can be satisfactorily welded, but the lower-carbon steels are the most readily welded. Nickel, chromium, and vanadium improve the welding qualities slightly. Since the weld metal is essentially cast steel, it follows that cast-steel parts are easily welded by either the gas or electrical processes.

The high strength and other desirable properties of alloy steels are chiefly due to their action on the carbon and to their response to heat treatment. The cast weld material is normally weaker than the heat-treated alloys, so that the weld is weaker than the base metal unless special precautions are taken. Special composition welding rods, usually of the shielded type, producing weld material of nearly the same analysis as the base metal, should be used, and the parts should be heat-treated after welding.

7-3   Strength of Welds.   The properties of weld metal deposited by the gas or arc process are shown in Table 7-1. With the shielded arc, the weld metal has better properties than for bare weld material.

Butt welds may be assumed to have approximately 80 per cent of the strength of the base metal if the welds are flush, and 100 per cent if they are reinforced 15 per cent, or if a reinforcing plate is used. Plug welds are as strong as the weld metal in shear.

ASME, API, governmental agencies, and other groups have various codes which cover the design of bridges, pressure vessels, buildings, ships, and other structures. In general, however, these codes do not cover most of the welding applications which confront the designer; hence the following basic elements of weld design may be useful.

In general, the designer is capable of estimating maximum stresses which exist in a weld. By comparing these with experimentally determined allowable values, he is able to determine whether the design is satisfactory. The major items which affect the strength of a weld are the soundness of the weld, determined mainly by the skill employed in making the weld; whether or not the loading is static or fatigue loading in nature; whether shielded or unshielded welding techniques are employed; the type of joint employed; and the type of stress existing within the joint.

Table 7-1   **Properties of weld metal**

Engineering property	Shielded-arc weld metal	Light-coated and bare weld metal
Ultimate strength, tension, psi.............	65,000–85,000	45,000–55,000
Yield stress, tension, psi.................	50,000–55,000	28,000–32,000
Ultimate strength, shear, psi..............	....	36,000–40,000
Endurance limit, reversed stress, psi........	28,000–30,000	12,000–16,000
Impact value, Izod, ft-lb.................	45–80	8–15
Per cent elongation in 2 in................	20–25	5–10

Table 7-2 **Design stresses s_{tab} for welds***

Type of welded joint and stress	Shielded welding		Unshielded welding	
	Static loading	Fatigue loading	Static loading	Fatigue loading
Fillet welds................	14,000	5,000	11,300	3,000
Butt welds:				
Shear loading............	10,000	5,000	8,000	3,000
Tension loading..........	16,000	8,000	13,000	5,000
Compression loading.....	18,000	8,000	15,000	5,000

* C. H. Jennings, Welding Design, *Trans. ASME*, vol. 58, p. 497, 1936.

In general, the allowable stress s_{all} in a weld is given by

$$s_{all} = K s_{tab} \qquad\qquad (7\text{-}1)$$

where K = a stress-concentration factor*
$\quad$ = 1.0 for static loading of any type of joint
$\quad$ = 1.2 for reinforced butt welds subjected to fatigue loading
$\quad$ = 1.5 for the toe of transverse fillet welds with fatigue loading
$\quad$ = 2.0 for T butt joints with sharp corners with fatigue loading
$\quad$ = 2.7 for the end of parallel fillet welds with fatigue loading
s_{all} = allowable value of stress, psi
s_{tab} = value of stress from Table 7-2, psi

The stresses in Table 7-2 have an approximate design factor of three without taking into account stress concentration.

7-4 Strength of Fillet Welds. The five most common types of welds are *fillet, lap, edge, butt,* and *plug welds;* these are illustrated in Figs. 7-1 through 7-5. Figure 7-6 depicts the two types of fillet welds,

* C. H. Jennings, Welding Design, *Trans. ASME*, vol. 58, p. 497, 1936.

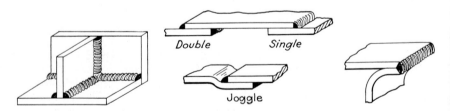

Fig. 7-1 Fillet welds.   **Fig. 7-2** Lap welds.   **Fig. 7-3** Edge weld.

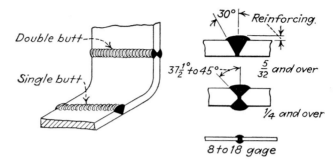

Fig. 7-4   Butt welds.

normal and *parallel welds.* Usually fillet welds are designed upon the basis of a certain allowable static load per linear inch of weld. These allowable loads depend upon the size of the fillet, the method of making the weld, the allowable design stress as indicated in Table 7-2, whether the weld is normal or parallel to the loading on the weld, and other items. Typical allowable loads per inch of weld are given in Table 7-3.

Table 7-3   **Allowable loads for mild-steel fillet welds***

Size of weld, in.	Allowable static load per linear inch of weld, lb			
	Bare welding rod		Shielded arc	
	Normal weld	Parallel weld	Normal weld	Parallel weld
$\frac{1}{8} \times \frac{1}{8}$	1,000	800	1,250	1,000
$\frac{3}{16} \times \frac{3}{16}$	1,500	1,200	1,875	1,500
$\frac{1}{4} \times \frac{1}{4}$	2,000	1,600	2,500	2,000
$\frac{5}{16} \times \frac{5}{16}$	2,500	2,000	3,125	2,500
$\frac{3}{8} \times \frac{3}{8}$	3,000	2,400	3,750	3,000
$\frac{1}{2} \times \frac{1}{2}$	4,000	3,200	5,000	4,000
$\frac{5}{8} \times \frac{5}{8}$	5,000	4,000	6,250	5,000
$\frac{3}{4} \times \frac{3}{4}$	6,000	4,800	7,500	6,000

* The values for fatigue loading are approximately 0.33 to 0.5 times those given in the table for static loading.

Fig. 7-5 Plug or rivet welds.

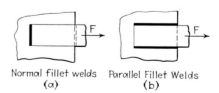

Normal fillet welds
(a)

Parallel Fillet Welds
(b)

Fig. 7-6 Fillet welds normal and parallel to the direction of loading.

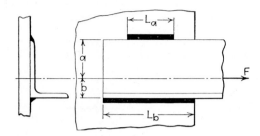

Fig. 7-7 Determining proper proportioning of parallel fillet welds.

7-5 Eccentric Loads. With axial loads on unsymmetrical sections such as angles, or channels welded on the flange edges, the weld lengths should be proportioned so that the sum of the resisting moments of the welds about the gravity axis is zero. In Fig. 7-7, if L is the total weld length, L_a and L_b the individual weld lengths, and a and b the distances of the welds from the gravity axis, then for equal stress s in both sections of the weld,

$$aL_a s = bL_b s \qquad (7\text{-}2)$$

and

$$L_b = L - L_a \qquad (7\text{-}3)$$

from which

$$L_a = \frac{bL}{a+b} \qquad (7\text{-}4)$$

and

$$L_b = \frac{aL}{a+b} \qquad (7\text{-}5)$$

A general case of eccentric loading is shown in Fig. 7-8 where the fillet welds are subjected to the action of a load F acting at a distance e from the center of gravity of the welds O. The welds are subjected to a primary shear stress s_1 and to a secondary shear stress s_2 which is proportional to the distance of the weld section from O and is maximum at the corners of the weld. If s_2 is the maximum secondary stress, then the

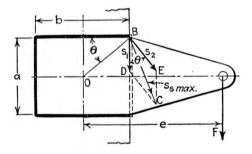

Fig. 7-8 Eccentrically loaded fillet welds.

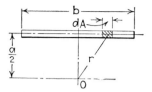

Fig. 7-9 Determining polar moment of inertia of the shear area of a fillet weld.

maximum total shear is the vector sum of s_1 and s_2. The vector sum BC is

$$s_{s\,max} = \sqrt{s_1^2 + s_2^2 + 2s_1s_2 \cos \theta} \qquad (7\text{-}6)$$

where

$$\cos \theta = \frac{b}{\sqrt{a^2 + b^2}} \qquad (7\text{-}7)$$

$$s_1 = \frac{F}{tL} \qquad (7\text{-}8)$$

and L is the total length of the welds. The value of the maximum secondary shear stress s_2 may be determined as follows:

It can be computed from Fig. 7-9 that the moment of the shear on a weld area dA is

$$dM = sr\,dA = \frac{2s_2r^2\,dA}{\sqrt{a^2 + b^2}} \qquad (7\text{-}9)$$

and

$$M = Fe = \frac{2s_2}{\sqrt{a^2 + b^2}} \int r^2\,dA \qquad (7\text{-}10)$$

The term $\int r^2\,dA$ is the polar moment of inertia J of the weld. Hence

$$s_2 = \frac{Fe\,\sqrt{a^2 + b^2}}{2J} \qquad (7\text{-}11)$$

Values of J for several combinations of eccentric-loaded fillet welds are given in Table 7-4.

7-6 Welded Pressure Vessels. The ASME Boiler Code contains rigid rules for the welding, inspection, and testing of vessels to be used as containers of gases and liquids under pressure. This code should be thoroughly studied before attempting the design of any pressure vessel.

7-7 Some Practical Welding Considerations. Welds should be placed symmetrically about the axis of the welded member unless the loading is unsymmetrical. For unsymmetrical members such as structural angles, the welded lengths are determined by the method previously outlined. When the strength of the weld is computed, $\frac{1}{2}$ in. should be subtracted from the length to allow for starting and stopping the weld, and

Table 7-4   **Values of J for fillet welds**

Type of weld	Moment of inertia J
	$\dfrac{tb^3}{12}$
	$\dfrac{ta^3}{12}$
	$\dfrac{tb(3a^2+b^2)}{6}$
	$\dfrac{ta(a^2+3b^2)}{6}$
	$\dfrac{t(a+b)^3}{6}$

welds having a length less than four times the width should not be considered. Intermittent welds should not be less than $1\frac{1}{2}$ in. long, and should be spaced not less than 16 times the plate thickness or more than 4 in. in the clear.

Fillet welds should be laid out so as to make it possible to obtain good fusion at the bottom of the weld. It is desirable that the weld should not be subjected to direct bending stresses; hence corner welds should be avoided unless the plates are properly supported independently of the weld. No special preparation of the plate edges is required for fillet and edge welds. The tee welds may be made with plain fillet welds, or the plate may be beveled on one or both sides.

Butt welds in plates of less than $\frac{1}{8}$ in. thickness do not require beveling of the plate edge, plates from $\frac{1}{8}$ to $\frac{1}{4}$ in. should be beveled or cut to form a U groove, and heavier plates should be cut to form a U groove on both sides. When plates of unequal thickness are butt-welded, the edge of the thicker plate should be reduced so that it is of approximately the same thickness as the thinner plate.

Unreinforced holes should not be located in a welded joint, and if they are located near a weld, the minimum distance from the edge of the hole to the weld should be equal to the plate thickness for plates from 1 to 2 in. thick, and never less than 1 in. For plates over 2 in. thick, the minimum distance should be 2 in.

Residual stresses resulting from the contraction of the weld metal upon cooling may be removed by annealing or by stressing the entire weld to the yield point. Residual stresses may be serious if the weld is subjected to repeated shock loads.

Example 1.   Two long plates $\frac{1}{2}$ in. thick and $6\frac{1}{2}$ in. wide are joined by a double lap joint having a total of two $\frac{1}{2}$-in. fillet welds normal to the direction of tensile loading. Determine whether or not the unshielded welds are overloaded when a tensile load of 22,000 lb is applied to the plates.

It is assumed that each fillet weld transmits the total load and has an effective length of 6 in. The average unit stresses on the weld faces are therefore

$$s_t = \frac{22,000}{6} = 3,667 \text{ lb/in.}$$

The allowable load given in Table 7-3 for a $\frac{1}{2}$- by $\frac{1}{2}$-in. bare normal weld is 4,000 lb/in.; hence the weld is not overloaded.

Example 2.   Determine the thickness required for a welded drum of a 60-in. boiler that is to operate at a steam pressure of 900 psi.

Since this is a Class 1 vessel, the longitudinal joint must be butt-welded, the joint efficiency must be 90 per cent, and the permissible stress at the temperature of the steam is 11,000 psi. The thickness of plate required is determined by Eq. (20-2), page 444, when this is rearranged so that

$$t = \frac{pd_i}{2s_t e} = \frac{900 \times 60}{2 \times 11,000 \times 0.90} = 2.73, \text{ say } 2\frac{3}{4} \text{ in.}$$

Example 3.   A 6- by 4- by $\frac{1}{2}$-in. angle is to be welded to a steel plate by fillet welds along the edges of the 6 in. leg. This angle is subjected to a tension load of 50,000 lb. Determine the unshielded weld lengths required if it is placed as shown in Fig. 7-7.

The line of action of the load may be assumed to be the gravity axis of the angle. From tables of the properties of structural shapes, the axis is 1.99 in. from the short leg. Hence the distance b in the figure is 1.99 in. and a is 4.01 in. Table 7-3 gives 3,200 lb as the permissible load per inch of a $\frac{1}{2}$-in. parallel unshielded weld. Hence the total length of weld required is

$$L = \frac{50,000}{3,200} = 15.62 \text{ in.}$$

The individual weld lengths are

$$L_a = \frac{15.62 \times 1.99}{6.00} = 5.18 \text{ in.}$$

and

$$L_b = \frac{15.62 \times 4.01}{6.00} = 10.44 \text{ in.}$$

If $\frac{1}{2}$ in. is allowed for starting and stopping the weld, a should be made $5\frac{3}{4}$ in. and b should be 11 in. Shorter lengths of shielded weld could be used.

Example 4.   A 6- by 4- by $\frac{1}{2}$-in. angle is welded to its support by two fillet welds. A load of 4,200 lb is applied normal to the gravity axis of the angle at a distance of 15 in. from the center of gravity of the welds. Determine the maximum shear stress in the welds, assuming each weld to be 3 in. in length and parallel to the axis of the angle.

The primary shear stress is

$$s_1 = \frac{4,200}{0.5 \times 3 \times 2} = 1,400 \text{ psi}$$

and the secondary shear is

$$s_2 = \frac{4,200 \times 15 \sqrt{36 + 9}}{2[0.5 \times 3(3 \times 36 + 9)/6]} = 7,230 \text{ psi}$$

Combining these stresses, $s_{s\,max}$ is

$$s_{s\,max} = \sqrt{1,400^2 + 7,230^2 + 2 \times 1,400 \times 7,230 \times 0.446}$$
$$= 7,950 \text{ psi}$$

7-8   Uses of Riveted Joints.   In the assembly of the individual members to form a complete machine or structure, some form of fastening such as riveting, welding, bolting, or keying must be used. Riveting makes a permanent joint that cannot be disassembled without destruction of the rivets. Tanks, pressure vessels, bridges, and building structures often are built of steel plates and rolled shapes riveted together; however, welding is rapidly displacing riveting as a fastening method.

Riveted joints may be roughly divided into three classes: those in which strength and rigidity are the chief requirements, those in which both strength and rigidity are required, and those in which sealing against fluid leakage as well as strength and rigidity are required.

7-9   Rivets.   Rivets are made of wrought iron or soft steel for most uses, but where corrosive resistance or light weight is a requirement, rivets of copper, aluminum, and other materials are used. In making up the joint, the rivet is inserted in a punched or drilled hole and the second head is formed by a die or rivet set; pressure is exerted by hand hammers, air hammers, or air- or hydraulic-pressure machines, the particular type of machine to be used depending upon the rivet size and the class of work desired. For the best class of work, the rivets are heated to a white heat before they are inserted in the holes and headed. Hot-driven rivets shrink on cooling and the resulting stress may be only approximately determined. Caution should be exercised where the design depends on the ability of a rivet to carry a tensile load. In some classes of work, the rivets are accurately sized, fitted in the holes, and inserted cold, and the heads are formed by peening or spinning the cold metal.

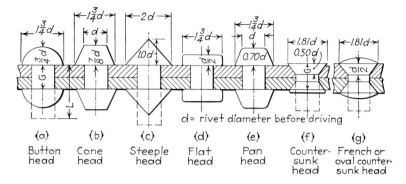

Fig. 7-10 Proportions of rivet heads in common usage.

The forms of heads generally used are shown in Fig. 7-10. The round or button, the pan, and the cone heads are the most commonly used. Countersunk heads are not so strong as the other forms and should not be used except where little clearance is available or where smooth outer surfaces are necessary.

For blind riveting the aircraft industry has introduced the draw-type rivet, known as the Huck rivet or the Cherry rivet, and the explosive rivet. The flush-type rivet, which reduces friction, is useful for ship hulls and aircraft structures.

Boiler and structural rivets are furnished commercially in diameters increasing in $\frac{1}{16}$-in. increments from $\frac{3}{8}$ in. to $1\frac{5}{8}$ in., the even $\frac{1}{8}$-in. sizes being the ones commonly carried in stock. In special cases, rivets up to 4 in. in diameter are used. The holes for the rivets should be approximately $\frac{1}{16}$ in. larger in diameter than the rivets.

7-10   Types of Riveted Joints.   When the two plates are simply laid over each other at the joint and riveted together as in Fig. 7-11, they

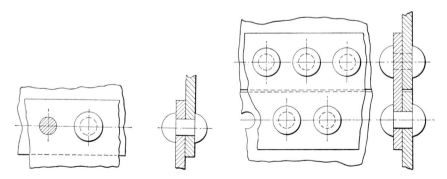

Fig. 7-11 Single-riveted lap joint.

Fig. 7-12 Single-riveted butt joint with single strap.

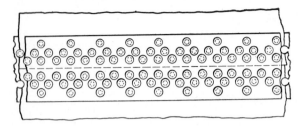

Fig. 7-13 Triple-riveted butt and double strap joint with straps of equal width, according to the ASME Boiler Code.

form a *lap* joint. Because the tension forces acting on the plates are not in the same plane they create a bending moment and thereby produce bending stresses in the plates and tension in the rivets. When the plates are placed end to end and connected by cover plates, as in Fig. 7-12, they form a *butt* joint. Because of the bending stresses, lap joints and single-cover-plate butt joints should not be used for high-pressure service. One type of joint used in boiler and tank design is illustrated in Fig. 7-13.

7-11 Assumptions in the Conventional Design of Riveted Joints. Riveted joints may fail in a number of different ways, as shown in Figs. 7-14 through 7-18, which illustrate failures by tension, shear, and crushing in various parts of the joint. The actual stresses set up in a riveted joint are complicated and cannot be simply computed. The usual methods employed in the design of such joints are based on the apparent direct stresses, and the unknown stresses are provided for in the factor of safety. Common assumptions may be stated as follows:

1. The load is distributed among the rivets according to the shear areas.

2. There is no bending stress in the rivets.

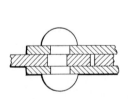

Fig. 7-14 Rivet failure by double shear of rivet.

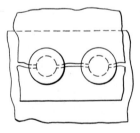

Fig. 7-15 Riveted joint failure by tearing in plate.

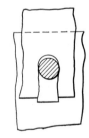

Fig. 7-16 Riveted joint failure by shearing of plate.

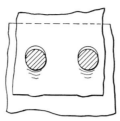 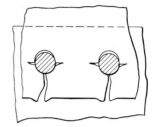

Fig. 7-17 Riveted joint failure by crushing of the plate.

Fig. 7-18 Riveted joint failure due to bending of plate.

3. The tensile stress is equally distributed over the section of metal between the rivets.

4. The crushing pressure is equally distributed over the projected area of the rivets.

5. In a rivet subjected to double shear, the shear is equally distributed between the two areas in shear.

6. The holes into which the rivets are driven do not weaken the member if it is in compression.

7. After they have been driven, the rivets completely fill the holes.

8. Friction between adjacent surfaces does not affect the strength of the joint.

It may be noted that the first six of the assumptions listed above are not conservative; i.e., the stresses actually developed in the joint are higher than would be indicated by calculations based on these assumptions. Assumption 8, however, is very conservative. The net effect of idealizing the strength of a riveted joint could not be determined without test. The experience of many years has shown that reasonable results are obtained when a boiler or structural joint is designed as indicated.

7-12 Notation Used with Riveted Joints. The following symbols are used in the discussion of riveted joints:

F = total load carried by any repeating group of rivets, lb

F_t, F_s, F_c = total load that may be carried in tension, shear, or crushing by a repeating group, lb

s_t, s_s, s_c = unit stress in tension, shear, or crushing, psi

t = main-plate thickness, in.

t_c = cover-plate thickness, in.

d = rivet-hole diameter, in.

p = pitch or center distance of rivet holes, in. In joints of more than one row of rivets, the pitch is measured in the outer row. Subscripts refer to the row, beginning with the inner row.

p_b = back pitch or distance between rows of rivets, in.

p_c = pitch on calking edge or outer row of cover plate, in.

n = number of rivet areas in any repeating group; when used with a subscript n refers to the row indicated by the subscript.

a = edge distance or distance from plate edge to center of nearest rivet, in.

e = joint efficiency; when used with the subscript t, s, or c it refers to the efficiency in tension, shear, or crushing only.

7-13 Efficiency of Riveted Joints.   Plates joined by riveting never are so strong as the original plates. The *efficiency of the joint* is defined as the ratio of the load that will produce the allowable stress in any part of the joint to the load that will produce the allowable tension stress in the unpunched plate.

The ideal joint would be one in which the allowable stresses in tension, shear, and crushing would all be produced by the same load. In practice, no joint is equally strong in every possible method of failure, and the efficiency is always less than that of the ideal joint. The average efficiency obtained in commercial boiler joints is shown in Table 7-5, the larger value in each case being that obtained with the thinner plates.

7-14 Rivet Diameters.   In boilers and high-pressure vessels, the rivet holes are always drilled or reamed with the plates and cover plates bolted in position, then the plates are separated and all fins and rough edges are removed from the holes. When the plates are reassembled and the rivets driven at high temperature, the rivets expand and completely fill the holes. Hence, in all strength calculations, the diameter of the hole is used and not the undriven diameter of the rivet. This is the practice in

Table 7-5 **Typical efficiency of commercial boiler joints**

Type of joint	Typical efficiency, %	Maximum efficiency, %*
Lap joints:		
Single riveted............	45–60	63.3
Double riveted..........	63–70	77.5
Triple riveted...........	72–80	86.6
Butt joints:		
Single riveted............	55–60	63.3
Double riveted...........	70–83	86.6
Triple riveted...........	80–90	95.0
Quadruple riveted........	85–94	98.1

* Maximum efficiencies are for ideal equistrength joints with s_t = 11,000, s_s = 8,800, and s_c = 19,000 psi.

Table 7-6   **Minimum cover-plate thickness and recommended rivet hole diameters, in.***

Thickness of shell plate	Thickness of cover plate	Diameter of rivet hole	Thickness of shell plate	Thickness of cover plate	Diameter of rivet hole
$\frac{1}{4}$	$\frac{1}{4}$	$\frac{11}{16}$	$\frac{17}{32}$	$\frac{7}{16}$	$\frac{15}{16}$
$\frac{9}{32}$	$\frac{1}{4}$	$\frac{11}{16}$	$\frac{9}{16}$	$\frac{7}{16}$	$1\frac{1}{16}$
$\frac{5}{16}$	$\frac{1}{4}$	$\frac{3}{4}$	$\frac{5}{8}$	$\frac{1}{2}$	$1\frac{1}{16}$
$\frac{11}{32}$	$\frac{1}{4}$	$\frac{3}{4}$	$\frac{3}{4}$	$\frac{1}{2}$	$1\frac{3}{16}$
$\frac{3}{8}$	$\frac{5}{16}$	$\frac{13}{16}$	$\frac{7}{8}$	$\frac{5}{8}$	$1\frac{5}{16}$
$\frac{13}{32}$	$\frac{5}{16}$	$\frac{13}{16}$	1	$\frac{11}{16}$	$1\frac{7}{16}$
$\frac{7}{16}$	$\frac{3}{8}$	$\frac{15}{16}$	$1\frac{1}{8}$	$\frac{3}{4}$	$1\frac{7}{16}$
$\frac{15}{32}$	$\frac{3}{8}$	$\frac{15}{16}$	$1\frac{1}{4}$	$\frac{7}{8}$	$1\frac{7}{16}$
$\frac{1}{2}$	$\frac{7}{16}$	$\frac{15}{16}$	$1\frac{1}{2}$	1	$1\frac{9}{16}$

* For shell plates over $1\frac{1}{2}$ in. thick the cover-plate thickness should be at least $\frac{2}{3}t$. Cover-plate thickness from ASME Boiler Code.

boiler design but does not apply to the calculation of structural joints where the holes are punched rather than drilled or reamed. A common rule is to make the rivet-hole diameter from $1.2\sqrt{t}$ to $1.4\sqrt{t}$ for rivets in single or double shear. This rule gives diameters as shown in Table 7-6.

7-15   Some Practical Rivet Considerations.   Tests indicate that when the rivets are placed at least $1\frac{1}{2}$ diameters from the plate edge, there is no danger of the plate shearing or tearing in front of the rivet. Ensuring a leakproof joint by calking will be difficult when this distance is too large. The ASME Boiler Code requires that the edge distance must be not less than $1\frac{1}{2}d$ or more than $1\frac{3}{4}d$.

In order to ensure a reasonable rigidity, and to allow for corrosion and unknown handling stresses, certain minimum thicknesses must be maintained as indicated in Table 7-7. When the plates are supported by stays, the minimum thickness is $\frac{5}{16}$ in. The minimum thickness of the cover plate to be used in double-strap joints is given in Table 7-6.

The longitudinal joints of shells in excess of 36 in. diameter must be of the double-strap butt construction. Smaller shells may be lap-welded if the pressure does not exceed 100 psi. The strength of circumferential joints, when the head is not stayed by tubes or through braces, must be at least 50 per cent that of longitudinal joints.

When the joint is subjected to shock such as is encountered in hydraulic work, the factor of safety should be increased above the usual value of 5.

7-16   Design of a Typical Boiler Joint.   A boiler is to be designed for a steam pressure of 350 psi. The inside diameter of the largest course of the drum is 54 in. The completed joint is shown in Fig. 7-19.

Table 7-7   **Minimum thickness of boiler plates***

Shell plates		Tube sheets of fire-tube boilers	
Diam. of shell, in.	Minimum thickness after flanging, in.	Diam. of tube sheet, in.	Minimum thickness, in.
36 and under	$\frac{1}{4}$	42 and under	$\frac{3}{8}$
36–54	$\frac{5}{16}$	42–54	$\frac{7}{16}$
54–72	$\frac{3}{8}$	54–72	$\frac{1}{2}$
72 and over	$\frac{1}{2}$	72 and over	$\frac{9}{16}$

* From ASME Boiler Code.

Table 7-8   **Suggested types of riveted joints**

Diam. of shell, in.	Thickness of shell, in.	Type of joint
24–72	$\frac{1}{4}$– $\frac{1}{2}$	Double riveted
36–84	$\frac{5}{16}$–1	Triple riveted
60–108	$\frac{3}{8}$–1$\frac{1}{4}$	Quadruple riveted

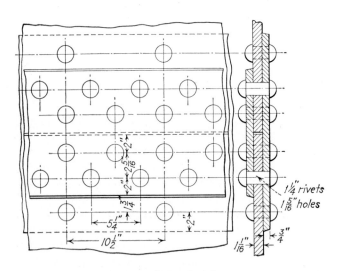

Fig. 7-19 Riveted joint to be analyzed.

The working stresses to be used are s_t equal to 11,000, s_s equal to 8,800, and s_c equal to 19,000 psi.

1. $t = \dfrac{PD}{2s_t e} = \dfrac{350 \times 54}{2 \times 11,000 \times e} = \dfrac{0.859}{e}$ in. [from Eq. (20-2)]

2. Table 7-8 indicates that a triple-riveted double-strap butt joint would be suitable, and the efficiency of this joint is shown in Table 7-5 to be 85 per cent.

$$t = \frac{0.859}{0.85} = 1.011, \text{ say } 1\tfrac{1}{16} \text{ in.}$$

Table 7-6 gives the cover-plate thickness as $\tfrac{3}{4}$ in.

3. The pitch p_c of the rivets of the row along the calking edge must be sufficiently close to permit adequate sealing against leakage of steam. Experience has indicated that the following* is useful:

$$p_c = d + 21.38 \sqrt[4]{\frac{t_c{}^3}{P}} \tag{7-12}$$

where d is the rivet diameter, p_c is the rivet pitch, and t_c is the cover-plate thickness, all in inches; thus

$$p_c = d + 21.38 \sqrt[4]{\frac{t_c{}^3}{P}} = d + 21.38 \sqrt[4]{\frac{0.75^3}{350}}$$
$$= d + 3.98 \text{ in.}$$

In the triple-riveted joint selected, the pitch in the outer row is twice that along the calking edge, and

$$p = 2p_c = 2d + 7.96$$

A repeating group of rivets will be equal in length to the pitch in the outer row. The tensile strength of the plate between rivets in the outer row is

$$F_t = (p - d)ts_t = (2d + 7.96 - d)1.0625 \times 11,000$$
$$= (d + 7.96)11,688 \text{ lb}$$

There are four rivets in double shear and one in single shear in each repeating group. Then the total shear strength of the rivets is

$$F_s = 9\frac{\pi d^2}{4} s_s = \frac{9\pi d^2}{4} 8,800 = 62,190 d^2 \text{ lb}$$

Equate F_t and F_s and solve for d.

$$(d + 7.96)11,688 = 62,190 d^2$$
$$d = 1.31, \text{ say } 1\tfrac{5}{16} \text{ in.}$$

* Perhaps first reported in G. B. Haven and G. W. Swett, "The Design of Steam Boilers and Pressure Vessels," John Wiley & Sons, New York, 1915.

The rivet diameter will be $\frac{1}{16}$ in. less, or $1\frac{1}{4}$ in.

4. $p_c = d + 3.98 = 1.3125 + 3.98 = 5.2925$ in., say $5\frac{1}{4}$ in., and

$p = 2 \times 5.25 = 10\frac{1}{2}$ in.

5. The actual efficiency of the joint can now be computed from the strengths of the joint in all the different modes of failure.

The strength of the unpunched plate for one repeating group is

$$F = pts_t = 10.5 \times 1.0625 \times 11,000 = 122,720 \text{ lb}$$

The joint may fail by tension in the plate between the rivets in the outer row; then

$$F_t = (p - d)ts_t = (10.5 - 1.3125)1.0625 \times 11,000 = 107,380 \text{ lb}$$

The joint may fail by shearing all rivets; then

$$F_s = \frac{9\pi d^2}{4} s_s = \frac{9\pi \times 1.3125^2}{4} \times 8,800 = 107,160 \text{ lb}$$

The joint may fail by crushing all rivets; then

$$F_c = (4dt + dt_c)s_c = (4 \times 1.0625 + 0.75)1.3125 \times 19,000$$
$$= 124,690 \text{ lb}$$

Since the last two equations show that the joint is stronger in crushing than in shear, it is useless to check any other type of failure involving crushing of the rivets.

The joint may fail by tension in the plate between the rivets in the second row and shearing the rivet in the outer row; then

$$F_t + F_s = (p - 2d)ts_t + \frac{\pi d^2}{4} s_s$$

$$= (10.5 - 2 \times 1.3125)1.0625 \times 11,000 + \frac{\pi 1.3125^2}{4} \times 8,800$$

$$= 103,950 \text{ lb}$$

The efficiency of the joint as designed is

$$e = \frac{103,950}{122,720} = 0.847 = 84.7 \text{ per cent}$$

which is only slightly lower than the efficiency assumed and may be considered to be satisfactory.

6. The design is now completed by determining the back pitches and the distance from the rivet holes to the plate edges.

In this joint there is a rivet in the inner row between each two rivets in the second row, and p/d equals 4. Hence the back pitch will be

$$p_b = 1\frac{3}{4}d = 1\frac{3}{4} \times 1.3125 = 2.297 \text{ or } 2\frac{5}{16} \text{ in.}$$

The rivets in the third or outer row must be placed to provide clearance between the rivet die and the edge of the cover plate. This distance must be

$$\frac{2 \times 1.25 + 0.75}{2} = 1.625, \text{ say } 1\tfrac{3}{4} \text{ in.}$$

The edge distance must be greater than $1\tfrac{1}{2}d$, i.e., 1.969, and less than $1\tfrac{3}{4}d$, or 2.297. Hence 2 in. is satisfactory.

7-17 Tank and Structural Joints. Ordinary tanks, coal bunkers, and similar structures where leakage is of minor importance may have proportions approaching the theoretical proportions of equal strengths in all methods of failure. The thicker plates, however, usually require excessive rivet sizes, and for practical reasons the rivet diameter is made approximately $1.2 \sqrt{t}$. The joints may be designed substantially as outlined in the preceding article, except that the pitch may be determined by equating the strength of a repeating group in tension to the strength of the rivets in shear, instead of by consideration of the calking requirements.

The permissible working stresses for the design of riveted connections in bridges, building structures, and machine frames are somewhat higher than those used in pressure-vessel design. The working stresses commonly used are given in Table 7-9.

7-18 Eccentric Loads on Structural Connections. The line of application of the load generally should pass through the center of gravity of the rivet areas. When it is necessary that the load line must be dis-

Table 7-9 **Permissible stresses for mild-steel structures**

Type of stress	Permissible unit stress, psi
Tension	20,000
Compression	20,000
Shear on power-driven rivets	15,000
Shear on hand-driven rivets	10,000
Crushing on power-driven rivets:	
In single shear	32,000
In double shear	40,000
Crushing on hand-driven rivets:	
In single shear	16,000
In double shear	20,000

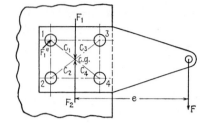

Fig. 7-20 Eccentrically loaded riveted joint to be analyzed.

placed from this position, the direct shear on each rivet will be supplemented by a secondary shear caused by the tendency of the force to twist the joint about the center of gravity. In Fig. 7-20 it is desired to determine the resulting forces and the resulting shear stresses in the rivets due to the application of the force F with an eccentricity of e. The actual forces resulting would be difficult to determine because of the influence of the stiffness of the bracket and the stiffness of the member to which the bracket is riveted. If, however, it is assumed that both these members are absolutely rigid and that all deformation or relative movement of the bracket to the members takes place as a result of shear deformation of the rivets, a ready solution is possible. The center of gravity of the rivet areas is located as shown, and equal and opposite forces F_1 and F_2, equal and parallel to F, are imagined acting at the center of gravity. The result is a couple, formed by F_2 and F, equal to Fe, that tends to twist the bracket, and a force F_1 that produces direct shear. The direct shear stress s'_s, called the primary shear stress, on the rivets produced by F_1 acts vertically downward, is the same for each rivet, and is

$$s'_s = \frac{F}{4A} \tag{7-13}$$

where A is the cross-sectional area of one rivet, in square inches.

The couple Fe will rotate the bracket around the point for which the required torque per unit of angular deformation is least, i.e., the center of gravity of the rivet cross-sectional areas. This is analogous to a beam bending around the neutral axis of its cross section. If it is assumed that the bracket rotates through an angle $d\theta$, the resulting shear deformation of rivet 1, at a distance of c_1 from the center of gravity, would be

$$\Delta = c_1 \, d\theta \tag{7-14}$$

and the unit shear deformation would be

$$\epsilon_s = kc_1 \, d\theta \tag{7-15}$$

where k is a constant that depends on the length of the section of the rivet subjected to the shear force.

The resulting stress, called the secondary shear stress, in rivet 1 is

$$s_{s1}'' = kGc_1 \, d\theta \tag{7-16}$$

The torque T_1 about the center of gravity as a result of the secondary stress in rivet 1 is

$$T_1 = F_1'' c_1 = s_{s1}'' A_1 c_1 = kGc_1^2 \, d\theta A_1 \tag{7-17}$$

where A_1 is the cross-sectional area of rivet 1, in square inches.

The total torque T about the center of gravity is found as the summation of the torques due to the stresses in all rivets.

$$T = kG \, d\theta \sum Ac^2 = \frac{s_s''}{c} \sum Ac^2 \tag{7-18}$$

since

$$kG \, d\theta = \frac{s_{s1}''}{c_1} = \frac{s_{s2}''}{c_2} \quad \text{etc.} \tag{7-19}$$

$$s_s'' = \frac{Tc}{\Sigma Ac^2} = \frac{Tc}{J} \tag{7-20}$$

where $\Sigma Ac^2 = J$, which is the polar moment of inertia of the rivet areas about the center of gravity, in inches fourth.

The secondary shear stresses are directed perpendicular to the lines through the rivet center and the center of gravity.

If it is desired to deal with forces instead of stresses, the primary force

$$F' = \frac{F}{n} \tag{7-21}$$

where n is the number of rivets.

The secondary forces

$$F'' = s_s'' A' = \frac{Tc}{J} A' \tag{7-22}$$

where A' is the area of rivet under consideration, in square inches, and J is the polar moment of inertia of all rivets in the connection, in inches fourth.

The above analysis offers a manner for handling joints when rivets of different diameter and material are used. In the latter case, all rivets are transformed to the same material.

Example.   Assume the arrangement shown in Fig. 7-21. All rivets are the same size.

Solution 1.   The loads F' and F'', equal to F, can be added to the system without changing the condition of equilibrium. The direct, or primary, shear force on each rivet $= 24,000/4 = 6,000$ lb acting parallel to F.

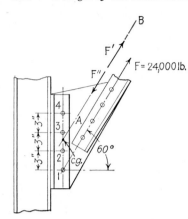

Fig. 7-21   Structural riveted joint to be analyzed.

The couple $(F$–$F'')$ produces a twisting moment about the center of gravity $= 24,000 \ (4.5 \sin 30) = 54,000$ lb-in. This moment is resisted by the combined moments of the forces acting on the rivets 1, 2, 3, and 4. Hence

$$54,000 = F''_1 \times 4.5 + F''_2 \times 1.5 + F''_3 \times 1.5 + F''_4 \times 4.5$$

If the angles and gusset plate are considered to be rigid, the forces F''_1, F''_2, F''_3, and F''_4 will be proportional to their distances from the center of gravity, and

$$F''_1 = F''_4$$

and

$$F''_2 = F''_3 = \frac{1.5}{4.5} F''_1$$

Substituting these values in the moment equation and solving,

$$F''_1 = F''_4 = 5,400 \text{ lb}$$

and

$$F''_2 = F''_3 = 1,800 \text{ lb}$$

These secondary shear forces acting normal to the lines joining the rivets and the center of gravity must be added to the direct shear forces by vector addition. Hence the actual load on rivet 1 is

$$F_1 = \sqrt{6,000^2 + 5,400^2 + 2 \times 6,000 \times 5,400 \times \cos 60}$$
$$= 9,880 \text{ lb}$$

Similarly,

$$F_2 = 7,070 \text{ lb}$$
$$F_3 = 5,330 \text{ lb}$$
$$F_4 = 5,720 \text{ lb}$$

It should be noted that the largest actual force acting on one of these rivets is nearly twice the apparent direct shear force.

Solution 2.   (Applying the method discussed in this article.) The direct, or primary, shear force on each rivet, as in solution 1, $= 24,000/4 = 6,000$ lb.   The secondary force on rivet 1:

$$F_1'' = \frac{Tc_1A_1}{J} = \frac{Tc_1A_1}{A_1\Sigma c^2} = \frac{Tc_1}{\Sigma c^2} = \frac{54,000 \times 4.5}{2(4.5)^2 + 2(1.5)^2} = 5,400 \text{ lb}$$

The secondary force on rivet 2:

$$F_2'' = \frac{54,000 \times 1.5}{2(4.5)^2 + 2(1.5)^2} = 1,800 \text{ lb}$$

From symmetry, $F_1'' = F_4''$ and $F_2'' = F_3''$. The resulting total loads on the rivets are determined as in solution 1.

8

Shafts

8-1 Shafts. A *shaft* is a rotating member supported by bearings and transmitting power. An *axle* is a stationary member, primarily loaded in bending, with gears, pulleys, and wheels rotating on it. Short axles and shafts are called *spindles*. Common usage does not make this distinction. The term shaft is used to designate all three forms.

Shafts are usually round and may be either solid or hollow. Shafts transmitting power by torsion may be divided into two general classes, transmission shafts and machine shafts. *Transmission shafts* are those used to transmit power between the source and the machines absorbing the power, and include countershafts, line shafts, head shafts, and all factory shafting. *Machine shafts* are those forming an integral part of the machine itself.

Shafts are sized for stresses and deformation, lateral deflection and torsional deflection, by applying the principles discussed in Chap. 3. The maximum shear theory is commonly used for shafts made of ductile materials and the maximum normal stress theory for shafts of brittle materials. The Mises-Hencky distortion energy theory may also be used.

8-2   Stresses in Shafts.   Shafts may be subjected to torsional, bending, or axial loads or a combination of these loads. If the load is torsional, the stress induced is shear; if bending, the stresses are tension and compression. When a shaft is subjected to a combination of loads, the maximum resulting stresses are determined by Eqs. (3-52) and (3-56). Since shafts are generally made of ductile materials, their diameters usually are obtained by the use of the maximum-shear stress theory, Eq. (3-52). However, for shafts of brittle material the maximum normal stress theory, Eq. (3-56), is used for shaft size determination.

When the loading is torsional only, the maximum stress and the angular deformation are

$$s_s = \frac{Tc}{J} \quad \text{and} \quad \theta = \frac{TL}{JG} \tag{8-1}$$

where s_s = torsional shear stress, psi
  T = torsional moment, in.-lb
  c = distance from neutral axis to outermost fiber, in.
  J = polar moment of inertia of cross-sectional area about axis of rotation, in.^{4}
  L = length of shaft, in.
  θ = angular deformation in length L, rad
  G = modulus of rigidity in shear, psi

For round hollow shafts, Eq. (8-1) becomes

$$s_s = \frac{16Td_o}{\pi(d_o^4 - d_i^4)} = \frac{16T}{\pi d_o^3} \frac{1}{1 - K^4} \tag{8-2}$$

where d_o = outside diameter, in.
  d_i = inside diameter, in.
  K = ratio of inside to outside diameter

When the shaft is subjected to bending only, the stress, given by the beam formula, is

$$s_t = \frac{32M}{\pi d_o^3} \frac{1}{1 - K^4} \tag{8-3}$$

where s_t = tensile or compressive stress, psi
  M = bending moment, in.-lb
  I = rectangular moment of inertia of cross-sectional area about neutral axis, in.^{4}

When the shaft is subjected to both torsional and bending loads, the stresses s_t and s_s from Eqs. (8-2) and (8-3), when substituted in Eqs. (3-52) and (3-56), become

$$s_{s\,\text{max}} = \frac{16}{\pi d_o^3} \sqrt{M^2 + T^2} \frac{1}{1 - K^4} \tag{8-4}$$

and

$$s_{t\,max} = \frac{16}{\pi d_o^3} (M + \sqrt{M^2 + T^2}) \frac{1}{1 - K^4} \tag{8-5}$$

These equations are applicable to solid shafts if K is made equal to zero. The term $\sqrt{M^2 + T^2}$ is often referred to as the *equivalent twisting moment*, and the term $\frac{1}{2}(M + \sqrt{M^2 + T^2})$ as the *equivalent bending moment*.

In certain installations the shaft may be subjected to an axial load F_a in addition to the torsional and bending loads. The average axial stress $4\alpha F_a/\pi(d_o^2 - d_i^2)$ must be added to the stress s_t from Eq. (8-3). In this case Eqs. (8-4) and (8-5) become

$$s_{s\,max} = \frac{16}{\pi d_o^3} \sqrt{\left(M + \frac{\alpha F_a d_o(1 + K^2)}{8}\right)^2 + T^2} \left(\frac{1}{1 - K^4}\right) \tag{8-6}$$

and

$$s_{t\,max} = \frac{16}{\pi d_o^3} \left[M + \frac{\alpha F_a d_o(1 + K^2)}{8} \right. $$
$$\left. + \sqrt{\left(M + \frac{\alpha F_a d_o(1 + K^2)}{8}\right)^2 + T^2} \right] \frac{1}{1 - K^4} \tag{8-7}$$

where d_o = shaft diameter, in.

F_a = axial tension or compression load, lb

K = ratio of inside to outside diameter of hollow shafts, and hence $K = 0$ for solid shafts

M = maximum bending moment, in.-lb

T = maximum torsional moment, in.-lb

$s_{s\,max}$ = maximum shear stress, psi

$s_{t\,max}$ = maximum tensile or compressive stress, psi

α = ratio of the maximum intensity of stress resulting from the axial load to the average axial stress.

The diameter of a shaft may be determined by substituting the design stress s_d for $s_{s\,max}$ or $s_{t\,max}$, depending upon which theory of failure is assumed to apply.

The value of α is obtained by considering the axial load, or thrust, as a load on a column of diameter d having a length equal to the distance between bearings. (See Arts. 3-12 to 3-15.) A straight-line formula commonly used for columns having a slenderness ratio less than 115 gives

$$\alpha = \frac{1}{1 - 0.0044(L/k)} \tag{8-8}$$

where L is the length between supporting bearings, in inches, and k is the radius of gyration of the shaft, in inches.

When the slenderness ratio is greater than 115, Euler's equation may be assumed to apply and gives

$$\alpha = \frac{s_y}{n\pi^2 E}\left(\frac{L}{k}\right)^2 \tag{8-9}$$

where s_y = yield stress in compression, psi
  n = constant for the type of column end support
   = 1 for hinged ends, 2.25 for fixed ends, and 1.6 for ends partly restrained, as in bearings
  E = modulus of elasticity, psi

Example. A hollow steel shaft is to transmit 19.8 hp at 250 rpm. The loading is such that the maximum bending moment is 10,000 in.-lb, the maximum torsional moment 5,000 in.-lb, and the axial compressive load 4,000 lb. The shaft is supported on rigid bearings 6 ft apart and is subjected to minor torsional loads suddenly applied. The maximum allowable shear stress is 3,100 psi. The ratio of the inside diameter to the outside diameter is 0.75.

Since the shaft diameter is not known, the values of k and α cannot be determined. It is necessary to find or assume a trial value of d, because d is included on the right-hand side of the equation and also affects the value of α. The value of α varies from 1 to 2.02 for slenderness ratios from 0 to 115, and in this case a value of 2.00, corresponding to a slenderness ratio of 115, can be assumed. The trial value of k is $L/115$ equal to $72/115$, or 0.625, and the trial value of d is $4k$ or 2.50 in. Then from Eq. (8-6):

$$d_o = \sqrt[3]{\frac{16}{\pi \times 3,100}} \sqrt{\left(10,000 + \frac{2.0 \times 4,000 \times 2.5(1 + 0.75^2)}{8}\right)^2 + (5,000)^2} \times \frac{1}{\sqrt[3]{1 - 0.75^4}}$$

from which d_o equals 2.93 in. By the use of this value of d_o, k is equal to 0.73, L/k equal to 98.6, and α equal to 1.76. By the substitution of these values in the equation and solving for d_o, the shaft diameter is found to be 2.94 in. This checks the last trial value of 2.93 and is the required diameter. For a transmission shaft, the next larger standard size is $2\frac{15}{16}$ in. For a machine shaft a diameter of $2\frac{15}{16}$ in. with an inside diameter of $2\frac{3}{16}$ in. is satisfactory, provided the deflections are within the specified limits.

8-3 Effect of Keyways. The keyway cut into the shaft materially affects the strength or load-carrying capacity of the shaft, since highly localized stresses occur at and near the corners of the keyway; the effect of these is more pronounced when shock and fatigue conditions prevail. Mathematical analyses of the stresses around the keyway are complex and are seldom made in design. Experimental work by H. F. Moore* indicates that the static weakening effect of the keyway is given by the

---

* H. F. Moore, *Univ. Illinois Eng. Exp. Sta. Bull.* 42.

formula

$$e = 1.0 - 0.2\frac{w}{d} - 1.1\frac{h}{d} \qquad (8\text{-}10)$$

where e = shaft strength factor or ratio of strength of shaft with keyway
to the same shaft without keyway
w = width of keyway, in.
h = depth of keyway, in.
d = shaft diameter, in.

The same experiments indicate that the angular twist of a shaft with a keyway is given by the formula

$$k = 1.0 + 0.4\frac{w}{d} + 0.7\frac{h}{d} \qquad (8\text{-}11)$$

where k is the ratio of the angular twist of the shaft with the keyway to the same shaft without the keyway. Since the twist is increased only in that part of the shaft containing the keyway, and since the hub of the mating member tends to stiffen the shaft, the increase in angular twist may be disregarded except for long keyways used with sliding or feather keys.

The high concentration of stresses at the bottom corners of the keyway may be reduced by the provision of filleted keyways. A fillet radius equal to one-half the keyway depth is considered good practice.* This type of keyway has not been standardized but the U.S. Navy and the American Bureau of Shipping use filleted keyways.

8-4 Transmission Shafts. Transmission shafts serve primarily to transmit power by torsion and are therefore subjected principally to shear stresses. Belt pulleys, gears, chain sprockets, and similar members carried by the shaft introduce bending loads, which cannot in general be determined, and it is customary to assume simple torsion; allowance for the unknown bending stresses is made by using lower design stresses for those shafts in which experience indicates that the bending stresses are severe.

Torsional deformation in transmission shafts should be limited to 1 deg in 20 diameters. Lateral deflection caused by bending should not exceed 0.01 in. per foot of length.

8-5 Commercial Shafting. Transmission shafts are commonly made of cold-rolled stock. For diameters over 3 in. many engineers prefer hot-rolled stock, and for diameters over 6 in. forged stock is generally used. Cold-rolled shafting is stronger than hot-rolled stock of the same analysis,

* *Inco*, vol. 23, no. 2, p. 10, 1949.

but the resistance to deformation is the same. Cold-rolled shafting is made in diameters increasing in steps of $\frac{1}{16}$ in. from $\frac{1}{2}$ to $2\frac{1}{2}$ in., of $\frac{1}{8}$ in. up to 4 in., and of $\frac{1}{4}$ in. up to 6 in. However, transmission shafts are standardized in $\frac{1}{4}$-in. steps from $1\frac{5}{16}$ in. to $2\frac{7}{16}$ in., and in $\frac{1}{2}$-in. steps up to $5\frac{15}{16}$ in. Standard stock lengths are 16, 20, and 24 ft.

8-6 Machine Shafts. The deflection of a machine shaft is usually as important as its strength, and in many cases more important. In most cases a shaft that is rigid enough is also strong enough, and the shaft should be designed for stiffness and checked for strength. The strength of the material does not affect the stiffness, which depends only on the dimensions and the modulus of elasticity of the material; so shafts of similar dimensions and loading have the same deflection, regardless of the kind of steel used. Hence, if the dimensions of a shaft are limited by the permissible deformations, a plain carbon-steel shaft may be just as satisfactory as a shaft of higher-cost alloy steel.

The permissible deformation of a machine shaft depends upon the service for which it is intended. The deflection of shafts carrying gears should be limited to $0.005/f$ in. at the gear, where f is the width of the gear face in inches. For very smooth-running gears, the deflection should be much smaller than this. Brown and Sharpe limit the relative slope of the shafts at the mesh to 0.0005 in./in. The deflection of any machine shaft supported on plain bearings should not exceed $0.00015L$, where L is the distance from the load point to the center of the bearing, in inches.

The angular twist of machine shafts should be limited to 6 min/ft for ordinary service, to $4\frac{1}{2}$ min/ft for variable loads, and to 3 min/ft for suddenly reversed loads and long feed shafts. The angular rigidity of machine-tool spindles should prevent a movement of more than $\frac{1}{64}$ in. at the circumference of the face plate. Milling cutter spindles should have an angular twist of less than 1 deg at the edge of the cutter.

8-7 Determination of Shaft Size for Strength. The shaft diameter required for strength can be found from Eq. (8-6) or (8-7) when the bending moment, the torsional moment, and the conditions of loading are known. The torsional moment is determined from the equation

$$T = \frac{hp \times 33,000 \times 12}{2\pi N} = \frac{63,000\ hp}{N} \tag{8-12}$$

where T = torsional moment, in.-lb
hp = horsepower transmitted
N = shaft speed, rpm

The torsional moment may differ at various points along the shaft if the power is taken off through gears, belts, or chains.

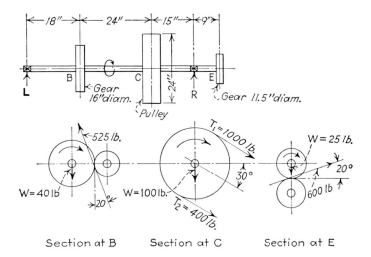

Section at B   Section at C   Section at E

Fig. 8-1

The bending moment is most easily found by replacing each individual load and bearing reaction by its vertical and horizontal components. All vertical components may then be considered as loads on a beam, and the vertical bending moments can be determined. Similarly, all horizontal components may be considered as loads on a beam, and the horizontal bending moments can be determined. The vector sum of the vertical and horizontal moments at any section of the beam is the total bending moment at that section.

Example. Find the diameter of the solid round shaft, $5\frac{1}{2}$ ft long, supported on two bearings, driven by a belt pulley, and driving two gears as shown in Fig. 8-1. The pulley is keyed to the shaft with a key. SAE 1045 annealed steel is used. The torsional moment is suddenly applied with moderate shock. The shaft is rough-finished.

Resolve each load into vertical and horizontal components as shown in Table 8-1.

Table 8-1

Load	lb	Vertical component	Horizontal component
Force B.........	525	$+493$	$+\ \ 179$
Weight B........	40	$-\ \ 40$	
Force C.........	1,400	-700	$-1,212$
Weight C........	100	-100	
Force E.........	600	$+205$	$-\ \ 564$
Weight E........	25	$-\ \ 25$	

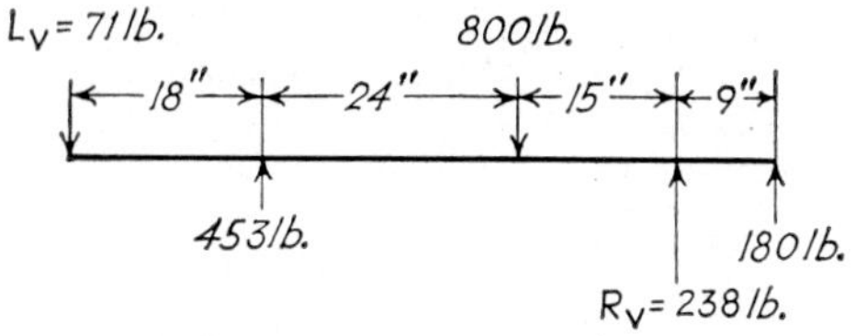

$L_V = 71\,lb.$ $800\,lb.$

Fig. 8-2   Vertical components.

$453\,lb.$ $180\,lb.$

$R_V = 238\,lb.$

The negative sign indicates forces acting vertically downward or horizontally backward.

If only the vertical forces are considered, a beam loaded as in Fig. 8-2 results. The vertical reactions at L and R are

$$L_v = 71 \text{ lb}$$
$$R_v = 238 \text{ lb}$$

The vertical bending moments at L, B, C, R, and E are

$$M_{Lv} = 0$$
$$M_{Bv} = -71 \times 18 = -1{,}278 \text{ in.-lb}$$
$$M_{Cv} = -71 \times 42 + 453 \times 24 = 7{,}890 \text{ in.-lb}$$
$$M_{Rv} = 180 \times 9 = 1{,}620 \text{ in.-lb}$$
$$M_{Ev} = 0$$

The horizontal forces acting on the beam are shown in Fig. 8-3 and the vertical and horizontal moments are given in Table 8-2. The total moments, found by combining the vertical and horizontal moments, are also tabulated.

The maximum bending moment, located at the pulley, is 11,814 in.-lb. The maximum torsional moment, also at the pulley, is $T = 493 \times 8 = 3{,}944$ in.-lb.

From Art. 2-10, $s_{es} = 0.55s_{ef}$; from Table 2-6, $s_{ef} = 42{,}000$ psi. Then $s_{es} = 0.55(42{,}000) = 23{,}100$ psi. Equation (2-1) gives

$$s_e = \frac{s_{es}}{flp} = \frac{23{,}100}{1.2 \times 1.0 \times 1.0} = 19{,}230 \text{ psi}$$

where f is 1.2 (Fig. 2-3 for rough-finished steel), s_{tu} is 85,000 psi (Table 2-6), l is 1.0 for a life corresponding to number of cycles for which s_{ef} was determined, and p is 1.0 for complete stress reversal.

Table 8-2

Section	Bending moments		
	Vertical	Horizontal	Total
L	0	0	0
B	1,278	1,926	2,311
C	7,890	8,790	11,814
R	1,620	5,076	5,336
E	0	0	0

Equation (4-5) gives

$$s_d = \frac{s_e}{aK(df)}$$

Using a service factor a of 2 for suddenly applied loads with minor shock, a stress-concentration factor K of 1.6 (from Table 3-2), and a design factor df of 1.5

$$s_d = \frac{19{,}230}{2.0 \times 1.6 \times 1.5} = 4000 \text{ psi}$$

Using the design stress s_d for $s_{s \text{ max}}$, Eq. (8-4) becomes

$$d = \sqrt[3]{\frac{16}{\pi s_d}} \sqrt{M^2 + T^2} = \sqrt[3]{\frac{16}{\pi \times 4000}} \sqrt{(11{,}814)^2 + (3{,}944)^2}$$
$$= 2.51 \text{ in., say } 2\tfrac{1}{2} \text{ in.}$$

If the maximum bending and torsional moments are not at the same section of the shaft, it is necessary to determine the diameter required by the combination of moments at each point of load application, and to use a shaft of varying diameters or select the largest diameter. If shoulders or other discontinuities causing stress concentrations are present the diameters must be calculated at these sections taking into account the stress-concentration factors. The use of the torsional endurance limit (when the torque was steady) instead of the flexural endurance limit is also debatable.

Since this calculation gives the diameter required for strength only, the shaft should now be checked for deflection and twist, if these have specified maximums.

8-8 Graphical Determination of Bending Moments. When there are many loads to be considered, the use of graphical methods for determining the bending moments will expedite the solution. The method is best explained by an illustrative solution, and for this purpose, the data in the preceding article are used. The shaft and the loads are shown in Fig. 8-1. The resultant loads are shown in Fig. 8-4 as forces acting through the center of the shaft.

In Fig. 8-4 the total belt load is shown to scale, by the vector F_C. If the weights of the pulley and belt are added, R_C is the resultant load. Similarly, if gear weights are added to each of the gear loads, R_B and R_E are the resultant loads. The resultant loads scale 1,453,488, and 591 lb respectively.

The total bending moment at any section of the beam is the vector sum of the bending moments produced at that section by the individual

Fig. 8-3 Horizontal components.

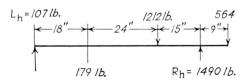

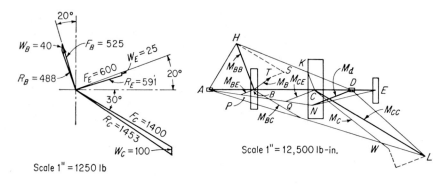

Fig. 8-4   Fig. 8-5

loads. Consequently, if the vectors representing the individual moments can be obtained graphically, they can be added graphically to determine the total moment at any desired section of the beam.

Consider all the loads removed except the load at the gear B. The bending moment under the gear will then be

$$M_{BB} = \frac{488 \times 39}{57} \times 18 = 6{,}010 \text{ in.-lb}$$

In Fig. 8-5 if BH is drawn parallel to R_B and equal to M_{BB}, i.e., 6,010 in.-lb, then $AHDE$ is the bending-moment diagram for the beam, and CK, parallel to BH, represents the bending moment at C produced by the load R_B acting alone.

Now consider the beam with the loads at C acting, the resultant of these loads being R_C. In a manner similar to that just used, CL is made equal to M_{CC} and parallel to R_C, and $ALDE$ becomes the bending-moment diagram of the beam for the load R_C. Then BG parallel to CL represents the bending moment at B produced by the load R_C.

Similarly, ANE is the bending-moment diagram for the load R_E, and BP and CQ represent the moments at B and C, respectively, produced by this load.

At the section B, there are three bending moments, M_{BB}, M_{BC}, and M_{BE}. If these vectors, as indicated by the vector polygon $BHST$, are added, then BT, their vector sum, represents to scale the bending moment at B, or M_B. Similarly, M_C and M_D, the bending moments at C and D, are found. By scaling the vectors, M_B is found to be 2,300 in.-lb, M_C 11,800 in.-lb, and M_D 5,330 in.-lb. The maximum bending moment having been found to be 11,800 in.-lb, the shaft diameter can be determined by substitution in Eq. (8-4), as in the preceding example.

8-9 Design of Shafts for Deflection. As previously stated, most machine shafts must be designed for a specified maximum deformation, and as the rigidity, rather than the strength, usually determines the shaft size, the deformation computations should be made first, and then the shaft should be checked for strength. When there are a number of loads acting, the principle of superposing deformations may be used to advantage, i.e., the total deformation at any point is found by adding the deformations produced at this point by each load acting separately.

Example. Assume the same shaft and loading as in the previous example and in Fig. 8-1. Determine the required shaft diameter if the deflection at any load is limited to 0.025 in.

Consider the shaft with the single vertical load at B. From the beam formulas, the vertical deflection at the gear B is found to be

$$y_{vBB} = \frac{Fb^2a^2}{3EIL} = \frac{453 \times 18^2 \times 39^2 \times 64}{3 \times 30,000,000 \times \pi d^4 \times 57} = \frac{0.890}{d^4}$$

and the deflection at the pulley C is

$$y_{vCB} = \frac{Fbx}{6EIL} (L^2 - b^2 - x^2)$$

$$= \frac{453 \times 18 \times 15 \times 64}{6 \times 30,000,000 \times \pi d^4 \times 57} (57^2 - 18^2 - 15^2)$$

$$= \frac{0.631}{d^4}$$

and the deflection at the gear E is

$$y_{vEB} = (\text{slope at } D) \times 9 = \frac{Fb}{6EIL} (L^2 - b^2 - 3x^2) \times 9$$

$$= \frac{453 \times 18 \times 64}{6 \times 30,000,000 \times \pi d^4 \times 57} (57^2 - 18^2 - 0) \times 9$$

$$= \frac{0.448}{d^4}$$

Note that the deflections y_{vBB} and y_{vCB} are up, whereas y_{vEB} is down.

Table 8-3

| Load | Deflection | | | | | |
| | Vertical | | | Horizontal | | |
	B	C	E	B	C	E
B.........	$+\dfrac{0.890}{D^4}$	$+\dfrac{0.631}{D^4}$	$-\dfrac{0.448}{D^4}$	$+\dfrac{0.352}{D^4}$	$+\dfrac{0.250}{D^4}$	$-\dfrac{0.018}{D^4}$
C.........	$-\dfrac{0.223}{D^4}$	$-\dfrac{1.265}{D^4}$	$+\dfrac{0.297}{D^4}$	$-\dfrac{1.510}{D^4}$	$-\dfrac{8.600}{D^4}$	$+\dfrac{2.020}{D^4}$
E.........	$-\dfrac{0.170}{D^4}$	$-\dfrac{0.200}{D^4}$	$+\dfrac{0.218}{D^4}$	$+\dfrac{0.532}{D^4}$	$+\dfrac{0.627}{D^4}$	$+\dfrac{0.684}{D^4}$
Total.......	$+\dfrac{0.497}{D^4}$	$-\dfrac{0.834}{D^4}$	$+\dfrac{0.067}{D^4}$	$-\dfrac{0.626}{D^4}$	$-\dfrac{7.723}{D^4}$	$+\dfrac{1.318}{D^4}$

In a similar manner, the deflections caused by the vertical components of the loads at C and E, and the deflections caused by the horizontal components, are found. The deflections are tabulated in Table 8-3. Evidently the maximum total deflection at a load point is at the pulley C, and by vector addition of the vertical and horizontal deflections

$$y_C = \frac{1}{d^4} \sqrt{0.834^2 + 7.723^2} = \frac{7.76}{d^4}$$

from which

$$d^4 = \frac{7.76}{y_C} = \frac{7.76}{0.025} = 310.2$$

and

$$d = 4.20 \text{ in.}$$

In this case, the diameter required for strength is less than that required for deflection, and so the shaft diameter should be made $4\frac{1}{4}$ in.

8-10 Graphical Determination of Deflection.   The mathematical determination of deflection becomes very tedious when there are many loads, and especially so when the shaft is made up of sections of different diameters. Graphical determination of the deflection can be conveniently applied with sufficient accuracy for most purposes.

Texts on mechanics show that the second derivative of the deflection equation of any shaft is expressed by the relation

$$\frac{d^2y}{dx^2} = \frac{M}{EI} \tag{8-13}$$

where y = deflection at a distance x from one end of shaft, in.
$\quad\quad M$ = bending moment at section x, in.-lb
$\quad\quad E$ = modulus of elasticity of material, psi
$\quad\quad I$ = rectangular moment of inertia of shaft area at same section, in.[4]

From the known loads and reactions acting on the shaft, the bending-moment diagram is plotted. When the moments are divided by the values of EI for the corresponding sections of the shaft and plotted, a diagram representing the variations of M/EI over the entire length of the shaft is obtained. Double integration of this diagram gives the deflection curve of the shaft. The process of graphical integration is best shown by an illustrative example.

Example.   Assume a shaft loaded as shown in Fig. 8-6. Determine the deflection curve.

The bending moments are found to be 192,000 in.-lb at the 10,000-lb load, and 224,000 in.-lb at the 16,000-lb load. These values determine the moment diagram, curve 1, Fig. 8-6. From the left reaction to the section B, the shaft has a diameter of 4 in., and the value of I is 12.57 in.[4] From curve 1, the moment

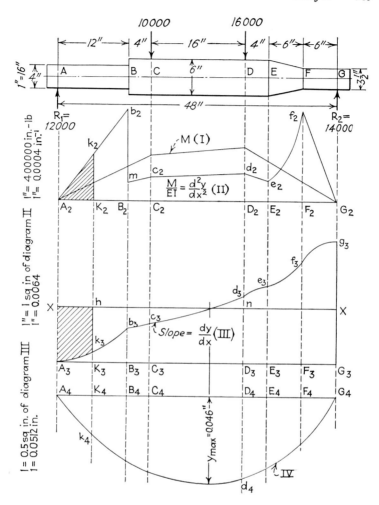

Fig. 8-6

at section B is found to be 144,000 in.-lb. Then, at this section

$$\frac{M}{EI} = \frac{144,000}{30,000,000 \times 12.57} = 0.000382 \text{ in.}^{-1}$$

The diameter changes at this section from 4 to 6 in. and M/EI changes. Hence

$$\frac{M}{EI} = \frac{144,000}{30,000,000 \times 63.62} = 0.000075 \text{ in.}^{-1}$$

Similarly, the values of M/EI at the sections C, D, E, F, and G are found. By plotting these to scale, curve 2 is obtained. To integrate this curve select any section such as K_2 and draw K_2k_2. By any means desired, measure the area $A_2k_2K_2$, and plot this area to scale in curve 3, locating the point k_3. Measure the area $A_2b_2B_2$ and plot as B_3b_3, locating point b_3. Measure the area $A_2b_2mc_2C_2$ and plot, locating the point c_3. Continue in this manner until the entire curve 3 has been plotted. This curve represents the first integral of curve 2, and therefore represents dy/dx, the slope of the deflection curve. The ordinates of curve 3 are measured in abstract units since areas in curve 2 are ordinates times abscissae, or inches^{-1} times inches. To eliminate the constant of integration, find the mean ordinate of curve 3 and plot this value as the line XX. This is the base line from which to measure values of dy/dx.

Integrate curve 3 by measuring the areas between the curve and the base line XX. Thus K_4k_4 represents to scale the area A_3k_3hX, and D_4d_4 represents the area $A_3b_3c_3d_3nX$. Remember that areas below the base line XX are negative and areas above are positive. The complete integration of curve 3 gives curve 4, which is the deflection curve for the shaft. Ordinates measured from the base line A_4G_4 represent deflections in inches, to the proper scale.

8-11   Critical Speeds of Shafts.

It is a commonly recognized fact that at certain speeds a rotating shaft becomes dynamically unstable, and that at these speeds excessive and even dangerous deflections may occur. The speeds at which a shaft becomes dynamically unstable are called the critical speeds and correspond to the speeds at which the number of natural vibrations, or natural frequency, equals the number of revolutions per minute.

Consider the shaft in Fig. 8-7 with the disk of mass M located between the supports. Also consider that the center of mass of the disk is at a distance e from the axis of rotation of the shaft. When the shaft is rotating with an angular speed ω, in radians per second, the disk rotates at the same angular speed; but since it is not rotating about an axis through its center of mass there is set up a centrifugal force Mv^2/r, which tends to deflect the shaft an amount y. The radius of the path of the center of mass of the disk is then $y + e$, and the centrifugal force becomes

$$F_c = \frac{Mv^2}{y + e} = M(y + e)\omega^2 \tag{8-14}$$

where M is in pounds-seconds squared per inch, V in inches per second, y and e are in inches, and ω is in rads per second. This force causes the deflec-

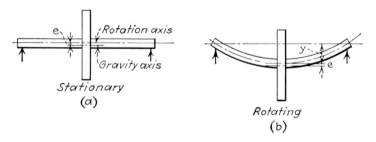

Fig. 8-7

tion y to increase until the flexural resisting moment of the shaft balances the bending moment produced by the centrifugal force. The relation between the deflection in any shaft and the force F producing that deflection is expressed by the equation

$$y = C\frac{FL^3}{EI}$$

or

$$F = \frac{EI}{CL^3}y \tag{8-15}$$

where C is a constant depending on the type of beam and the method of loading and I is the rectangular moment of inertia, in.[4]

When the condition of equilibrium is reached,

$$F_c = F$$

or

$$M(y + e)\omega^2 = \frac{EI}{CL^3}y$$

from which

$$y = \frac{e\omega^2}{EI/CML^3 - \omega^2} \tag{8-16}$$

Examination of this equation shows that the deflection becomes infinite when the angular velocity is such that

$$\omega_{cr}^2 = \frac{EI}{CML^3}$$

and

$$\omega_{cr} = \sqrt{\frac{EI}{CML^3}} \tag{8-17}$$

From Eq. (8-15), $F/y = EI/CL^3$, and if this ratio is called K, Eq. (8-17) becomes

$$\omega_{cr} = \sqrt{\frac{K}{M}} = \sqrt{\frac{Kg}{W}} \tag{8-18}$$

where g is the acceleration due to gravity, in inches per second per second.

If the equation for the deflection of the beam is known, the critical speed ω_{cr} can be computed.

For a central disk of mass M on a shaft of negligible weight, rotating on antifriction bearings, the shaft is assumed to be simply supported, and

$$y = \frac{FL^3}{48EI}$$

and

$$\frac{F}{y} = \frac{48EI}{L^3} = K$$

Hence

$$\omega_{cr} = \sqrt{\frac{48EIg}{L^3W}} \tag{8-19}$$

For a central disk of mass M on a shaft of negligible weight, rotating in sleeve bearings, the shaft is assumed to be fixed-ended, and

$$y = \frac{FL^3}{192EI}$$

and

$$\omega_{cr} = \sqrt{\frac{192EIg}{WL^3}} \tag{8-20}$$

For a noncentral disk of mass M at a distance a from the left support and at a distance b from the right support, a shaft of negligible weight rotating in rolling-element bearings, the deflection under the load is

$$y = \frac{a^2b^2F}{3EIL}$$

and

$$\omega_{cr} = \sqrt{\frac{3EILg}{Wa^2b^2}} \tag{8-21}$$

For the same condition as above except using sleeve bearings, the deflection is

$$y = \frac{Fa^3b^3}{3EIL^3}$$

and

$$\omega_{cr} = \sqrt{\frac{3EIL^3g}{Wa^3b^3}} \tag{8-22}$$

8-12 Critical Speed of a Shaft with Several Disks. When there are several disks or masses M_1, M_2, M_3, . . . M_n, on the shaft, the critical speed of the shaft may be found approximately by the equation

$$\omega_{cr} = \frac{\omega_1\omega_2\omega_3 \cdots \omega_n}{\sqrt{(\omega_1\omega_3\omega_4 \cdots \omega_n)^2 + (\omega_1\omega_2\omega_4 \cdots \omega_n)^2 + \cdots + (\omega_2\omega_3\omega_4 \cdots \omega_n)^2}} \tag{8-23}$$

where ω_1 is the critical speed of the shaft and a mass M_1, ω_2 is the critical speed of the shaft and a mass M_2, etc.

8-13 Critical Speed of a Uniform Shaft. The critical speed of a shaft supporting a uniform load, or of a shaft that supports no loads except its own weight, is found by considering the shaft to be made up of a number of short lengths of known mass. The critical speeds of the weightless shaft carrying these individual masses can then be computed and combined by means of Eq. (8-23) to determine the critical speed of the entire shaft. By this procedure, the critical speed of a shaft of uniform section, simply supported in end bearings and not supporting any concentrated masses, is found to be

$$\omega_{cr} = \sqrt{\frac{\pi^4 E I g}{W L^3}} = \sqrt{\frac{1.275g}{y}} \tag{8-24}$$

where W = total uniform load or shaft weight, lb
 L = distance between supports, in.
 y = deflection produced in a simple uniformly loaded beam by the distributed weight W, in.

The speed determined by this equation is the lowest of a series of critical speeds. Other critical speeds occur at 4, 9, 16, 25, etc., times the lowest critical speed.

A shaft of uniform section, uniformly loaded, and rigidly supported in end bearings has for its lowest critical speed

$$\omega_{cr} = \sqrt{\frac{5.0625\pi^4 E I g}{W L^3}} = \sqrt{\frac{1.275g}{y}} \tag{8-25}$$

and others at $(\frac{5}{3})^2$, $(\frac{7}{3})^2$, $(\frac{9}{3})^2$, $(\frac{11}{3})^2$, etc., times this speed.

8-14 Critical Speeds in Terms of the Weight and Deflection. In cases where the total deflection of the shaft at each supported disk is known, or can be easily determined by computation or graphical analysis, it is more convenient to state the critical speeds in terms of the disk weights and the deflections. In Eq. (8-17), change rads per second to revolutions per minute, and substitute the deflection y for its equivalent CFL^3/EI. Then, for a shaft with a single disk,

$$N_{cr} = \frac{60\omega_{cr}}{2\pi} = \frac{30}{\pi}\sqrt{\frac{F}{My}} = \frac{30}{\pi}\sqrt{\frac{g}{y}} \tag{8-26}$$

For a shaft supporting n disks

$$N_{cr} = \frac{30}{\pi}\sqrt{\frac{g(W_1 y_1 + W_2 y_2 + \cdots + W_n y_n)}{W_1 y_1^2 + W_2 y_2^2 + \cdots + W_n y_n^2}} \tag{8-27}$$

where W_1, W_2, etc., are the weights of the disks, and y_1, y_2, etc., are the deflections at the respective disks.

8-15 Operating Speeds. Damping effects in the shaft will reduce the total deflection when the shaft is rotating at a critical speed, but large deflections will occur that may cause serious trouble and possibly structural damage to the machine. This is particularly true if the operating speed is alternately above and below the critical speed. For this reason, the operating speed should be well removed from the neighborhood of any of the series of speeds at which these extreme deformations and vibrations may occur. In some machines regular impulses may be transmitted to the shaft and if the timing of these impulses approximates the natural frequency, or critical speed of the shaft, trouble may be expected although the shaft speed may be safely removed from the critical speed. This often happens in shafts connected to multiple-cylinder internal-combustion engines.

In order to avoid vibration troubles, some machines are mounted with very flexible shafts that, when rotated at high speeds, allow the disks to rotate on their own centers of gravity. De Laval used this principle in his early high-speed turbines. Centrifugal separators that operate at speeds of from 20,000 to 50,000 rpm also employ these flexible shafts.

In general, if the operating speed of any shaft is removed at least 20 per cent from any critical speed, there will be no vibration troubles.

⑨
Rolling-element Bearings

9-1 Definition of a Bearing. Whenever there is relative motion between two machine members, that member which controls the motion of, guides, or supports the other is known as a bearing. This motion may be rotational, translational, or a combination of the two.

9-2 Bearing Classification According to Load Application. Bearings may be classified as radial, thrust, or guide bearings. The common ball bearings or oil-lubricated cylindrical bearings used to support shafts are *radial bearings* since the load application onto these bearings is radial to the axis of rotation of the supported shafts; radial bearings often are referred to as *journal bearings* since the rotating shafts they support are called journals. A *thrust bearing* carries a load collinear to the axis of possible rotation of the supported member; bearings which carry the thrust from the propeller of an outboard motor or the weight of a phonograph turntable are thrust bearings. A *guide bearing* exists primarily to guide the motion of a machine member without specific regard to the direction of load application; the cylinders of reciprocating pumps and the crosshead bearings of double-acting engines are guide bearings. Often a bearing is a combination of two types; the common ball or tapered roller bearing which radially supports a portion of the shaft weight and its applied radial loads as well as a thrust component caused by a helical gear on the shaft is a combination of a radial and a thrust bearing.

9-3   Bearing Classification According to Type of Friction.   Depending upon whether primarily sliding or rolling friction exists within the bearing, bearings may be classified as sliding bearings and rolling-element bearings. Oil-lubricated journal bearings and reciprocating engine cylinders are *sliding bearings* since the action within the lubricant and between the mating surfaces is essentially a sliding or shearing action. Ball, roller, and needle bearings are *rolling-element bearings* since the bearing elements essentially are in rolling contact. Both sliding and rolling-element bearings have applications in which they may be used advantageously over the other type; there are many applications in which either sliding or rolling-element bearings may be employed to equal advantage. As a rule of thumb, sliding bearings are used wherever minimum initial cost, quietness of operation, and minimum running friction are desired; rolling-element bearings are used where minimum starting friction is desired, under conditions of intermittent or no lubrication during operation of the bearing, and in some applications where adequate bearing inspection and maintenance cannot be assured.

9-4   Classification of Rolling-element Bearings.   Rolling-element bearings may be classified as *ball* or *roller bearings* depending upon whether the primary rolling elements are spheres or rollers. Although sometimes *needle bearings* are listed as a third type of rolling-element bearings, they really are roller bearings.

The ball bearing consists of an outer race and an inner race held concentric with each other by several spherical balls circumferentially equally spaced. Usually a ball retaining cage is employed to assure the equal spacing of the balls. In some instances, one or both of the races may not exist, and hardened shafts and/or hardened bored holes may be substituted. In general, ball bearings may be divided into single-row radial bearings without loading grooves, single-row radial bearings with loading grooves (also called filling slots), single-row angular contact bearings, double-row bearings, bearings derived from the basic types with added modifications (such as shields, seals, and snap-ring grooves), and special-purpose bearings. These typical bearings are shown in Fig. 9-1.

There are many types of ball bearings providing for different radial- and thrust-load characteristics, mounting methods, and lubrication requirements. *Single-row radial bearings* without loading grooves contain as many balls as can be introduced by eccentric displacement of the races; a ball cage or retainer is employed to keep the balls equally spaced after

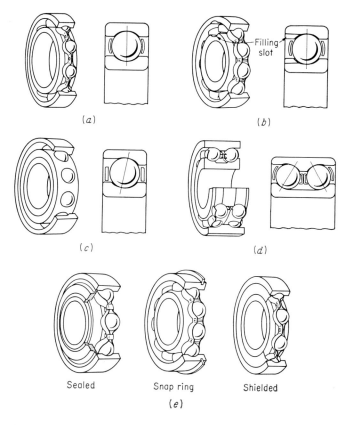

Fig. 9-1 Types of ball bearings. (*a*) Single-row radial (without filling slot). (*b*) Single-row radial with filling slot. (*c*) Single-row angular contact. (*d*) Double-row angular contact. (*e*) Derived bearings.

the races have become concentric. These bearings have high radial-load capacity with essentially an equal thrust-load capacity for loading in either direction. Single-row radial bearings provided with loading grooves or filling slots in each race permit the balls to be inserted into the bearing when the races are concentric; cages are employed to retain the balls. These bearings have very high radial-load capacities because of the large number of balls; on the other hand, because of the loading grooves in the races, they are capable of resisting only moderate thrust loads in either direction. Single-row *angular contact bearings* have a high race shoulder on one side of both races and are open on the opposite side except for a small shoulder which serves to hold the bearing together after assembly (effected by cooling the inner race and heating the outer race). Some angular contact bearings have shoulders on both sides of the inner race and on one side of the outer race; the outer race is completed by a cylindrical race opposite the shoulder, thereby permitting the outer race to be removed. Angular contact bearings have a comparatively high radial load capacity and a high thrust capacity in one direction. To permit taking thrust loads in either direction, angular contact bearings often are mounted in pairs, either side by side or at opposite ends of a shaft. Duplex bearings are two angular contact bearings that have been ground with offset race rings to permit them to be clamped together with an axial thrust preload existing between the two bearings; duplex bearings can carry both heavy radial loads and heavy thrust loads from either direction. *Double-row bearings* are essentially two angular contact bearings manufactured with a common inner race and a common outer race; they are capable of supporting high radial loads and high thrust loads from either direction. The *derived bearings* are any of the previously mentioned types of ball bearings with one or more modifications, such as the addition of shields, seals, and snap rings. Many special types of ball bearings exist, such as self-aligning, thrust, conveyor, wheel, adapter, pump-shaft, and instrument bearings. Their construction and capacity vary with the various manufacturers. Some self-aligning bearings are made with the outer race a spherical surface centered on the axis of the shaft so that the bearing is free to adjust itself to angular misalignment of the shaft. A ball bearing capable of carrying only thrust loads consists of two concentric thrust collars separated by several balls with or without a ball retainer; the balls are constrained to roll along circular retaining grooves in each collar. Several commercially available special-purpose ball bearings are depicted in Fig. 9-2.

A *roller bearing* consists of an outer race, an inner race, and a set of rollers with or without a roller cage or retainer. In some instances, either or both of the races may be absent with a hardened shaft and/or a bored hole serving for the missing race. If the axes of the rollers are parallel to the rotational axis of the supported shaft, the bearing usually is a *cylindrical* (or straight) *roller bearing*.* If the axes are inclined to and intersect the rotational axis of the shaft at a common point, the bearing is a *tapered* (or conical) *roller bearing*. If the length-to-diameter ratio of the rollers is large (say 6 or more), the bearing often is called a *needle* (or quill) *bearing;* in this case a roller cage usually is not used, only retaining lips on the races. Bearings with smaller length-to-diameter ratios are referred to simply as roller bearings. In the simplest type of roller bearings, no retainer is used. The loose rollers tend to twist in the races, thus increasing the friction, wearing the rollers cigar-shaped, and causing a bending action which may break the rollers. Consequently, retainer lips on the races and retaining cages on the rollers are used to confine, space, and guide the rollers.

---

* The rollers also may be barrel or hourglass in shape.

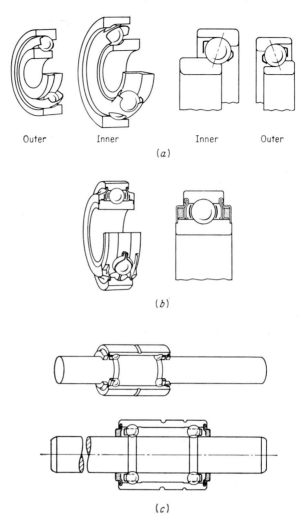

Outer   Inner   Inner   Outer

(*a*)

(*b*)

(*c*)

Fig. 9-2 Several special-purpose ball bearings. (*a*) Automotive front wheel bearing. (*b*) Automotive rear wheel bearing. (*c*) Pump and fan shaft bearing.

Some forms of roller bearings have rollers formed of helically wound steel strips, which are especially suitable for shock loads. Since adjacent rollers are wound in opposite directions, they aid in sweeping the lubricant across the bearing surface. The rollers and races of conical roller bearings are truncated cones having a common apex on the shaft center to ensure true rolling contact. The taper usually is rather small (6 to 7 deg) to prevent imposed radial loads from causing excessive end thrust. Roller bearings with barrel- and hourglass-shaped rollers are made with one spherical race

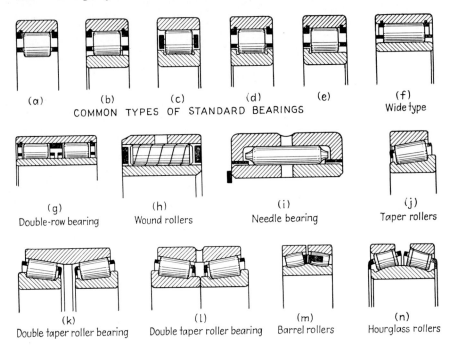

(a)

(b)

(c)

(d)

(e)

(f)
Wide type

COMMON TYPES OF STANDARD BEARINGS

(g)
Double-row bearing

(h)
Wound rollers

(i)
Needle bearing

(j)
Taper rollers

(k)
Double taper roller bearing

(l)
Double taper roller bearing

(m)
Barrel rollers

(n)
Hourglass rollers

Fig. 9-3   Types of roller bearings.

to provide self-alignment with the shaft. Each manufacturer has its own classification of roller bearing, and hence it is difficult to classify them further in a general way. Typical roller bearings are shown in Fig. 9-3.

9-5   Bearing Series.   Almost all ball bearings and many roller bearings are made to standard dimensions of outside diameter, bore, and width. For each standard bore, the bearings are usually made in four different widths: the *extra-light, light, medium,* and *heavy series,* sometimes referred to as the *100, 200, 300,* and *400* series respectively. Sometimes an *extra-heavy-* or *500*-series bearing is manufactured. The common series of ball bearings are illustrated in Fig. 9-4. The selection of the proper series of bearing depends upon the load, shaft size, housing space limitations, and other design considerations. The extra-light-series bearings are used mainly where the loads involved are so small that the major purpose of the bearing is to hold the supported shaft in a desired position. Light-series bearings are used when loads are moderate, shafts are relatively large in order to achieve stiffness and are hollow, and housing space limitations demand that the outer race be small in outside diameter. Medium-series bearings are wider and have larger outside diameters than light-series bearings of the same bore; they have approximately one-third larger radial-load capacity, and they are used when loads are relatively heavy in comparison to shaft diameter. Heavy-series bearings are used when loads are heavy in comparison to shaft size; they are approximately one-fourth to one-third stronger than medium-series bearings of the same bore, and they are used mainly in connection with specially proportioned shafts for heavy loads. Extra-heavy-series bearings are used to carry unusually heavy loads in comparison to shaft size. More and more roller bearings are replacing extra-heavy ball bearings, hence extra-heavy ball bearings have limited usage in special applications.

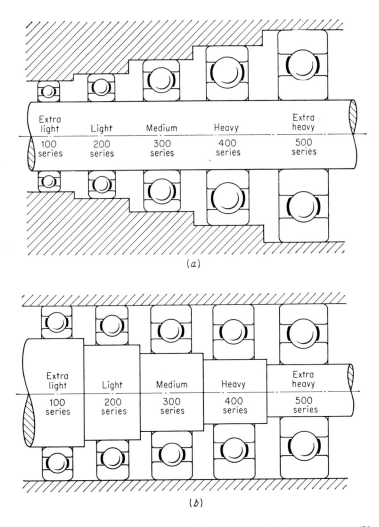

Fig. 9-4 Ball-bearing series. (*a*) Five series with the same bore. (*b*) Five series with the same outside diameter.

9-6 Standard Indication of Bearing Bore. The majority of ball bearings and many roller bearings are manufactured to standard increments of bore variation, either in millimeters or fractions of an inch, known respectively as the *metric-series* and the *inch-series* bearings. For the metric series, the *International Standard Number* (same as the SAE number) usually is used to indicate the bore.

Each manufacturer has its own bearing numbering system which, in addition to indicating the bore and series of the bearing, may also indicate preloading information, operational noise, tolerance of manufacture, special additions such as seals or snap-ring grooves, lubricant, and other items.

9-7 Mechanics of Rolling Friction. Although in rolling-element bearings the theoretical action between the balls or rollers and the races is rolling contact, this is not completely correct. Instead of point or line contact between the element and the races, because of elastic and plastic deformations, the areas of contact are instead circular or oblong. In addition, some sliding action occurs in most cases; this is particularly noticeable for bearings with loose fits with the shafts and/or the housings into which they are pressed. The effect of a roller pressed against a flat supporting surface is shown in exaggerated form in Fig. 9-5. Because of deformations of the roller and its contacting surface, when the roller is pulled horizontally to the right, it must climb out of the indentation formed. When moments are taken about the effective forward point of contact P, it is found that

$$F_f R \cos \theta = F_n R \sin \theta \qquad (9\text{-}1)$$

and

$$\tan \theta = \frac{F_f}{F_n} = f_r \qquad (9\text{-}2)$$

where F_f = frictional resisting force, lb
$\quad F_n$ = normal force between the roller and the flat surface, lb
$\quad f_r$ = coefficient of rolling friction
$\quad R$ = radius of the roller, in.
$\quad \theta$ = friction angle, rad

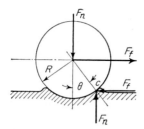

Fig. 9-5   Roller on flat elastic surface.

Table 9-1   Average values of coefficients of rolling friction f_r (R is radius of roller, in.)

Commercial radial ball bearings............................	$0.0008/R$ to $0.0012/R$
Commercial ball thrust bearings.........................	$0.0032/R$ to $0.0036/R$
Commercial roller bearings.............................	$0.0010/R$ to $0.0015/R$
Steel rollers and plates with well-finished clean surfaces without lubricant......................................	$0.0005/R$ to $0.001/R*$
Steel rollers and plates with well-finished and lubricated surfaces..	$0.001/R$ to $0.002/R*$
Steel on steel.......................................	$0.002/R$ (average)*
Iron on iron..	$0.002/R$ (average)*
Hard polished steel on hard polished steel...............	$0.0002/R$ to $0.0004/R*$

* Lionel S. Marks, "Mechanical Engineer's Handbook," 5th ed., pp. 222–223, McGraw-Hill Book Company, New York, 1951.

As in sliding friction, the *coefficient of rolling friction* f_r is defined as the tangent of the friction angle between the normal force and the frictional resisting force. Data on rolling friction are scarce and highly variable. Such friction depends upon the radii and the deflections of the rolling surfaces, the roughnesses of the mating surfaces, the conditions of lubrication, and other items. Keeping all items constant but the radius R, f_r varies inversely as R. The coefficient of rolling friction decreases as the hardness and Young's modulus of the materials increase and apparently is almost completely independent of velocity and temperature. Typical values of f_r are given in Table 9-1. The usually published coefficients of rolling friction for rolling-element bearings are effective coefficients of friction at the bore of the bearing.

9-8 Bearing Capacity Based Upon Stresses. The balls, rollers, and races of loaded bearings deform, causing circular or oblong area contact as indicated in Fig. 9-6. As previously discussed in Chap. 3, the maximum Herzian* contact stress occurs at the center of each contact area. For a Poisson's ratio of 0.30 and for the same Young's modulus for all materials in rolling contact, the maximum Herzian contact stresses are as follows:

$$s_{\max} = 0.418 \sqrt{\frac{FE(R_1 + R_2)}{LR_1R_2}} \quad \text{for two rollers} \tag{9-3}$$

$$s_{\max} = 0.418 \sqrt{\frac{FE}{LR}} \quad \text{for a roller and a plane} \tag{9-4}$$

$$s_{\max} = 0.388 \sqrt[3]{\frac{FE^2(R_1 + R_2)^2}{R_1^2 R_2^2}} \quad \text{for two balls} \tag{9-5}$$

$$s_{\max} = 0.388 \sqrt[3]{\frac{FE^2}{R^2}} \quad \text{for a ball and a plane} \tag{9-6}$$

* Also called Hertzian contact stresses.

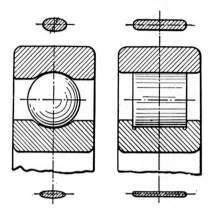

Fig. 9-6   Typical contact areas for ball and roller bearings.

where E = Young's modulus, psi
F = force between one ball or roller and mating race or plane, lb
L = length of roller, in.
R, R_1, R_2 = radius of ball or roller, in.
s_{max} = maximum Herzian contact stress, psi

Because the radii of the races are large compared with those of the balls or rollers, Eqs. (9-4) and (9-6) respectively are suitable for roller and ball bearings. Unless the ball or roller diameters are known, these equations are not very useful. Timoshenko[*] points out that the maximum shear stress occurs below the surface at the point of maximum contact stress and is equal to

$$s_{s\,max} = 0.31 s_{max} \qquad \text{for balls} \qquad (9\text{-}7)$$
$$s_{s\,max} = 0.304 s_{max} \qquad \text{for cylinders} \qquad (9\text{-}8)$$

For ductile materials, this $s_{s\,max}$ usually is the critical stress. For brittle materials, s_{max} usually is critical.

The fundamental theory applying to roller and ball bearings is based on the work of Striebeck,[†] who developed a mathematical treatment based on his own experiments and the earlier ones of Herz. From static tests, Striebeck found that the compressive breaking load F_c of a single ball of diameter D is expressed by the formula

$$F_c = K_1 D^2 \qquad (9\text{-}9)$$

where K_1 is a constant varying with the material and with the shape of the supporting races. At the breaking load, K_1 is about 100,000 for carbon-

---

[*] S. Timoshenko, "Strength of Materials," 2d ed.. part II, pp. 355–361, D. Van Nostrand Co., Inc , Princeton, N.J., 1951.

[†] Striebeck, "*Kugellager für beliebige Belastung,*" vol. 45, 1901, p. 75; vol 29, 1902, p. 1421, D.I.V. Translation by H. Hess, *Trans. ASME*, vol. 29, p. 367, 1907.

steel balls, and 125,000 for hardened alloy-steel balls. A factor of safety of at least 10 should be used with these values.

For the average radial bearing containing n balls, the maximum load per ball is

$$F_{max} = \frac{4.37F_r}{n} \tag{9-10}$$

where F_r is the total radial load on the bearing, in pounds.

Hence the radial load capacity of a ball bearing is

$$F_r = \frac{F_{max}n}{4.37} = \frac{K_1nD^2}{4.37} \quad \text{or approximately} \quad \frac{K_1nD^2}{5} \tag{9-11}$$

Safe working values of K_1 for average bearing life are 550 for unhardened steel, 700 for hardened carbon steel, 1,000 for hardened alloy steel on flat races, 1,500 for hardened carbon steel, and 2,000 for hardened alloy steel on grooved races having a radius equal to $0.67D$. Modern commercial bearings, using the higher-strength alloy steels and races having nearly the same curvature as the balls, have much higher values of K_1. These values of K_1 are for a life of 3,000 hr at 100 rpm.

For roller bearings, Eq. (9-11) takes the form

$$F_r = \frac{K_2nLD}{5} \tag{9-12}$$

where L is the roller length, in inches, and D is the roller diameter, in inches. The value of K_2 is 7,000 for hardened carbon steel and 10,000 for hardened alloy steel.

Table 9-2 gives the basic standard dimensions for many of the SAE and International Standard ball and roller bearings. There are sizes other than those listed, e.g., an extra-light metric series, an inch series, miniature bearings with a bore as small as 0.04 in., and other special bearings.

The actual radial- and/or thrust-load capacity of ball and roller bearings is determined mainly by the fatigue life of the bearing materials. This in turn is a function of the rotational speeds of the elements, the loads applied onto the bearing, the expected life of the bearing, cleanliness and lubrication of the bearing, mounting of the bearing, and other items. The useful life of rolling-element bearings usually is ended by pitting or spalling of the rolling surfaces, a typical fatigue failure. Equation (9-12) indicates that the capacity of any roller bearing depends upon the number, length, and diameter of the rollers together with the materials used. The capacity also depends upon the expected life in hours, the speed of rotation, the type of bearing, and the type of application. Proper selection of a rolling-element bearing depends upon all of these items as well as the acceptable probability of the bearing not failing before its expected life.

Table 9-2   **SAE and International Standard dimensions for ball and roller bearings**

SAE and IS numbers	Bore, all series		Extra-light (100) series		Light (200) series		Medium (300) series		Heavy (400) series	
	mm	in.	O.D., in.	Width, in.	O.D., in.	Width, in.	O.D.. in.	Width, in.	O.D., in.	Width, in.
00	10	0.3937	1.0236	0.3150	1.1811	0.354	1.3780	0.433		
01	12	0.4724	1.1024	0.3150	1.2598	0.394	1.4567	0.472		
02	15	0.5906	1.2598	0.3543	1.3780	0.433	1.6535	0.512		
03	17	0.6693	1.3780	0.3937	1.5748	0.472	1.8504	0.551	2.4409	0.669
04	20	0.7874	1.6535	0.4724	1.8504	0.551	2.0472	0.591	2.8346	0.748
05	25	0.9843	1.8504	0.4724	2.0472	0.591	2.4409	0.669	3.1496	0.827
06	30	1.1811	2.1654	0.5118	2.4409	0.630	2.8346	0.748	3.5433	0.906
07	35	1.3780	2.4409	0.5512	2.8346	0.669	3.1496	0.827	3.9370	0.984
08	40	1.5748	2.6772	0.5906	3.1496	0.709	3.5433	0.906	4.3307	1.063
09	45	1.7717	2.9528	0.6299	3.3465	0.748	3.9370	0.984	4.7244	1.142
10	50	1.9685	3.1496	0.6299	3.5433	0.787	4.3307	1.063	5.1181	1.220
11	55	2.1654	3.5433	0.7087	3.9370	0.827	4.7244	1.142	5.5118	1.299
12	60	2.3622	3.7402	0.7087	4.3307	0.866	5.1181	1.220	5.9055	1.378
13	65	2.5591	3.9370	0.7087	4.7244	0.906	5.5118	1.299	6.2992	1.457
14	70	2.7559	4.3307	0.7874	4.9213	0.945	5.9055	1.378	7.0866	1.654
15	75	2.9528	4.5276	0.7874	5.1181	0.984	6.2992	1.457	7.4803	1.772
16	80	3.1496	4.9213	0.8661	5.5118	1.024	6.6929	1.535	7.8740	1.890
17	85	3.3465	5.1181	0.8661	5.9055	1.102	7.0866	1.614	8.2677	2.047
18	90	3.5433	5.5118	0.9449	6.2992	1.181	7.4803	1.693		
19	95	3.7402	5.7087	0.9449	6.6929	1.260	7.8740	1.772		
20	100	3.9370	5.9055	0.9449	7.0866	1.339	8.4646	1.850		
21	105	4.1339	6.2992	1.0236	7.4803	1.417	8.8583	1.929		
22	110	4.3307	6.6929	1.1024	7.8740	1.496	9.4488	1.969		
24	120	4.7244	7.0866	1.1024	8.4646	1.574	10.2362	2.165		
26	130	5.1181	7.8740	1.2992	9.0551	1.574	11.0236	2.284		
28	140	5.5118	8.2677	1.2992	9.8425	1.654	11.8110	2.441		

In Chap. 2 it was shown that if the relation between imposed stress and the number of stress repetitions at failure is plotted to logarithmic scales, the curve is a straight line so long as the imposed stress is greater than the endurance limit.* This relation can be represented by the equation

$$N_f = K_3 s^a \tag{9-13}$$

where a = experimentally determined exponent

K_3 = experimentally determined material strength factor, in.²/cycles-lb

---

* Imposed contact stresses of 120,000 to 160,000 psi and higher are common in rolling-element bearings.

N_f = number of stress cycles at failure, cycles

s = maximum imposed contact stress, psi

For a particular size and type of rolling-element bearing, this reduces to

$$N_f = \frac{K_4}{F_r^b} \tag{9-14}$$

or

$$F_r = \sqrt[6]{\frac{K_4}{N_f}} \tag{9-15}$$

from which

$$F_r = \frac{K_5}{\sqrt[b]{60 H_a N_a}} \tag{9-16}$$

where b = experimentally determined exponent, approximately 3

F_r = radial-load capacity of the individual bearing, lb

H_a = life expectancy of bearing, under conditions of actual usage, hr

K_4 = experimentally determined material strength factor, rev-lb

K_5 = experimentally determined material strength factor, lb-sec-hr-rev/min²

N_a = actual rotational speed of the inner race relative to the outer race, rpm

N_f = number of stress cycles at failure, rev

9-9 Determining Required Radial-load Catalog Capacities. Ball and roller bearing manufacturers usually tabulate the capacities of their bearings in terms of a *catalog rated radial-load capacity* F_c in pounds, a specified catalog speed N_c in rpm, and a specified catalog life H_c in hours.

Bearing manufacturers do not agree as to the basis of determining expected life. Some use the minimum life expectancy; i.e., 90 per cent of any large number of bearings tested will have a life greater than the rated life, whereas 10 per cent will fail before the rated life is reached. Other manufacturers use the average life; i.e., 50 per cent will have a life longer than the rated life. Various manufacturers use 500, 1,000, 1,500, 3,000, 3,500, 3,800, and 10,000 hr as the basic life rating on which catalog rated capacities are based. Catalog ratings are also based on different speeds in rpm. It is therefore important that when comparing the rated capacities of bearings made by different manufacturers, capacities are all reduced to the same life expectancy and speed.

If a large group of ball or roller bearings as nearly identical as possible are manufactured of the same material, to the same dimensions, and by the same production methods, and if each of these bearings is tested until failure under conditions as nearly identical as possible, there will be a wide variation in the lives of the bearings. A few bearings will fail almost immediately, whereas some will have an apparently endless life. The various catalog capacities for ball and roller bearings usually are based upon 90 per cent of the bearings not failing before reaching their catalog life, corresponding to a *reliability factor* K_{rel} of unity. Typical values of K_{rel} are given in Table 9-3.

Table 9-3 Typical reliability factors K_{rel} for rolling-element bearings

Per cent bearings which will not fail during catalog rated life	K_{rel}
100	0
90	1.0
80	2.0
70	2.9
60	4.0
50	5.0
40	6.4
30	8.0
20	10.5
10	14.5
0	∞

In order for a designer to be able to make a proper selection of a ball or roller bearing, he must consider actual radial and thrust loads, rotational speeds and whether or not these speeds are constant, desired life, probability of the bearing failing before the end of its desired life, pre-

Table 9-4 Typical values of application factor K_a for roller and ball bearings*

Type of service	K_a values for Eq. (9-17)	
	Ball bearings	Roller bearings
Uniform and steady load......................	1.0	1.0
Light shock load............................	1.5	1.0
Moderate shock load........................	2.0	1.3
Heavy shock load...........................	2.5	1.7
Extreme and indeterminate shock load.........	3.0	2.0

* Courtesy Norma-Hoffman Corporation.

loading conditions, application of the bearing, mounting and lubrication of the bearing, and other items. The following equation is useful in assisting him to make this selection:

$$F_c = (K_a K_l) K_o K_p K_s K_t F_r \qquad (9\text{-}17)$$

where F_c = catalog rating of bearing, lb (see Tables 9-7 and 9-8)

F_r = actual radial load on the bearing, lb

H_a = desired life of bearing, hr of use

H_c = catalog rated life of bearing, hr

K_a = application factor taking into account the amount of shock (see Table 9-4)

$K_l = \sqrt[3]{H_a/H_c K_{rel}}$, the life factor

K_o = oscillation factor

= 1.0 for constant rotational speeds of the races

= 0.67 for sinusoidal oscillations of the races

= from 0.5 to 1.5 for all other nonconstant rotational conditions

K_p = preloading factor

= 1.0 for nonpreloaded ball bearings and straight roller bearings

= 1.05 to 1.20 for preloaded duplex ball bearings

= 1.20 or more for three or more adjacent preloaded ball bearings and for preloaded tapered roller bearings (manufacturers' literature gives specific preloading recommendations)

K_r = rotational factor

= 1.0 for bearings with fixed outer races and rotating inner races

= 1.4 for 100-series bearings with fixed inner races and rotating outer races

= 1.5 for 200-series bearings with fixed inner races and rotating outer races

= 1.6 for 300-series bearings with fixed inner races and rotating outer races

= 1.7 for 400-series bearings with fixed inner races and rotating outer races

= 2.0 for bearings with inner and outer races rotating in opposite directions

K_{rel} = reliability factor from Table 9-3

$K_s = \sqrt[3]{K_r N_a/N_c}$, the speed factor

K_t = thrust factor

= 1.0 if there is no thrust-load component

= 1.0 for untapered roller bearings since they are incapable of carrying thrust

= 1.0 to 11.0 for ball bearings (see Table 9-5) (See manufacturers' recommendations for tapered roller bearings.)

Table 9-5   Typical thrust factors K_t for ball bearings

$\dfrac{F_t}{F_r}$	Single-row deep groove type (no filling groove)	Single-row angular contact type	Double-row spherical and angular types
0	1.00	1.00	1.00
0.10	1.00	1.05	1.10
0.15	1.00	1.08	1.15
0.20	1.00	1.10	1.20
0.30	1.00	1.15	1.30
0.40	1.10	1.20	1.40
0.50	1.25	1.25	1.50
0.60	1.40	1.30	1.60
0.70	1.55	1.35	1.70
0.80	1.70	1.40	1.80
0.90	1.85	1.45	1.90
1.00	2.00	1.50	2.00
1.25	2.38	1.63	2.25
1.50	2.75	1.75	2.50
1.75	3.13	1.88	2.75
2.00	3.50	2.00	3.00
3.00	5.00	2.50	4.00
4.00	6.50	3.00	5.00
5.00	8.00	3.50	6.00
10.00	15.00	6.00	11.00

N_a = rotational speed of inner race for fixed outer-race bearings, rpm

   = rotational speed of outer race for fixed inner-race bearings, rpm

   = difference of the rotational speeds of the races for both races rotating in the same direction, rpm

   = sum of the rotational speeds of the races for opposite rotational directions of the two races, rpm

N_c = catalog rated rotational speed, rpm

Often it is difficult to determine proper values of K_a and K_l; from experience, however, it may be simpler to determine a combined value of K_a times K_l. Table 9-6 lists typical values of combined application and life factors.

9-10   Ball and Roller Thrust Bearings.   Several types of rolling-element thrust bearings are shown in Fig. 9-7. Since all the balls or rollers support their share of the load, the thrust capacity of a pure thrust bearing is much greater than that of a pure radial bearing. However, at the higher speeds of rotation, centrifugal force throws the balls against the

Table 9-6 **Typical combined application-life factors**
$K_a \times K_l{}^*$

Type of equipment	$K_a \times K_l$
Agricultural equipment	0.65–1.00
Air compressors	1.50–2.00
Belt conveyors	1.00
Belt drives	1.50
Blowers and fans	1.00
Cars, industrial	1.00
mine and mill	1.50
Centrifugal extractors	1.00–1.40
Cranes and hoists,	
hand operated	0.80
power operated	1.00
powerhouse	0.80
steel mill	1.50
Crushers, pulverizers	1.40–1.8
Foundry equipment	1.00
Gear drives	1.75
Glassmaking equipment	1.00
Industrial locomotives	1.50–1.75
Machine tools, except spindles	1.50
Mining machinery	1.50–2.00
Oil-field equipment	2.00
Pumps, centrifugal	1.00
reciprocating	1.40–2.00
dredge, sludge, line	2.00
Refrigerating machinery	1.70
Road machinery	1.00
Rock crushers	2.00–3.00
Steam shovels	1.00
Textile machinery	1.00–1.50
Tractors, general	1.00
crawler tracks	3.00–4.00
crawler wheels	2.20
Transmission machinery	1.70
Turbines	2.00
Woodworking and sawmill equipment	1.00–1.50

* These suggested factors are for fixed inner-race bearings. For fixed outer-race and rotating inner-race bearings, these factors should be multiplied by the K_r values given for Eq. (9-17).

outer side of the race, causing a wedging action. For this reason, the speed limit for pure thrust bearings is much lower than for radial bearings. For the higher speeds, the deep-groove type of angular contact bearing should be used. In the following paragraphs are some examples of ball and roller bearings selection.

Example 1. Select a single-row ball bearing to support a $2\frac{1}{2}$-in.-diameter shaft rotating at 1500 rpm for an estimated 10 hr/day for 5 years. The bearing is

Table 9-7 Typical radial capacity F_c of ball bearings (based on average life H_c of 10,000 hr and speed N_c of 500 rpm)*

Radial-load capacity F_c, lb

SAE No.	Bore mm	Bore in.	Deep-groove 100	Deep-groove 200	Deep-groove 300	Deep-groove 400	Filling-slot 200	Filling-slot 300	Filling-slot 400	Angular* 200	Angular* 300	Angular* 400	Two-row spherical 200	Two-row spherical 300	Two-row spherical 400	Two-row angular 200	Two-row angular 300	Two-row angular 400
00	10	0.3937	167	190	330	....	....	....	....	265	330	....	205	310	....	335	....	....
01	12	0.4724	185	220	380	....	....	....	....	330	410	....	250	360	....	470	....	....
02	15	0.5906	200	260	470	....	....	....	....	415	515	....	290	430	....	515	....	....
03	17	0.6693	214	330	560	910	....	....	....	455	610	1,025	360	505	770	....	930	1,690
04	20	0.7874	366	400	650	1,380	....	970	....	615	735	1,550	495	640	1,050	770	1,020	2,080
05	25	0.9843	395	500	910	1,610	....	1,200	....	660	1,025	1,900	610	770	1,140	840	1,530	2,610
06	30	1.1811	544	610	1,190	1,945	940	1,460	....	1,020	1,335	2,290	760	1,050	1,370	1,160	2,020	3,090
07	35	1.3780	677	850	1,500	2,290	1,270	1,820	....	1,270	1,685	2,690	880	1,240	1,740	1,550	2,520	4,010
08	40	1.5748	717	1,000	1,840	2,630	1,560	2,200	....	1,675	2,070	3,285	1,000	1,460	1,960	1,660	3,120	4,980
09	45	1.7717	920	1,110	2,220	3,070	1,660	2,880	....	1,775	2,245	3,615	1,210	1,870	2,360	1,820	3,740	5,640
10	50	1.9685	972	1,220	2,490	3,510	1,840	3,350	....	1,950	2,905	4,270	1,300	2,120	2,870	2,120	4,450	6,780
11	55	2.1654	1,320	1,540	2,940	4,000	2,300	3,850	....	2,250	3,430	4,870	1,590	2,530	3,100	2,540	5,950	7,490
12	60	2.3622	1,390	1,640	3,050	4,510	....	....	....	2,520	4,000	5,595	1,850	2,900	3,390	3,110	6,320	8,770
13	65	2.5591	1,457	1,890	3,990	5,050	....	....	....	2,935	4,595	6,145	1,970	3,110	3,920	3,220	7,650	9,650
14	70	2.7559	1,840	2,070	4,500	6,210	....	....	....	3,330	5,240	7,550	2,250	3,650	4,800	3,640	8,240	11,000
15	75	2.9528	1,925	2,280	4,800	6,820	....	....	....	3,740	5,610	8,305	2,360	3,920	5,080	3,980	8,700	12,400
16	80	3.1496	2,280	2,490	5,070	7,470	....	....	....	4,050	6,180	9,080	2,570	4,330	5,350	4,430	10,300	14,900
17	85	3.3465	2,485	2,940	5,680	8,130	....	....	....	4,550	7,090	9,895	2,880	4,780	6,700	5,150	10,500	16,000
18	90	3.5433	2,750	3,140	6,200	8,825	....	....	....	5,140	7,560	10,740	3,340	5,480	7,980	5,900	12,400	16,600
19	95	3.7402	2,880	3,620	6,980	10,280	....	....	....	5,905	8,750	12,510	3,740	6,100	8,500	6,700	12,900	20,300
20	100	3.9370	3,000	3,920	7,680	11,825	....	....	....	6,740	9,290	14,390	4,100	7,100	10,600	7,160	16,100	22,700
21	105	4.1339	3,380	4,120	8,400	....	....	....	....	7,115	10,480	....	4,460	7,450	....	8,500	17,500	....
22	110	4.3307	3,760	4,500	9,400	....	....	....	....	7,485	11,170	....	4,800	7,970	....	9,300	18,900	....

* Thrust capacity of angular-type bearings is approximately twice the radial capacity. These illustrative values are for comparison only. For the capacity of **any** specific bearing, consult the manufacturer's catalog.

Table 9-8 **Typical radial capacity F_c of straight cylindrical roller bearings (based on average life H_c of 10,000 hr and speed N_c of 500 rpm)***

SAE No.	Bore		Radial-load capacity F_c, lb		
	mm	in.	200 series	300 series	400 series
00	10	0.3937	370	615	
01	12	0.4724	415	795	
02	15	0.5905	460	895	
03	17	0.6693	595	1,090	2,040
04	20	0.7874	745	1,190	2,680
05	25	0.9843	895	1,760	3,370
06	30	1.1811	1,320	2,450	4,150
07	35	1.3780	1,540	2,660	5,000
08	40	1.5748	1,840	3,220	6,000
09	45	1.7717	2,150	4,050	7,070
10	50	1.9685	2,450	4,980	8,240
11	55	2.1654	2,610	5,390	9,480
12	60	2.3622	2,760	6,000	10,810
13	65	2.5591	3,220	7,000	12,200
14	70	2.7559	3,620	7,500	13,650
15	75	2.9528	4,050	9,610	15,170
16	80	3.1496	4,550	10,300	16,770
17	85	3.3465	5,600	12,610	18,430
18	90	3.5433	5,910	13,510	20,180
19	95	3.7402	6,750	14,420	21,960
20	100	3.9370	7,500	15,920	23,840
21	105	4.1339	8,240	18,190	
22	110	4.3307	9,270	19,560	
24	120	4.7244	12,160	22,350	
26	130	5.1181	23,720	25,210	
28	140	5.5118	26,120	29,800	

* These illustrative values are for comparison only. For the capacity of any specific bearing, consult the manufacturers' catalog.

to carry an estimated radial load of 1000 lb with moderate shock plus a 500 lb thrust load. Assume a 90 per cent chance that the bearing will meet its desired life, therefore $K_{rel} = 1.0$. Also assume no preloading, hence $K_p = 1.0$.

Solution. From Table 9-4, $K_a = 2.0$.

$$H_a = 10 \times 5 \times 365 = 18,250 \text{ hr}$$

$$K_l = \sqrt[3]{\frac{H_a}{H_c \times K_{rel}}} = \sqrt[3]{\frac{18,250}{10,000 \times 1}} = 1.22$$

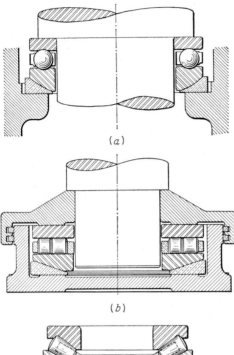

(a)

(b)

Fig. 9-7 Rolling-element thrust bearings. (a) Ball thrust bearing. (b) Tapered roller thrust bearing. (c) Spherical roller thrust bearing.

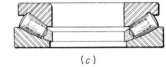

(c)

For constant shaft rotation, $K_o = 1.0$; $K_r = 1.0$ since the outer race is presumed to be fixed. Since there is no preloading, $K_p = 1.0$.

$$K_s = \sqrt[3]{\frac{K_r N_a}{N_c}} \approx \sqrt[3]{\frac{1.0 \times 1500}{500}} = 1.44$$

$F_t/F_r = 500/1000 = 0.5$, from Table 9-5, and $K_t = 1.25$. Using Eq. (9-17), the minimum required radial load catalog capacity at 500 rpm and 10,000 hr is

$$\begin{aligned} F_c &= K_a K_l K_o K_p K_s K_t F_r \\ &= 2 \times 1.22 \times 1 \times 1 \times 1.44 \times 1.25 \times 1000 \\ &= 4390 \text{ lb minimum} \end{aligned}$$

From Table 9-7, the following data are obtained:

Bearing type	Bearing number	F_c, lb	Bore, in.
Single-row deep groove........	314	4,500	2.7559
Single-row deep groove........	412	4,510	2.3622
Single-row with filling slot.....	None has an F_c as large as 4,390 lb		
Single-row angular...........	313	4,595	2.5591
Single-row angular...........	411	4,870	2.1654

If the diameter of the shaft at the location of the bearing could be reduced to approximately (not exactly, since pressing the bearing onto the shaft must be considered) 2.36 or 2.17 in., then either the No. 412 single-row deep groove or the No. 411 single-row angular bearing could be used. If the shaft diameter could not be reduced below the $2\frac{1}{2}$ in., and if the shaft diameter could be increased (either by use of an integral increase in shaft diameter or the use of a sleeve pressed onto the shaft) to approximately 2.76 or 2.56 in., then either No. 314 single-row deep groove or No. 313 single-row angular contact bearing could be used. No single-row bearing with filling slot could be used since no bearing of this type has the minimum required radial capacity of 4390 lb. Which of these four possible bearings actually would be used by a designer would depend upon limitations of the shaft diameter, maximum permissible outside diameter of the outer race, and other considerations not given in the problem statement.

Example 2. Select a two-row angular contact ball bearing suitable for mounting a wire cable idler pulley on a stationary shaft of a steel mill hoist. The estimated maximum radial load on the pulley is 3230 lb; no thrust is anticipated. From other design considerations, it was determined that the minimum diameter of the shaft is $2\frac{1}{8}$ in. and the maximum bore of the pulley where it presses onto the bearing is $5\frac{1}{4}$ in. Assume an effective pulley rotational speed of 1,000 rpm.

Solution. From Table 9-6, the combined value of $K_a \times K_l$ is found to be 1.50. From data given for Eq. (9-17), a trial K_r value of 1.6 is assumed to take into account the fixed inner race. Considering the usage of the hoist, the bearing is assumed to rotate intermittently in both directions, therefore a K_o value of 0.8 is assumed. Without preloading, $K_p = 1.0$. Since there is no thrust load, $K_t = 1.0$.

$$K_s = \sqrt[3]{\frac{K_r N_a}{N_c}} = \sqrt[3]{\frac{1.6 \times 1000}{500}} = 1.47$$

Thus $F_c = (K_a \times K_l)K_o K_p K_s K_t F_r$
$= 1.5 \times 0.8 \times 1 \times 1.47 \times 1.0 \times 3230$
$= 5,700$ lb minimum required catalog rated radial-load capacity

From Tables 9-2 and 9-7, the following are tabulated:

Two-row angular contact bearing number	F_c, lb	Bore, in.	O.D. of outer race, in.
218	5900	3.5433	6.2992
311	5950	2.1654	4.7244
410	6780	1.9685	5.1181

Each of the three bearings listed meets the minimum required F_c even after corrected to the proper K_r value for the particular series. The No. 218 bearing has too large an outside diameter to fit into the pulley, and the No. 410 bearing has too small a bore to fit onto the minimum required shaft diameter; hence the No. 311 two-row angular contact ball bearing would be selected for mounting the pulley.

Example 3. Select a straight roller bearing for the low-speed shaft of a gear-reduction unit operating at 65 rpm with an applied radial load of 2,850 lb. The shaft size has been determined and is $2\frac{1}{4}$ in.

Solution. The combined life and application factor from Table 9-6 is 1.75. The speed ratio is 65/500; hence

$$K_s = \sqrt[3]{\tfrac{6\,5}{5\,0\,0}} = 0.51$$

The required catalog capacity at 500 rpm is

$$F_c = (K_a \times K_l)K_oK_pK_sK_tF_r = 1.75 \times 1 \times 1 \times 0.51 \times 1 \times 2,850$$
$$= 2,544 \text{ lb minimum}$$

From Table 9-8, if the shaft can be turned down slightly at the bearing seat, bearing No. 211 with a bore of 2.1654 in. and a capacity of 2,610 lb could be used. If the shaft diameter cannot be reduced, then bearing No. 212 with a bore of 2.3622 in. could be used.

9-11   Mounting Rolling-element Bearings.

Major considerations in mounting a ball or roller bearing are keeping the inner race concentric with the shaft on which it is mounted, the outer race concentric with the housing or bored hole into which it fits, the races each axially located in their proper position, and the bearing properly preloaded if preloading is employed.

Keeping the inner race concentric with the shaft is almost always effected by pressing the inner race onto the shaft. For a similar reason, the outer race is pressed into its housing. Ball bearings usually are installed with the shaft and inner race rotating, but in some cases the outer race rotates. It is general practice to install the rotating race with a tight press fit interference of approximately 0.0025 in. per each inch bore of the bearing if the inner race rotates, or per each inch of outside diameter of outer race if the outer race rotates. The stationary race usually is installed with a wringing press fit with an interference somewhat less than that indicated for the rotating race. This method of mounting permits the stationary race angularly to creep around gradually in its supporting housing or on its supporting shaft, presenting new parts of the stationary race to the load and thereby distributing the wear throughout the entire race. This rate of creep is very slow, and care should be exercised to prevent looseness and its consequent slippage and excessive wear on the supporting housing.

In order to keep a shaft properly located axially, the stationary race of one bearing on the shaft should be clamped in position with a press fit permitting the slight angular creeping action. All other races should be mounted in a manner permitting their free axial movement within their housing or on their shaft; this is to prevent cramping of the bearings from changes in the axial dimensions of the shaft due to temperature changes or inaccuracies in machining. Many single-row ball bearings

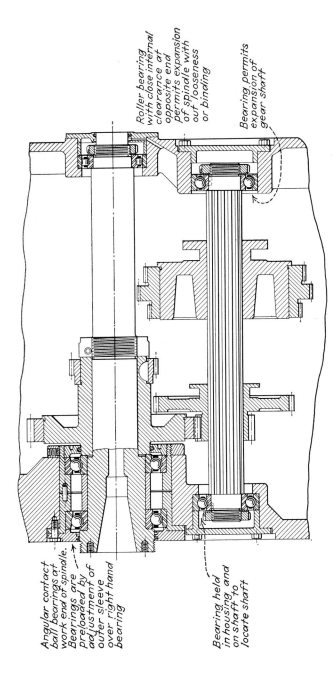

Roller bearing
with close internal
clearance at
opposite end
permits expansion
of spindle with
out looseness
or binding

Bearing permits
expansion of
gear shaft

Angular contact
ball bearings at
work end of spindle.
Bearings are
preloaded by
adjustment of
outer sleeve
over right hand
bearing

Bearing held
in housing and
on shaft to
locate shaft

Fig. 9-8 Use of ball and roller bearings to support shafts.

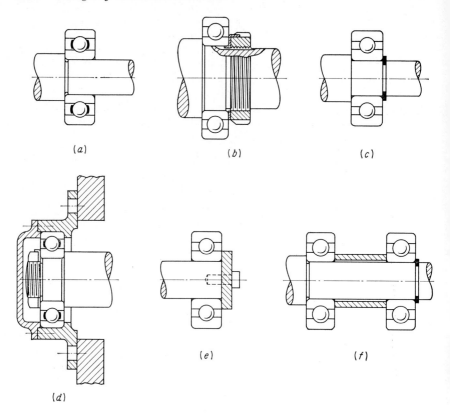

Fig. 9-9   Methods of axially locating ball bearings on a shaft. (*a*) Press fit against a shoulder. (*b*) Locknut with lock washer. (*c*) Snap ring and groove. (*d*) Bearing cartridge. (*e*) Disk on end of shaft. (*f*) Cylindrical sleeve to separate bearings.

permit an axial movement from 0.002 to 0.010 in. When straight roller bearings are used, usually no steps are necessary to permit slight axial movement of the shaft, thereby preventing cramping of the bearings. The proper method of mounting bearings permitting axial shaft motion is indicated in Fig. 9-8. The proper diameters of retaining shoulders on the shafts and in bearing housings usually are available in the bearing information bulletins of the various manufacturers.

The inner race may be held in its correct axial position on the shaft simply by use of a press fit only, often in conjunction with a locating shoulder on the shaft. Methods for installing ball bearings are indicated in Fig. 9-9.

9-12   Preloading of Bearings.   When a rolling-element bearing is placed under load, a certain amount of deformation takes place in the rolling members and the races they contact. This deformation results in

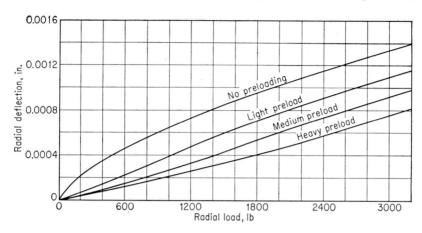

Fig. 9-10 Typical effect of preloading upon the radial deflection of a radially loaded angular contact ball bearing.

an unwanted movement or change of position of the supported machine parts. For many applications, this deflection or change of position may be of little consequence. Where machine members must be accurately located under load, however, it often becomes desirable to adopt preloading to reduce deflective motion within the bearings. Preloading is accomplished by the application onto the bearing (or between two or more bearings on a shaft) of a sufficient axial load to produce a predetermined amount of bearing deflection before the work load is imposed upon the bearing. This effect of preloading upon a particular single-row angular contact bearing is illustrated in Figs. 9-10 and 9-11.

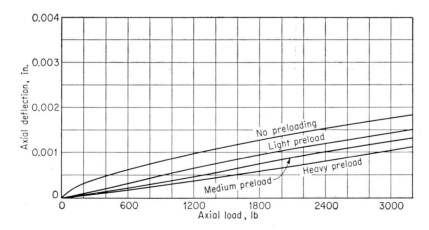

Fig. 9-11 Typical effect of preloading upon the axial deflection of an axially loaded angular contact ball bearing.

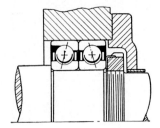

Fig. 9-12 Pair of duplex bearings mounted together.

The preloading of two or more opposed ball bearings requires a short axial movement of either the inner or outer race of at least one bearing. Sufficient looseness of fit either on the shaft or in the housing to allow this axial movement is usually easily accomplished. Bearings usually are preloaded by means of preloading nuts, closure caps in bearing cartridges, and other positive means.

The amount of preload employed in a bearing generally is only a small percentage of the capacity of the bearing; in these cases, the effect of preloading upon bearing performance is negligible, and in some cases preloading improves bearing performance. When heavy preloading is employed, the capacity of the bearing usually is noticeably reduced. Examples of preloading are shown in Figs. 9-12 through 9-14.

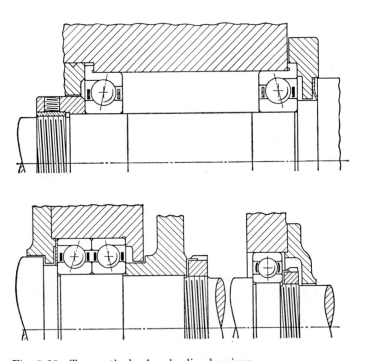

Fig. 9-13   Two methods of preloading bearings.

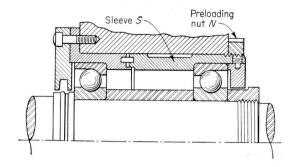

Fig. 9-14   Bearings with provision for preloading.

9-13   Bearing Shields, Seals, and Lubrication.   The presence of dirt, grit, metallic chips, and other undesirable foreign matter within a bearing is one of the major causes of noise and premature failure. To prevent this, many types of shields and/or seals are used. These seals and shields are available as integral parts of ball bearings, or they are available as separate items for roller and ball bearings. Applications are indicated in Figs. 9-15 through 9-17.

The use of oil and grease seals not integral with bearings is illustrated in Fig. 9-18. It almost always is desirable to provide lubrication in ball and roller bearings to reduce friction occurring between the rolling elements and the retaining cages, to reduce friction occurring where the contact between the rolling elements and the races is not pure rolling contact, to protect the bearing elements from dirt and corrosive action, and to assist in the dissipation of heat generated in the bearing. Light mineral oils and greases are used rather than animal oils or greases since the latter may corrode the elements of the bearings.

Ball bearings with two seals are prelubricated by the manufacturer with a light grease, usually requiring no further lubrication throughout

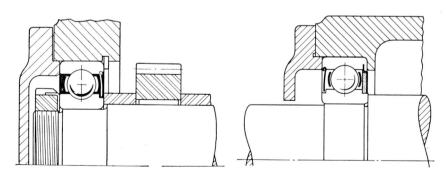

Fig. 9-15   Bearing with single shield toward inside of shaft.

Fig. 9-16   Bearing with outer shield and inner seal.

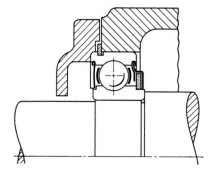

Fig. 9-17   Snap-ring-located ball bearing with outer shield and inner seal.

the useful life of the bearing. A special variety is available with plastic elements in the seals; the bearings can be relubricated after their initial period of usage by injecting oil or grease through the self-sealing plastic by means of a hypodermic needle. All roller bearings and ball bearings without two seals require lubrication during their usage.

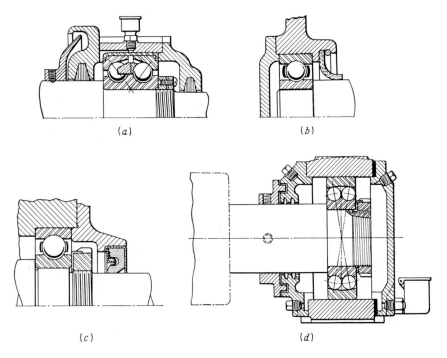

Fig. 9-18   Bearings with nonintegral oil seals. (*a*) Felt oil seal. (*b*) Cover and separate leather oil seal. (*c*) Separate shaft seal. (*d*) Gasketed cover plate.

10

Sliding-element Bearings

10-1 Types of Sliding-element Bearings. The ball, roller, and needle bearings of Chap. 9 are known as rolling-element bearings since essentially rolling friction exists within these bearings. Where essentially sliding friction exists, the bearing is known as a *sliding bearing* or *sliding-element bearing*. Depending upon whether the primary purpose of the bearing is to support a radial load, axial load, or simply to guide the translational motion of a machine member, the bearings are known respectively as *radial bearings, thrust bearings,* and *guide bearings*. Radial bearings support rotating shafts, often referred to as journals, and hence are called *journal bearings;* however, they also are known as *plain bearings* and *cylindrical bearings*.

10-2 Definition and Purposes of a Lubricant. A lubricant is any substance that will form a film between the two surfaces of a bearing. The major purposes of a lubricant are to cool a bearing by the absorption of heat by the lubricant, to reduce the coefficient of friction between the bearing surfaces and hence reduce the heat generation and corresponding power loss, and to prevent rapid wear due to contact of the bearing surfaces.

10-3   Types of Lubricants.   Depending upon their state of matter, lubricants may be liquids, solids, gases, or a combination of two or all three. By far the most commonly used lubricants are liquids, and include oil, grease, water, silicones, acids, bases, and other fluids. Solid lubricants in general use are graphite, molybdenum disulfide, lead oxide, soap, mica, soapstone, powdered glass, and plastics. Air, nitrogen, and hydrogen are the generally employed gas lubricants, though other gases sometimes are used to an advantage.

10-4   Methods of Providing Lubricant.   Bearings may be lubricated during usage by providing lubricant continuously or intermittently, or they may be prelubricated and no additional lubricant provided during the normal usage of the bearing. Methods of lubricating during usage consist of oiling intermittently by hand, providing drop feed of the lubricant at a rate usually determined by the flow of oil through a variable valve, bottle lubrication, wick lubrication, capillary lubrication, employing ring oilers which dip into a lubricant and cause oil to be conducted into the bearing, immersing the bearing in a bath of lubricant, causing the machine member to splash into a sump of oil and so deliver oil to the bearing, providing a continuous supply of oil from a pressurization system usually employing a pump, and providing grease fittings to be used when the machine is not running or, in some cases, when the machine is in operation. Bearings are prelubricated either by providing volumes within the bearing assembly into which a quantity of lubricant can be placed with the hope that it will suffice for the life of the bearing, or by the selection of mating bearing and journal materials which possess sufficient inherent lubricating properties. The voids in the material of sintered bearings can be filled with lubricant with the hope that during usage of the bearing, adequate lubrication will occur. Grooves or other voids prelubricated with grease when the bearing is installed often give suitable results.

The use of steel or cast iron journals in graphite bearings is an example of the selection of materials which possess inherent lubricating properties. Sometimes the journal and/or bearing surface are coated with a solid lubricant, such as molybdenum disulfide, the use of which may provide an adequately low coefficient of friction so that no additional lubrication is required. The method of lubrication chosen by the designer depends upon the type and application of the particular bearing, the amount and type of service which is expected during the life of the bearing, the amount of heat generated within the bearing, the method and quantity of heat dissipation, and other items.

10-5   Viscosity of Lubricants.   *Viscosity* of a lubricant is that property which resists shearing of the lubricant. Viscosities which are determined by direct measurement of shear resistance of the lubricants are known as *absolute viscosities*. Those which are not determined by direct measurements of shear resistance but by other means are known as *secondary viscosities*. *Kinematic viscosity* is absolute viscosity divided by the specific gravity of the lubricant.

Absolute viscosities usually are measured as the force required on a given area of a differential cube of the lubricant to give a specified difference in velocity (or shear rate) between the two parallel surfaces of the cube parallel to the force. A *poise* is the unit of viscosity resulting when a force of 1 dyne acting on 1 cm^2 surface of 1 cm^3 of lubricant causes a relative shear rate of 1 cm/sec, thus 1 poise = 1 dyne-sec/cm^2. A poise is a very large unit of viscosity; hence the unit *centipoise* was devised and assigned the symbol Z; 1 centipoise = 0.01 poise = 0.01 dyne-sec/cm^2. The unit of absolute viscosity in the English system is the *reyn;* 1 reyn results when a 1-lb force on one surface of a 1-in. cube of lubricant causes a relative shear rate of 1 in./sec; thus 1 reyn = 1 lb-sec/in.^{2} The symbol μ is used to indicate viscosity in reyns. Conversion between viscosities in centipoises and reyns may be accomplished by

$$Z = 6,895,000\mu \tag{10-1}$$
$$\mu = 0.000,000,145Z \tag{10-2}$$

Viscosities of lubricating oils decrease with an increase of temperature, and increase with increases of pressure. For the normal pressures encountered, the variation of viscosity with pressure is negligible in comparison to the temperature effects; hence designers usually neglect the effect of pressure upon viscosity. Typical viscosities in reyns on various classified oils are tabulated in Table 10-1. These values are approxi-

Table 10-1 **Average viscosity-temperature data of classified oils**

Temperature, °F	(Approximate viscosity in reyns) $\times$ 10^6 for:							
	AN-06	SAE 10	SAE 20	SAE 30	SAE 40	SAE 50	SAE 60	SAE 70
80	3.5	9.5	13.5	23.5	35.0	59.0	80.0	117.0
90	3.0	7.0	11.0	17.5	25.0	42.5	58.0	87.0
100	2.5	5.5	8.5	13.5	19.0	32.0	43.0	61.5
110	2.0	4.0	6.5	10.0	14.5	24.5	32.5	45.0
120	1.5	3.5	5.0	7.5	11.0	19.0	25.0	33.5
130	1.0	2.5	4.0	6.0	8.5	14.0	18.5	24.5
140	1.0	2.0	3.5	5.0	6.5	10.5	13.5	18.5
150	1.0	1.5	2.5	4.0	5.0	8.5	10.5	14.0
160	1.0	1.0	2.0	3.0	4.0	6.5	8.0	11.0
170	1.0	1.0	1.5	3.0	4.0	5.0	6.5	9.0
180	1.0	1.0	1.0	2.0	3.0	4.0	5.5	6.5

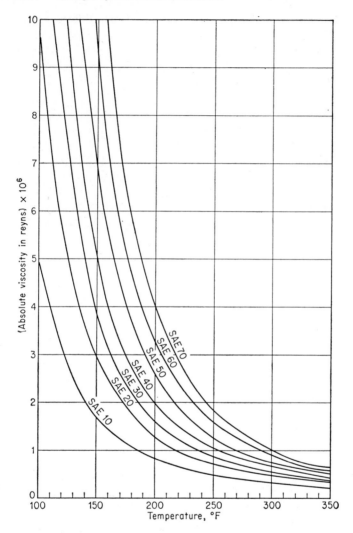

Fig. 10-1 Average viscosities of SAE-classified oils.

mate mid-range values for each viscosity classification listed and have been rounded off to the nearest 0.000,000,5 reyn. Viscosity data for the SAE oils are shown approximately in Fig. 10-1. Absolute viscosity Z divided by specific gravity ρ gives the kinematic viscosity ν in centistokes; thus

$$\nu = \frac{Z}{\rho} \text{ centistokes} = \frac{6,895,000\mu}{\rho} \qquad (10\text{-}3)$$

Perhaps the most common method of obtaining secondary viscosity is by use of the Saybolt universal viscosimeter. This method determines the time required for a given quantity of oil at a specified temperature to drop through the calibrated orifice of the standardized viscometer. The resulting viscosity S is in Saybolt universal seconds.

$$Z = \rho \left(0.22S - \frac{180}{S} \right) \tag{10-4}$$

$$\mu = 0.000,000,145\rho \left(0.22S - \frac{180}{S} \right) \tag{10-5}$$

10-6 Viscosity Index. The viscosity index of an oil is a measure of its change of viscosity with temperature relative to standard oil which exhibits the least change of viscosity with temperature. A sampling was made of the viscosity-temperature relationships for essentially all American crude oils. A Pennsylvania oil was found to exhibit the most nearly constant viscosity-temperature data, while a Gulf Coast oil exhibited the greatest change of viscosity with temperature. Arbitrarily this Pennsylvania oil was called a 100 *viscosity index* oil, while the Gulf Coast oil was called a zero viscosity index oil. Figure 10-2 depicts these two oils in terms of their Saybolt universal seconds viscosity S. The line between the 100 and the zero viscosity index oils represents the oil for which a viscosity index determination is to be made. The viscosity index VI is defined by

$$VI = \frac{100(H-U)}{H - L} \tag{10-6}$$

Fig. 10-2 Determination of viscosity index.

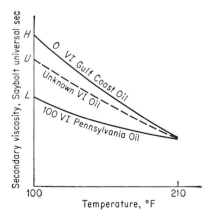

where H = viscosity of the zero viscosity index oil at 100°F, Saybolt universal seconds

L = viscosity of the 100 viscosity index oil at 100°F, Saybolt universal seconds

U = viscosity of the unknown oil at 100°F, Saybolt universal seconds

VI = viscosity index of the unknown oil

Oils with high viscosity indexes are useful in internal-combustion engines in which it is desired to have a large load-carrying capacity (essentially a linear function of viscosity) when the engine is hot without having an excessively high viscosity when the engine is cold.

10-7   Specific Gravity of Oils.   The absolute viscosity of any oil varies with its specific gravity, which is a function of temperature. The specific gravity ρ of a liquid is given by

$$\rho = \rho_0 + C_1(t - t_0) + C_2(t - t_0)^2 \tag{10-7}$$

where C_1 = a constant, 1/deg F

C_2 = a constant, 1/deg² F

t = temperature at which the specific gravity is desired, °F

t_0 = temperature corresponding to ρ_0, °F

ρ = specific gravity desired at t°F

ρ_0 = known specific gravity at temperature t_0

For lubricating oils, Peffer* indicated that C_1 varies between $-0.000,35$ and $-0.000,38$, with $-0.000,365$ as an average, and that C_2 is essentially zero. Thus for oil, Eq. (10-7) becomes

$$\rho = \rho_0 - 0.000,365(t - t_0) \tag{10-8}$$

The specific gravity of oils varies according to its geographic source as indicated in Table 10-2.

10-8   Newtonian and Non-Newtonian Lubricants.   A lubricant for which the resisting shear force varies linearly with the shear rate, all other items being held constant, is known as a *Newtonian lubricant*. Although there are several oils which exhibit essentially Newtonian properties, most of the lubricating oils in current usage are only approximately linear in their shear force–shear rate curve. Some deviate substantially from the linear condition. These non-Newtonian lubricants often may be employed to an advantage, as in automotive engines where a low shear resistance is desired when the engine is being started and where

---

* H. W. Bearce, and E. L. Peffer, Density and Thermal Expansion of American Petroleum Oils, *Bur. Standards, Tech. Paper 77*, Aug. 26, 1926.

Table 10-2 **Typical specific gravities of various crude oils***

Crude oil	Range of specific gravity ρ at 60°F
Pennsylvania crude............	0.86–0.88
Mid-Continent crude...........	0.87–0.88
California crude...............	0.89–0.91
Gulf Coast crude..............	0.91–0.94

* Note: The specific gravities of refined oils do not materially vary from the specific gravity of the crude stock. For most lubrication calculations, the above specific gravities are adequate.

a high load capacity (usually accomplished with higher-friction torques) is desired under highway conditions corresponding to high shear rates within the lubricant film.

10-9 Types of Journal Bearings.   Essentially there are two types of journal bearings, circular and other than circular. As indicated in Fig. 10-3, the circular journal bearing may be *full bearings* enclosing 360 deg of the journal, or *partial bearings* enclosing less than 360 deg of the journal. In both of these types, the diameter of the journal is less than the diameter of the bearing. If the diameters of the journal and partial bearing are equal, a *fitted bearing* results. In recent years, non-circular journal bearings have been employed to an advantage with circular journals. Since at present they have very little usage, they shall not be discussed in this book. The remainder of this discussion of journal bearings relates to full journal bearings.

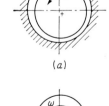

(*a*)

Fig. 10-3   Two types of journal bearings. (*a*) Full bearing. (*b*) Partial bearing.

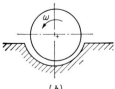

(*b*)

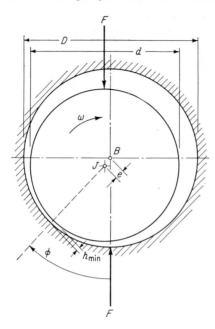

Fig. 10-4 Typical hydrodynamic running conditions for a journal bearing.

10-10　Bearing Dimensions, Ratios, Modulus, and Other Terms.
A journal bearing under typical conditions of operation is indicated in
Fig. 10-4. All linear dimensions are assumed to be in inches. For the sake
of clarity in the illustration, the oil film which occupies all or a portion
of the void between the journal and the bearing is not shown. The bore
of the bearing is the diameter D. The diameter of the journal is d. The
diametral clearance C_d and the *radial* clearance C_r in inches are defined by

$$C_d = D - d \qquad\qquad (10\text{-}9)$$

$$C_r = \frac{C_d}{2} = \frac{D - d}{2} \qquad\qquad (10\text{-}10)$$

The C_d/D ratio is called the *diametral clearance ratio*. The axial length L
of the journal contained inside the bearing is defined as the length of the
journal and bearing. The *projected area* of the bearing is the product of
L and D. A *square bearing* is one in which the length ratio L/D is unity;
long bearings have length ratios over 1 and *short bearings* have ratios less
than 1. F is the radial load in pounds carried by the journal and the bearing. The unit loading p in pounds per square inch (sometimes called the
bearing pressure) is the load per projected unit area of the bearing as
indicated by

$$p = \frac{F}{LD} \qquad\qquad (10\text{-}11)$$

It should be pointed out that p does not necessarily represent the pressure

within the oil film at any given location. The *eccentricity e* is the radial distance between the center B of the bearing and the displaced center J of the journal. The radial dimension h_{min} of the oil film at the minimum thickness is called the *minimum oil-film thickness*. The *eccentricity ratio* ϵ is the ratio of the actual eccentricity e to the maximum possible value C_r which occurs when the journal and bearing touch. Thus

$$\epsilon = \frac{e}{C_r} = \frac{C_r - h_{min}}{C_r} = 1 - \frac{h_{min}}{C_r} \tag{10-12}$$

An eccentricity ratio of zero indicates that the journal and bearings are collinear, whereas a ratio of unity indicates contact between the mating surfaces of the journal and the bearing. The *attitude angle* ϕ is measured from the line of radial loading on the bearing to the position of the minimum oil-film thickness. The rotational speed of the journal relative to the bearing is indicated by either N in rpm or n in rps. For a stationary bearing, N or n is simply the rotational speed of the journal. The *bearing modulus* is the product of absolute viscosity and rotational speed divided by the unit loading. Depending upon the viscosity and rotational speed used, the bearing modulus may be

$$\text{Bearing modulus} = \frac{ZN}{p} \tag{10-13}$$

or

$$\text{Bearing modulus} = \frac{\mu n}{p} \tag{10-14}$$

It can be shown that

$$\frac{ZN}{p} = 413,400,000 \frac{\mu n}{p} \tag{10-15}$$

and

$$\frac{\mu n}{p} = 0.000,000,002,42 \frac{ZN}{p} \tag{10-16}$$

10-11 Bearing and Journal Friction Torques. The frictional torque T_b in inch-pounds on the bearing is given by

$$T_b = \frac{F f_b D}{2} \tag{10-17}$$

where f_b is the coefficient of bearing friction. The curve in Fig. 10-5 of bearing coefficient of friction versus bearing modulus is typical of a journal bearing in operation. A voltage potential can be placed across the lubricant film and an oscilloscope used to monitor whether or not the journal and bearing touch. For bearing moduli of M_3 and higher, no contact ever occurs between the bearing, and the lubrication procedures are referred to as *hydrodynamic film lubrication* or *hydrodynamic lubrication*. For moduli of M_1 and less, continuous contact occurs between the journal and bearing, and *marginal lubrication* is said to exist. Intermittent contact occurs at least once each journal rotation for moduli between M_1 and M_3, and the lubrication is in a transitional zone, often included in the marginal lubrication zone. On the basis of minimum energy loss due to friction, it would be desirable to operate a bearing with a modulus of M_2, corresponding to operation with a minimum coefficient of bearing friction f_{min}. In practice this is difficult to accomplish since the moduli M_1 and M_3 are comparatively close together, and if an attempt is made to operate at modulus M_2, deviation of actual conditions from those desired could easily cause the bearing to operate well into the marginal range or the hydrodynamic range.

So long as the bearing operated hydrodynamically with a modulus of M_3 or higher, the coefficient of bearing friction f_b may be obtained from*

$$f_b = 4.73 \times 10^{-8} \frac{ZN}{p} \frac{D}{C_d} + k = 19.55 \frac{\mu n}{p} \frac{D}{C_d} + k \qquad (10\text{-}18)$$

where the factor k is a function of the length ratio as indicated in Fig. 10-6. Since many bearings operate with length ratios between 0.8 and 1.5, some designers use a value of 0.002 for k rather than actually determining its value from a curve similar to that in Fig. 10-6. For bearings operating marginally, analytical determination of f_b is much more difficult. In fact, no good equation exists for this case. Some designers use

* S. A. McKee and T. R. McKee, Friction of Journal Bearings as Influenced by Clearance and Length, *Trans. ASME*, vol. 51, APM-51-15, 1929.

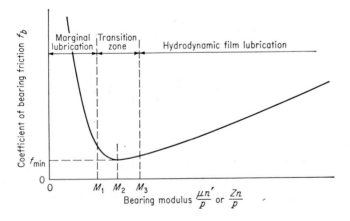

Fig. 10-5 Typical hook curve of coefficient of bearing friction versus bearing modulus.

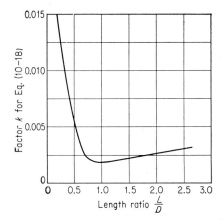

Fig. 10-6 Correction factor k for Eq. (10-18).

the following equation for marginal lubrication*:

$$f_b = \frac{C_1 C_2}{250} \sqrt[4]{\frac{p_a}{V}} \tag{10-19}$$

where C_1 and C_2 are constants determined respectively from Tables 10-3 and 10-4, V is the rubbing velocity in feet per minute $= \pi DN/12$, and p_a is the absolute value of the average unit loading on the bearing throughout one cycle. When p varies throughout the cycle, the value used for p_a should be either one-half the maximum instantaneous value of p or the absolute value of the mean value of p, whichever is the larger. Since the values of C_1 and C_2 vary widely, values of f_b obtained from Eq. (10-19) cover a wide range; this is typical of experimentally determined coefficients of bearing friction under marginal lubrication conditions.

Under conditions of hydrodynamic film lubrication, the friction torque T_j on the journal is larger than the bearing friction torque T_b as

* Louis Illmer, High Pressure Bearing Research, *Trans. ASME*, vol. 46, 1924.

Table 10-3 **Values of the circumstance constant C_1 for Eq. (10-19)**

Method of lubrication	Workman- ship	Attendance	Location and cleanliness	Constant C_1
Oil bath or flooded.....	High-grade	First-class	Clean and protected	1
Oil, free drop (constant feed)	Good	Fairly good	Favorable (ordinary conditions)	2
Oil cup or grease (intermittent feed)	Fair	Poor	Exposed to dirt, grit, or other unfavorable conditions	4

Table 10-4   Values of the type constant C_2 for Eq. (10-19)

Type of Bearing	Constant C_2
Rotating journals, such as rigid bearings and crankpins.....................	1
Oscillating journals, such as rigid wrist pins and pintle blocks.............	1
Rotating bearings lacking ample rigidity, such as eccentrices..............	2
Rotating flat surfaces lubricated from the center to the circumference, such as annular step or pivot bearings..	2
Sliding flat surfaces wiping over the guide ends, such as reciprocating cross-head shoes; use 2 for long guides and 3 for short guides.................	2–3
Sliding or wiping surfaces lubricated from the periphery or outer wiping edge, such as marine thrust bearings and worm gears.........................	3–4
Long power-screw nuts and similar wiping parts over which it is difficult to distribute lubricant or load uniformly..................................	4–6

indicated by

$$T_j = T_b + Fe \sin \phi = T_b + F\left(\frac{C_d}{2} - h_m\right) \sin \phi \qquad (10\text{-}20)$$

where ϕ is the attitude angle and h_m is the minimum oil-film thickness as indicated in Fig. 10-4 and as determined by the methods of Art. 10-13.

Under marginal lubrication conditions, the frictional torques on the journal and bearing are essentially equal and Eq. (10-20) is not applicable.

10-12   Hydrodynamic Load Capacity of Journal Bearings.   An exact solution of the Navier-Stokes relations yielding the relationship between the radial-load capacity of hydrodynamically lubricated journal bearing and the system parameters appears impossible. Various analytic approximations have been devised. Some of the more recent analytical and experimental investigations[*,†] have yielded results useful to the designer. Various dimensionless quantities called *load numbers* have been devised which relate unit loading to other parameters. Some of these are as follows:

$$N_L = \frac{p}{\mu n}\left(\frac{D}{L}\right)^2 \qquad (10\text{-}21)$$

$$N_c = \frac{p}{\mu n}\left(\frac{C_d}{D}\right)^2\left(\frac{D}{L}\right)^2 = \frac{p}{\mu n}\left(\frac{C_d}{L}\right)^2 \qquad (10\text{-}22)$$

$$N_c = \frac{p}{\mu n}\left(\frac{C_d}{D}\right)^2\left(\frac{D}{L}\right)^{1.728} \qquad (10\text{-}23)$$

---

* G. B. DuBois, F. W. Ocvirk, and R. L. Wehe, Experimental Investigation of Eccentricity Ratio, Friction, and Oil Flow of Long and Short Journal Bearings with Load-number Charts, *NACA Tech. Note* 3491, 1955.

† L. F. Kreisle, Very Short Journal-bearing Hydrodynamic Performance under Conditions approaching Marginal Lubrication, *Trans. ASME*, vol. 78, no. 5, pp. 955–963, July, 1956.

$$N_h = \frac{p}{\mu n}\left(\frac{h_m}{D}\right)^2\left(\frac{D}{L}\right)^2 = \frac{p}{\mu n}\left(\frac{h_m}{L}\right)^2 \tag{10-24}$$

$$N_h = \frac{p}{\mu n}\left(\frac{h_m}{D}\right)^2\left(\frac{D}{L}\right)^{1.728} \tag{10-25}$$

in which N_L is called a load number, N_c a clearance load number, and N_h an oil-film thickness load number. Perhaps the simplest load number to use and one of the most useful is that value of N_L given in Eq. (10-21) and indicated graphically in Fig. 10-7 by the cross-plotting of the three

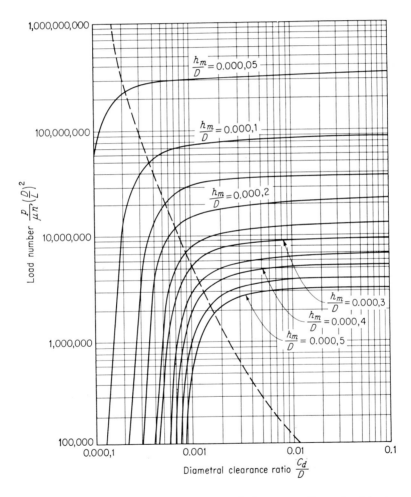

Fig. 10-7 Load number versus diametral clearance ratio and minimum oil-film thickness ratio for hydrodynamically lubricated short journal bearings (with length-to-diameter ratios below 1). For long bearings, useful results often may be obtained by substituting 1.0 for the actual length-to-diameter ratio in determining the load number.

dimensionless parameters

$$N_L = \frac{p}{\mu n}\left(\frac{D}{L}\right)^2, \qquad \frac{C_d}{D}, \qquad \text{and} \qquad \frac{h_m}{D}$$

If any two of the dimensionless quantities are known, the third is readily obtained by use of Fig. 10-7. For example, for $N_L = 30,000,000$ and $C_d/D = 0.001$, h_m/D is found from the plot to be approximately $0.000,15$. If D is known, then the minimum oil-film thickness h_m is readily calculated as $0.000,15D$ in. It should be pointed out that the use of Fig. 10-7 is valid only if hydrodynamic film lubrication exists; this figure is not applicable to marginal lubrication conditions. As a rough rule of thumb, conditions to the right of the dashed line in Fig. 10-7 have a fairly good possibility of being in the hydrodynamic range. Points lying to the left of the dashed line frequently correspond to marginal lubrication conditions. To be more nearly correct, the test for hydrodynamic lubrication given in Art. 10-13 should be made and then Fig. 10-7 used if hydrodynamic conditions exist.

Once the minimum oil-film thickness is known under hydrodynamic conditions by use of Fig. 10-7, the eccentricity ratio ϵ may be computed by use of Eq. (10-12). By use of Fig. 10-8, the attitude angle ϕ may be determined and hence the angular location of the minimum oil-film thickness. The friction torques on the bearing and journal now may be calculated by use of Eqs. (10-17) and (10-20).

10-13   Test for Hydrodynamic Lubrication.   None of the previously presented information concerning hydrodynamic lubrication of sliding-element bearings is valid if hydrodynamic lubrication conditions are nonexistent. Various tests, equations, maximum unit loadings, critical oil-film rupturing equations, and other methods have been devised in an attempt to prove or disprove the presence of hydrodynamic conditions. A recently developed criterion[*] states that hydrodynamic lubrication ceases and marginal lubrication procedures begin to exist when the asperities of the surfaces of the mating journal and bearing touch each other, thereby rupturing the hydrodynamic oil film. This is a function of the roughness of the mating bearing and journal surfaces measured in a circumferential direction. The average peak-to-valley surface roughness of metallic surfaces is defined and discussed in Arts. 19-8 and 19-9 along with a method of obtaining this roughness for bearing and journal surfaces. The criterion states that hydrodynamic lubrication exists if the minimum oil-film thickness h_{min} obtained by use of Fig. 10-7 equals or exceeds the sum of the two average peak-to-valley roughnesses of the

---

[*] L. F. Kreisle, Predominant-peak Surface Roughness, a Criterion for Minimum Hydrodynamic Oil Film Thickness, *Trans. ASME*, vol. 79, no. 6, pp. 1235–1241, August, 1957.

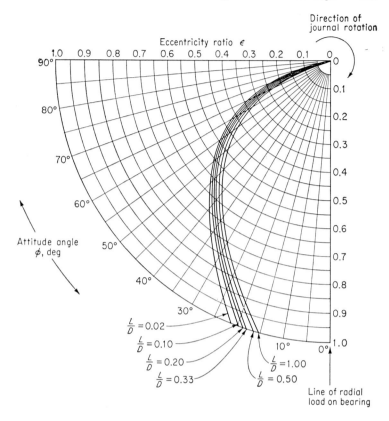

Fig. 10-8 Eccentricity ratio versus attitude angle for hydrodynamically lubricated short journal bearing.

mating journal and bearing surfaces. If $h_{\min}$ obtained from Fig. 10-7 is less than the sum of the two average peak-to-valley roughnesses of the bearing and journal, then marginal lubrication exists and the use of hydrodynamic lubrication considerations is invalid.

10-14 Typical Parameters for Hydrodynamically Lubricated Journal Bearings. Recent trends in the design of journal bearings have resulted in the use of short bearings rather than those in which the length exceeds the diameter. The length ratio L/D typically varies between 0.25 and 1.0, with perhaps 0.8 being an average value for many present-day applications. Diametral clearance ratios C_d/D of 0.001 in./in. are almost standard for any bearing; however, for bearings that may be subject to a temporary interruption of lubricant supply, ratios up to 0.002 in./in. may be used. Unit loadings p frequently vary between 100 and 1,500 psi; however, loadings as high as 3,500 psi sometimes are used

Table 10-5 **Typical loading conditions for journal bearings***

Bearing type	Unit loading p, psi	Maximum pressure-velocity relation pV, f-lb/(min)(in.²)	Viscosity Z	Bearing modulus ZN/p	Diametral clearance ratio C_d/D	Length ratio L/D
Axles:						
Locomotive...........	550	120,000	600	30–50	0.001 max	1.6–1.8
Railway car..........	300–500	120,000	500	50–100	0.001 max	1.8–2
Crankpins:						
Aircraft..............	750–5,000	2,500,000	8	10	....	0.5–1.5
Automobile...........	1,500–3,500	....	8	10	0.001 max	0.5–1.4
Diesel...............	1,500–4,000	....	20–60	5–10	0.001	0.8–1.5
Gas.................	1,200–1,800	400,000	20–60	10–15	0.001	0.8–1.5
Steam (high speed).....	400–1,200	400,000	30	6–15	0.001 max	1
Steam (low speed)......	800–1,500	....	80	6–8	0.001 max	1–1.25
Shears, punches.......	5,000–8,000	....	100	....	0.001	1–2
Crankshaft and main bearings:						
Aircraft..............	600–1,800	2,000,000	25–30	15–25	0.001 max	0.6–1.5
Automobile...........	300–1,800	....	25–30	20–30	0.001 max	0.8–1.5
Diesel...............	350–1,200	1,000,000	20–60	15–20	0.001 max	0.8–1.5
Gas.................	500–1,000	....	20–60	20–30	0.001 max	0.4–2.0
Steam (high speed).....	60–500	....	15–30	25–30	0.001 max	1.3–1.7
Steam (low speed)......	80–400	....	70	20–30	0.001 max	1.2–1.6
Shears, punches.......	2,000–4,000	....	100	....	0.001	1–2
Wrist pins:						
Aircraft..............	3,000–10,000	....	7–8	8–15	0.001 max	0.6–1.5
Automobile...........	1,500–5,000	....	7–8	8–15	0.001 max	0.8–1.5
Diesel...............	1,200–1,800	....	20–60	20–25	0.001 max	0.8–1.54
Gas.................	1,200–2,000	....	20–60	5–10	0.001 max	0.8–1.25
Steam (high speed).....	1,500–1,800	....	25	5	0.001	1.4–1.6
Steam (low speed)......	1,000–1,500	....	70	5	0.001	1.2–1.5
Generators, motors......	100–200	50,000	25	200	0.001	1–2
Hoisting machinery......	70–90					
Line shafts.............	15–150	25,000	25–60	30–100	0.001	2.5–3
Reducing gears.........	80–250	100,000	30–50	40–300	0.001	2–4
Machine tools..........	50–300	10,000	40	2–10	0.001	1.3
Steam turbines.........	75–300	1,000,000	10–20	100–200	0.001	1–2

* Note: Individual applications may vary widely from the above.

with success. Bearing moduli ZN/p between 5 and 200, or $\mu n/p$ values between 0.000,000,012 and 0.000,000,485 are common. Operating temperatures of the oil vary considerably; however, temperatures of 140 to 170°F are common in many applications, and temperatures as high as 240°F and higher frequently occur in internal-combustion engines. For bearings with a pressurized oil supply, it is common for the oil temperature to rise between 10 and 30°F as it flows through the bearing. Other information concerning design parameters in current usage is tabulated in Table 10-5.

An example of the design of a hydrodynamic journal bearing follows.

Example. A 1.9967-in.-diameter hardened-and-ground steel journal rotates uniformly at 1,800 rpm in a lathe-turned stationary bronze journal bearing which is 2.000 in. in diameter and 1.000 in. long. The effective temperature of the SAE 20 oil in the oil-film is 150°F. Assuming no shock conditions, determine the maximum radial load F_{max} that this journal can carry under these conditions and still operate hydrodynamically. While operating under these conditions, what is the power loss in the bearing?

$$C_d = D - d = 2.000 - 1.996 = 0.004 \text{ in.}$$
$$\frac{C_d}{D} = \frac{0.004}{2.000} = 0.002 \text{ in./in.}$$

From Table 10-1, for SAE 20 oil at 150°F, $\mu = 0.000,002,5$ reyn. Since $N = 1,800$ rpm,

$$n = \frac{1,800}{60} = 30 \text{ rps}$$

By use of Tables 19-2 and 19-3, typical values of the average peak-to-valley roughnesses are

For the bearing: $30 \times 5.0 = 150 \mu\text{in.} = 0.000,150 \text{ in.}$
For the journal: $10 \times 4.5 = 45 \mu\text{in.} = 0.000,045 \text{ in.}$

For hydrodynamic lubrication to exist, by use of the criterion of Art. 10-13, the smallest permissible value of the minimum oil-film thickness is

$$h_{min} = 0.000,150 + 0.000,045 = 0.000,195 \text{ in.}$$
$$\frac{h_{min}}{D} = \frac{0.000,195}{2.000} = 0.000,097,5$$

From Fig. 10-7, for the computed values of C_d/D and h_{min}/D,

$$N_L = 85,000,000 \quad \text{approx}$$

Hence

$$\frac{p_{max}}{\mu n}\left(\frac{D}{L}\right)^2 = \frac{F_{max}D}{\mu n L^3} = 85,000,000$$

Therefore

$$F_{max} = \frac{85,000,000 \times 0.000,002,5 \times 30 \times (1.000)^3}{2.000} = 3,185 \text{ lb}$$

the maximum radial load the bearing can carry and still operate hydrodynamically. This corresponds to a unit loading p of 1,593 psi. Equation (10-18) may be used to obtain the coefficient of bearing friction.

$$\frac{L}{D} = \frac{1.000}{2.000} = 0.5$$

therefore from Fig. 10-6, $k = 0.005$.

$$f_b = 19.55\left(\frac{\mu n}{p}\right)\left(\frac{D}{C_d}\right) + k = \frac{19.55 \times 0.000,002,5 \times 30 \times 2.0}{1,593 \times 0.004} + 0.005$$
$$= 0.005,5$$
$$V = \frac{\pi \, dN}{12} = \frac{\pi \times 1.996 \times 1800}{12} = 940 \text{ fpm}$$
$$\text{Power loss} = \frac{F_{max}f_b V}{33,000} = \frac{3,185 \times 0.005,5 \times 940}{33,000} = 0.498 \text{ hp}$$

10-15   Heat Generation in Journal Bearing Oil Films.   Due to frictional shearing of the oil film and possibly to the rubbing of bearing surfaces if marginal lubrication exists, energy is lost in the form of heat released within the oil film and the bearing surfaces. The heat generated H_{gen} is given by

$$H_{gen} = \frac{Ff_b\pi\, dN}{12 \times 778} = 0.000,336Ff_b\, dN \qquad \text{Btu/min} \qquad (10\text{-}26)$$

where f_b is obtained from Eq. (10-18) or (10-19) depending upon whether hydrodynamic or marginal lubrication exists.

10-16   Heat Dissipation Capacity of Journal Bearings.   The heat dissipation capacity of journal bearings depends upon the size of the bearing, whether or not pressurized lubrication exists, the temperature difference between the oil film and the ambient conditions, heat conduction conditions of the bearing support, and other items.

The natural heat dissipation capacity H_{diss} of a journal bearing based upon its projected area and ventilation of the bearing is given by

$$H_{diss} = \frac{Q_dLD}{778} \qquad \text{Btu/min} \qquad (10\text{-}27)$$

where Q_d is the heat dissipation coefficient of the bearing in ft-lb/(min) (in.²) of projected area. Typical values of Q_d are presented graphically in Fig. 10-9 as a function of the type of ventilation, temperature of the bearing t_b, and the ambient temperature t_a. For most bearings, natural dissipation of heat occurs mainly by convection, partly by conduction, and only slightly by radiation. The values of Q_d of Fig. 10-9 include the combined effects of convection, conduction, and radiation.

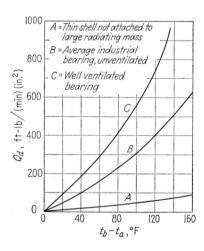

Fig. 10-9 Heat dissipation coefficient Q_d versus difference between bearing and ambient temperatures.

Table 10-6 **Specific heat c of petroleum oils, Btu/(lb)(°F)***

Temperature, °F	Specific gravity at 60/60°F							
	1.000	0.934	0.876	0.825	0.780	0.739	0.702	0.670
0	0.388	0.401	0.415	0.427	0.439	0.451	0.463	0.474
100	0.433	0.448	0.463	0.477	0.490	0.504	0.517	0.529
200	0.478	0.495	0.511	0.526	0.541	0.556	0.570	0.584
300	0.523	0.541	0.559	0.576	0.592	0.609	0.624	
400	0.568	0.588	0.607	0.625	0.643	0.661		
500	0.613	0.634	0.655	0.675	0.694			
600	0.658	0.681	0.703	0.724				
700	0.703	0.727	0.751					
800	0.748	0.774	0.799					

* J. C. Hunsaker and B. G. Rightmire, "Engineering Applications of Fluid Mechanics," p. 285, McGraw-Hill Book Company, New York, 1947.

For forced lubrication where oil under pressure is provided to the bearing, usually the heat absorbed by the oil is many times that dissipated by other means. The heat absorbed by the oil H_{oil} is

$$H_{oil} = w(h_2 - h_1) = wc_{av} \Delta t \qquad \text{Btu/min} \qquad (10\text{-}28)$$

where c_{av} = average specific heat of the oil flowing through the bearing, Btu/(lb)(°F)

h_1 = enthalpy of oil entering the bearing, Btu/(lb)(°F)

h_2 = enthalpy of oil leaving the bearing, Btu/(lb)(°F)

Δt = temperature rise of the oil as it flows through the bearing, °F

w = rate of lubricant flow, lb/min

Typical values of the specific heat c of lubricating oils as a function of the temperature and specific gravity of the oil are given in Table 10-6. Most designers determine the value of c_{av} corresponding to the average temperature of the oil flowing through the bearing.

An example of the calculation of bearing operating temperature follows.

Example. A 1.875-in.-diameter steel journal rotating 800 rpm is to carry a radial load of 700 lb. The journal bearing is 2.5625 in. long and has a 0.002,2-in. diametral clearance. An h_{min}/D operating ratio of 0.000,15 in./in. is assumed for the SAE 20 lubricating oil. Assuming 100°F ambient air temperature t_a around the bearing with average, unventilated industrial conditions, determine the probable operating oil film temperature t_o.

No direct method exists which in one calculation will lead to t_o. The procedure to be followed is to assume various operating temperatures and to calculate the heat generated H_{gen} and the natural heat dissipation capacity H_{diss} for each temperature. A plot will be made of H_{gen} versus bearing temperature and

Table 10-7   **Tabular data for calculation of bearing operating temperature**

Item	Assumed t_o, °F		
	170	175	180
μ	0.000,001,5	0.000,001,25	0.000,001,0
t_b, °F	165	170	175
f_b	0.004,28	0.003,90	0.003,52
H_{gen}, Btu/min	1.51	1.38	·1.24
H_{diss}, Btu/min	1.08	1.24	1.36

H_{diss} versus bearing temperature. Where these two curves intersect represents the condition where $H_{gen} = H_{diss}$, and the temperature corresponding to this intersection is the actual natural operating temperature of the oil film. For each assumed value of t_o, the bearing temperature t_b is assumed to be 5°F lower, a typical value for journal bearings. Oil viscosities are obtained from Table 10-1, and the coefficients of bearing friction f_b are computed by use of Eq. (10-18), assuming that hydrodynamic lubrication exists. Natural heat dissipation coefficients Q_d are read from the middle curve of Fig. 10-9. Equations (10-26) and (10-27) are used respectively to obtain the heat generated H_{gen} and natural heat

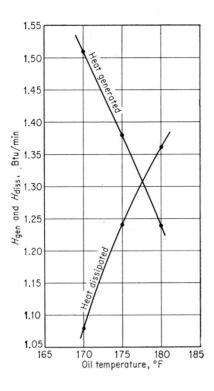

Fig. 10-10   Plot of computed data for example in Art. 10-16.

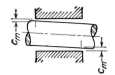

Fig. 10-11   Bearing clearance c_m at the ends of the journal.

dissipation capacity H_{diss} corresponding to each assumed oil-film operating temperature t_o. For t_o values of 170, 175, and 180°F, the data of Table 10-7 and Fig. 10-10 result.

The data of Table 10-7 are plotted in Fig. 10-10, where the heat generated and heat dissipation curves are seen to intersect at oil operating temperature t_o of approximately 177.6°F, assuming that the bearing operates hydrodynamically without pressurized lubrication.

Since no surface roughness data are given, it is impossible to use the criterion of Art. 10-13 to test for hydrodynamic conditions; however, the magnitude of the load number N_L at $t_o = 177.6°F$ may be useful.

$$N_L = \frac{p}{\mu n}\left(\frac{D}{L}\right)^2 = \frac{FD}{\mu n L^3} = \frac{700 \times 1.8772}{0.000,001,38 \times 13.33(2.5625)^3}$$

$$= 4,230,000$$

$$\frac{C_d}{D} = \frac{0.002,2}{1.8772} = 0.001,18 \text{ in./in.}$$

The point represented by the above values of N_L and C_d/D lie almost exactly on the dashed line of Fig. 10-7, indicating that this bearing may lie on the verge of marginal lubrication. With sufficiently smooth mating surfaces of the journal and bearing, hydrodynamic conditions can exist.

10-17   Journal Deflection and Clearance.   Clearance should be such that when the bearing is loaded the ends of the journal will not rub against the bearing surface. The clearance, indicated by c_m in Fig. 10-11, may be calculated as follows. For a shaft supported as a simple beam as shown in Fig. 10-12,

$$c_m = 2.55 \frac{pL}{E}\left(\frac{L}{d}\right)^3 + 3.4 \frac{pdL^2}{ED_s^4}(2C - a)a \qquad (10\text{-}29)$$

where the first term of the equation represents the deflection of the

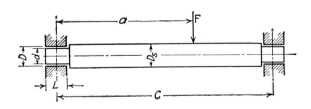

Fig. 10-12   Simply supported shaft without overhang.

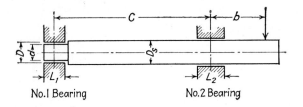

No.1 Bearing   No.2 Bearing

Fig. 10-13   Shaft with overhung load.

journal and the second term the shaft deflection, and p is the unit loading at the bearing under inspection.

When the load overhangs one bearing, as in Fig. 10-13, the minimum clearance is

$$c_{1m} = 2.55\,\frac{p_1 L_1}{E}\left(\frac{L_1}{d_1}\right)^3 + 3.4\,\frac{p_1 d_1 L_1^2}{E D_s^4}\,C^2 \qquad \text{for bearing No. 1}\quad(10\text{-}30)$$

$$c_{2m} = 2.55\,\frac{p_2 L_2}{E}\left(\frac{L_2}{d_2}\right)^3$$

$$\qquad\qquad + 6.8\,\frac{p_2 d_2 L_2^2}{E D_s^4}\left(\frac{C^2 b}{C+b}\right) \qquad \text{for bearing No. 2}\quad(10\text{-}31)$$

where a, b = distance of load from bearing, in.
$\qquad\quad C$ = axial distance between centers bearing, in.
$\qquad\quad D_s$ = shaft diameter, in.
$\quad d, d_1, d_2$ = journal diameter, in.
$\qquad\quad E$ = modulus of elasticity of shaft material, psi
$\quad L, L_1, L_2$ = bearing length, in.
$\quad p, p_1, p_2$ = unit loading on bearings, psi

Obviously the bearing clearances from Eqs. (10-29) through (10-31) should be positive values. For hydrodynamic lubrication to exist, some designers state that the computed value of c_m must be at least 0.8 times the sum of the two average peak-to-valley roughnesses of the mating journal and bearing surfaces as measured in a circumferential direction after breaking of the bearing.

10-18   Materials for Journal Bearings.   When a complete oil film is maintained between the journal and bearing surfaces, the materials used have little effect on the power loss and wear. With marginal lubrication, during the starting and stopping periods and when the lubricant supply fails, the surfaces come into contact; the materials therefore must be selected to resist wear and to provide a low coefficient of friction. Some bearing materials are desirable because they absorb some oil, providing lubrication during the starting period. Others are desirable because they are plastic enough to conform to slight irregularities of the journal. In general, it is claimed that unlike materials for the journal and bearing give the best results. However, for light pressures hardened steel on hardened steel, and cast iron on cast iron give excellent results. Good bearing materials should have a comparatively low coefficient of friction with the journal material, have good compressive strength, have good corrosion resistance to withstand any adverse compounds in the lubricant, have sufficient ductility to withstand fatigue and to allow for pounding and irregular loading, have a high thermal conductivity, have sufficient hardness to carry the load safely but not too high a hardness to prevent the solid impurities in the oil from being embedded in the comparatively soft bearing material rather than scoring the harder journal, and should have a modulus of elasticity permitting elastic deformation at high unit loadings. Other desirable characteristics of bearings are antiseize properties under conditions of lubricant failure and/or startup from rest under load, bearing wetted by the lubricant, adequate plasticity to enable the bearing to align itself (particularly during conditions of running-in), adequate compressive bearing yield strength at normal operating temperatures, high softening temperature for the alloy, and good wear resistance. It is highly probable that no one journal bearing will ever have the majority of these desirable characteristics, hence the designer must choose a material which best meets the requirements most important to any particular application.

The more common bearing materials are the babbitts, bronzes, leads, silver and its alloys, and various nonmetallics such as graphite, rubber, plastic, fabric, and wood.

10-19 Thrust Bearings. Thrust bearings are used to absorb the axial loads from shafts. They may be classified according to their type of lubrication as marginally lubricated, hydrodynamically lubricated, or hydrostatically pressurized externally.

10-20 Marginally Lubricated Thrust Bearings. The simplest type of thrust bearing for horizontal shafts is the *single collar* or *multiple collar* thrust bearings shown in Fig. 10-14. Such bearings may be oiled by reservoirs in the top of the bearings, by wick oilers, through hollow shafts, or by immersion in an oil bath. Automatic oil feeding devices are preferable since these bearings usually operate at rather high unit loadings and high rotational speeds. Coefficients of friction between 0.03 and 0.05 are typical. Typical ratios of outside to inside diameters of the annular bearing area are from 1.4 to 1.8. Unit loadings of 50 to 800 psi are common. After the bearing has been used for some time and the bearing surfaces have become worn, the friction torque is approximated by

$$T_f = 0.5 f F_a (r_o + r_i) \qquad\qquad (10\text{-}32)$$

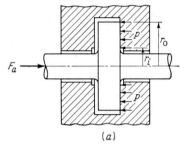

(a)

Fig. 10-14   Collar thrust bearings. (a) Single collar. (b) Multiple collar.

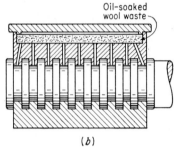

Oil-soaked
wool waste

(b)

where f is coefficient of friction, F_a is total axial thrust applied onto the bearing by the shaft, in pounds, r_i and r_o are respectively inside and outside radii of the bearing contact areas, in inches, and T_f is friction torque, in inch-pounds.

A simple *pedestal* or *vertical-thrust bearing* is shown in Fig. 10-15. Usually two or more washers are provided, with or without spherical faces to provide for self-alignment. The oil level should be high enough to completely submerge the thrust washers, and openings should be provided so that the oil can enter at the center and move outward between the bearing surfaces. Coefficients of friction between 0.01 and 0.02 are common.

A *suspension bearing*, of the type used in the upper end of vertical hydraulic turbines, is shown in Fig. 10-16. This bearing is enclosed in a housing that holds sufficient oil to submerge the bearing surfaces and provide oil-bath lubrication. Radial grooves on the upper side of the

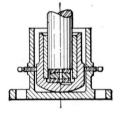

Fig. 10-15   Pedestal or vertical thrust bearing.

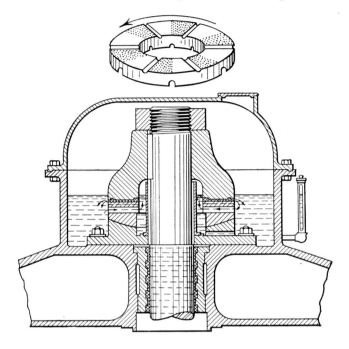

Fig. 10-16   Suspension bearing for hydraulic turbine.

bearing shoe distribute the oil over the bearing surfaces. Friction coefficients between 0.01 and 0.02 and outside to inside diameters ratios of the thrust area of 2 to 3 are conventional.

10-21   Hydrodynamic Thrust Bearings. In an attempt to substitute hydrodynamic lubrication for marginal lubrication, tapered lands or tilting pads may be used instead of the solid support pad, permitting the formation of a wedge-shaped hydrodynamic supporting oil film at each tapered land or tilted pad. These *tapered-land* and *tilting-pad* thrust bearings are illustrated schematically in Fig. 10-17. Usually the annular thrust area is broken into six or more lands or tilting pads with a net bearing support area of 0.75 to 0.85 times the gross annular area.

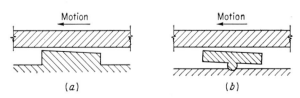

Fig. 10-17   Types of hydrodynamic thrust bearings. (*a*) Tapered land. (*b*) Tilting pad.

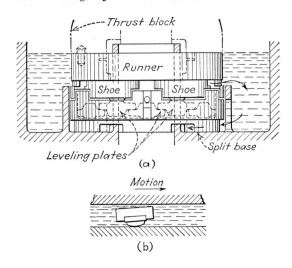

Fig. 10-18   Kingsbury thrust bearing. (*a*) Diagram
of six-shoe vertical bearing. (*b*) Diagram of one shoe
tilted under hydrodynamic pressure.

One of the most interesting applications is the *Kingsbury thrust
bearing* shown in Figs. 10-18 and 10-19. This bearing consists of a plain
collar supported by a number of pivoted segments, or shoes; the pivoted
shoes are mounted so that they are free to tilt radially and tangentially
under the hydrodynamic oil pressure. When the thrust collar begins to
move, oil adhering to it is carried between the collar and shoes, building
up a pressure that causes the shoes to tilt slightly, as shown in Fig. 10-18*b*.
When the bearing surfaces are completely submerged in oil, this wedge-
shaped oil film is continuously maintained; the bearing surfaces are completely separated, and true hydrodynamic film lubrication is maintained.
The amount of the tilting of the shoes, although very small, varies with

Fig. 10-19   Kingsbury combination thrust
and journal bearing. (*Allis-Chalmers Manufacturing Company.*)

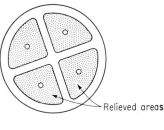

Fig. 10-20 Externally pressurized hydrostatic thrust bearing.

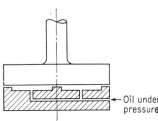

the oil viscosity, the thrust load, and the speed, so that the oil film automatically adjusts itself to operating conditions corresponding to minimum
friction torque.

Standard Kingsbury thrust bearings are rated at unit thrusts up
to 400 psi when using an oil of SAE 30 viscosity; however, unit loadings
up to 1,000 psi may be carried on specially built bearings. The coefficient
of friction ranges from 0.001 to 0.003 after running conditions are reached.
During the first revolution when starting, the coefficient is much higher.
The collar diameter is approximately $2\frac{1}{2}$ times the diameter of the shaft.
The bearing area is about 85 per cent of the collar area.

10-22 Externally Pressurized Hydrostatic Bearings. Upon occasion the unit loadings of journal or thrust bearings may be so high that
when the loaded journal is started from rest it is impossible for an oil
film to build up. By having relieved areas in the bearing and supplying
oil to these areas under external pressurization, a hydrostatic oil film
is formed before the journal is started in rotation and remains during
rotation. Figure 10-20 depicts an externally pressurized thrust bearing,
also called a hydrostatic bearing. Dead weights in excess of 1,000,000 lb
have been successfully supported on such a hydrostatic bearing, and it
rotates with only a fractional horsepower motor.

10-23 Lubrication by Solids. If bearing and journal surfaces of
dissimilar or similar material are completely cleaned of all contamination
whatsoever, including metal oxides (such as by employing out-gasing
under heat in a high vacuum), very high coefficients of friction result.
After these metallic surfaces have been permitted to oxidize by exposure

Table 10-8 Typical coefficients of boundary friction for various mating materials*

Mating materials	Coefficients of starting friction			Coefficients of running friction	
	Lubricated except for metallic oxides	Liquid-lubricated with indicated fluid	Dry-lubricated with molybdenum disulfide	Liquid-lubricated after wear-in with the liquid indicated	Dry lubricated after wear-in with molybdenum disulfide
Steel journal on steel bearing	0.4–0.8	0.1–0.2 with rape oil, castor oil, and mineral oil	0.05 for a 10,000-psi unit loading	0.10–0.15 with fatty acids and paraffin oil	0.07 for a 25,000-psi unit loading and a 50-fpm velocity; 0.03 for an 80,000-psi unit loading and a 50-fpm velocity
Steel journal on bronze bearing	0.22–0.45	0.12–0.16 with rape oil, castor oil, and mineral oil	0.078 for a 3,400-psi unit loading	0.16 with an unspecified liquid lubricant	0.075 for a 3,400-psi unit loading and a 6-ipm velocity
Steel journal on tungsten bearing	0.4–0.6	0.1–0.2 with an unspecified liquid lubricant	0.034 for a 10,000-psi unit loading	. . .	0.031 for a 100,000-psi unit loading and a 6-ipm velocity

* A. Sonntag, Lubrication by Solids as a Design Parameter, *Electro-Technol.*, p. 111, December, 1960.

to air, the resulting coefficients of friction usually are one-tenth or less of the previous values. Thus a metallic oxide and other contaminants on the mating bearing and journal surfaces act as a lubricant under non-hydrodynamic lubrication conditions where marginal processes prevail. In an attempt further to reduce the coefficients of friction, various solids have been bonded to the metallic surfaces or dispersed into the ordinary oil or grease lubricant. These include molybdenum disulfide, graphite, tungsten disulfide, titanium disulfide, lead oxide, lead chloride, tellurium disulfide, selenium disulfide, barium hydroxide, and many other materials, organic and otherwise. Typical effects of solid lubricants upon the coefficient of boundary friction are presented in Table 10-8.

Molybdenum disulfide has proved to be one of the best solid lubricants available, either bonded to surfaces or dispersed in oils or greases. It operates well over a temperature range of $-300°F$ to over $1500°F$, has a coefficient of running friction varying from approximately 0.10 for low unit loadings to about 0.023 for 400,000-psi loadings, and exhibits good anti-stick-slip characteristics, protecting the bearing against seizure. It has a high natural tendency to cling to the surfaces of metals.

Graphite, both natural and manufactured, is useful as a solid lubricant, either to replace the metallic surface of the journal or bearing or as a colloidal additive to oils or greases. It exhibits somewhat the same properties as molybdenum disulfide but usually to a lesser degree of desirability. When subjected to temperatures of approximately 750°F or above, the graphite begins to burn in the presence of the atmosphere. The characteristics of graphite as a lubricant depend mainly upon the graphitic carbon content rather than total carbon content of the solid lubricant, the kind of associated minerals, the particle size of the graphite, the kind of graphite, the effects of other additives or carriers used with the graphite, and the environmental atmosphere and temperature.

Typical coefficients of running and static friction of the solid lubricants in more common usage are presented in Table 10-9 together with whether or not essentially stick-slip (grab-release) action occurs. This information was obtained for hardened steel journals on hardened steel bearings with unit loadings between 8,000 and 15,000 psi.

10-24 Grease Lubrication of Journal Bearings. Calcium-base, lithium-base, or sodium-base greases, either with or without solid lubricant additives, have long been accepted as standard lubricants for rolling-element bearings. In these applications, it is usually desirable that the bearing be packed between one-third and one-half full of grease.

Grease lubrication of journal bearings has a long standing; however, because of the highly non-Newtonian nature of all greases, a nearly correct analysis of grease-lubricated bearings is almost impossible. In general, journal bearings which have been designed to operate successfully with a

Table 10-9   **Typical coefficients of boundary friction for solid lubricants***

Item	Solid lubricant	Coefficient of friction		Is stick-slip action typical?
		Running	Starting	
	None (dry steel on steel).........	0.40	0.4–0.8	yes
Layer lattice inorganics	Molybdenum disulfide, natural....	0.050	0.053	no
	Molybdenum disulfide, synthetic	0.091	0.106	no
	Molybdenum disulfide paste-concentrate..................	0.093	0.096	no
	Tungsten disulfide..............	0.090	0.098	no
	Titanium disulfide..............	0.25†	.....	yes
	Tellurium disulfide..............	0.25†	.....	yes
	Selenium disulfide..............	0.25†	.....	yes
	Graphite, natural..............	0.25†	.....	no
	Graphite, colloidal (22% in H_2O)	0.100	.....	no
	Boron nitride..................	0.25†	.....	yes
	Barium hydroxide..............	0.151	0.163	no
	Lead chloride..................	0.191	0.214	no
	Mica.........................	0.25†	.....	yes
	Silver iodide..................	0.231	0.245	yes
	Talc.........................	0.25†	.....	yes
Other inorganics	Borax........................	0.210	0.226	no
	Kaolin.......................	0.25†	.....	yes
	Rottenstone..................	0.189	0.195	no
	Lead oxide, zinc oxide...........	0.25†	.....	yes
	Vermiculite...................	0.160	0.167	no
Chemical conversion layers	Iron–manganese–phosphate layer......................	0.213	0.218	no
	Iron–manganese–phosphate layer with molybdenum disulfide rubbed on top................	0.067	0.074	no
	Sulfide melt (570°C) Sulfinuz.....	0.242	.....	no
Organics (with melting points)	Spermaceti wax (43–49°C)........	0.048	0.062	no
	Beeswax (60–63°C).............	0.050	0.055	no
	Bayberry wax (44–49°C).........	0.054	0.070	no
	Synthetic wax (comm. average) (93–96°C)...................	0.061	0.062	no
	Polyethylene glycol (high mol. wt.) (53–56°C)...............	0.077	0.078	no
	Hydrogenated tallow glyceride (59–63°C)....................	0.082	.....	no
	Candelilla wax (67–70°C)........	0.099	0.113	no
	Paraffin (49–77°C)..............	0.104	0.112	yes
	Montan wax (78–90°C).........	0.108	0.120	yes

Table 10-9 **Typical coefficients of boundary friction for solid lubricants.*** (*Continued*)

Item	Solid lubricant	Coefficient of friction		Is stick-slip action typical?
		Running	Starting	
Organics (with melting points)	Carnauba wax (83–86°C).........	0.143	0.169	no
	Diethylene glycol stearate (54–59°C).......................	0.083	0.089	no
	Calcium stearate (157–163°C).....	0.107	0.113	no
	Aluminum stearate (129–160°C)...	0.114	0.119	no
	Sodium stearate (198–210°C).....	0.164	0.192	yes
	Lithium-12-hydroxy stearate (210–215°C)...................	0.211	0.218	no

* A. Sonntag, Lubrication by solids as a Design Parameter, *Electro-Technol.*, p. 109, December, 1960.

† Because of limitations of the test equipment, maximum determinable values of the coefficient of friction were 0.25; hence whenever a value of 0.25 appears in the table it is probable that an unknown higher coefficient of friction exists.

high-viscosity oil will perform well when grease-lubricated, although usually with higher heat generation rates.

If properly carried out, grease lubrication of sleeve bearings offers many features for reliable lubrication of machinery. Equipment can be greased and is ready for immediate use, but there is no harm if a considerable period passes before the equipment is used.

The overload capacity of grease-lubricated journal bearings is far higher than that of oil-lubricated bearings of comparable radial-load capacity. Two of the major difficulties encountered in grease lubrication are excessive wear due to the admission of dirt and other abrasive particles, and the adverse conditions that result from providing too much grease and/or too frequently. Neither of these should be any real problem with proper maintenance conditions.

10-25 Mechanical Design Details of Journal Bearings. If the radial load on a bearing is more or less constant, the lubricating oil should be introduced through a radial hole on the unloaded portion of the bearing, essentially opposite to the location of the minimum oil-film thickness, and axially centered in the bearing. If a bearing has a circumferential oil-distributing groove, this groove divides the bearing into two sections, each somewhat less than one-half the total bearing length. In general, no oil grooving is necessary to properly distribute the lubricant throughout the bearing. If an oil-distributing groove is employed, it may be of the axial type, somewhat shorter than the bearing length, and connected to the radial hole supplying the oil.

When small diametral clearance ratios are employed, proper alignment of all bearings supporting the shaft must be assured or else the bearing must be mounted in somewhat flexible supports to permit self-alignment. Of course self-aligning bearings also can be used. In general, employing larger diametral clearance ratios increases the oil flow rate, decreases the necessity for careful alignment of bearings and shaft, but does not appreciably decrease the load-carrying capacity of an oil-lubricated journal bearing.

10-26 Bearing Caps. When split bearings are used, generally the bearing cap does not support the applied load but may be subjected to considerable pressure. When split connecting-rod ends are used in double-acting engines, the cap must carry the full load. In single-acting engines, the cap is subjected to heavy inertial loads under starting conditions.

The cap usually is checked for strength by assuming it to be a simple beam loaded at the center and supported at the bolt centers. In many cases, the requirements of forging or casting require greater thickness than does strength. For the bearing cap

$$s = \frac{Mc}{I} = \frac{3Fa}{2Lt^2} \tag{10-33}$$

for which

$$t = \sqrt{\frac{3Fa}{2Ls}} \tag{10-34}$$

where a is the distance between bolt centers, in inches, and t is the cap thickness, in inches.

When oil holes are provided in the cap, the length is the bearing cap length less the diameter or length of the oil hole.

The deflection of the cap is

$$y = \frac{1}{48}\frac{Fa^3}{EI} = \frac{Fa^3}{4ELt^3} \tag{10-35}$$

for which

$$t = 0.63a \sqrt[3]{\frac{F}{ELy}} \tag{10-36}$$

Most designers limit this deflection to 0.001 in. The bolts holding the cap in place are usually designed to carry 1.3 to 1.5 times their proportionate share of the load.

11

Spur Gears

11-1 Introduction. Belts, friction pulleys, and other types of power transmission that depend upon friction are subject to slippage and hence do not transmit a definite and invariable speed ratio. Chains and gears are used when positive drives are necessary, and, when the center distances are relatively short, toothed gears are preferred.

Although large speed ratios have been obtained, it is customary to limit the reduction to 6:1 for spur gears, and 10:1 for helical and herringbone gears. For larger reductions two or more pairs of gears are used. Before undertaking a study of gear design, the student should thoroughly understand gear nomenclature,* gear standards, and gear geometry.† Familiarization with gear fabrication would also be helpful. Essential principles and terminology will be reviewed in this treatise.

11-2 Spur Gear Nomenclature. Power may be transmitted by friction between two rotating cylinders or cones when they are pressed

---

* Gear Nomenclature, AGMA Standard, 112.03, November, 1954.

† V. L. Doughtie and W. H. James, "Elements of Mechanism," pp. 252–317, John Wiley & Sons, Inc., New York, 1954.

Fig. 11-1 Spur gear, welded construction. (*Courtesy Lukenweld, Inc.*)

Fig. 11-2 Internal gear. (*Courtesy Lukenweld, Inc.*)

together, but positive driving action without slippage will only be obtained if teeth are built upon these surfaces to form gears. The surfaces upon which the teeth are built are called the *pitch surfaces*, and the intersections of these surfaces with planes perpendicular to the axes of rotation are called the *pitch lines*, or *pitch circles*. The diameter of the pitch circle designates the size of the gear.

Spur gears (Fig. 11-1) are cylindrical in form, operate on parallel axes, and have straight teeth parallel to the axis. A gear may be either *external* (Fig. 11-1), with the teeth formed on the outer surface of a cylinder or cone, or *internal* (Fig. 11-2), with the teeth formed on the inner surface of a cylinder or cone. An internal gear can mesh only with an external pinion. A *rack* is a spur gear of infinite diameter, i.e., its pitch surface is a plane surface, and the pitch line a straight line. The common parts of spur gears are defined in Fig. 11-3. The *circular pitch p* of the gear is the distance from one face of a tooth to the corresponding face of the next adjacent tooth, measured along the pitch circle. The *diametral pitch P* is the ratio of the number of teeth to the pitch diameter, i.e., the number of teeth per inch of diameter. The *base pitch p_b* (sometimes called *normal pitch*) of involute gears is the distance between corresponding tooth profiles of adjacent teeth measured along the line of action, or along the circumference of the base circle from which the involute-tooth outline is generated. The *arc of action* is the arc traversed by a point on the pitch circle while any tooth is in contact with its mating tooth, and the *angle of action* is the subtending angle. The *addendum* is the height of the tooth outside of the pitch line and the *dedendum* is the depth of the tooth inside of the pitch line. The sum of the addendum and dedendum

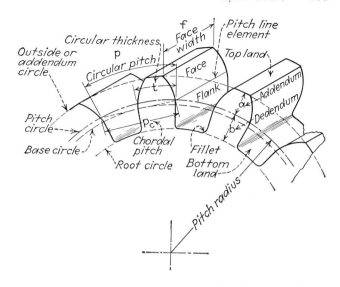

Fig. 11-3 Representation of terms used with gear teeth.

is the *whole depth*. *Clearance* is the amount by which the dedendum in a given gear exceeds the addendum of its mating gear. *Working depth* is the depth of engagement of two gears and is the sum of their addenda.

11-3 Tooth Profiles. In order that the teeth of two mating gears will transmit uniform angular velocity, the common normal to the mating tooth surfaces at their point of contact must always pass through the same *pitch point*, i.e., the point where the line of centers intersects the pitch circles. Tooth profiles fulfilling this requirement may be generated by rolling a template, with a suitable tracing point, on the outside of one pitch circle and on the inside of the mating pitch circle as in Fig. 11-4.

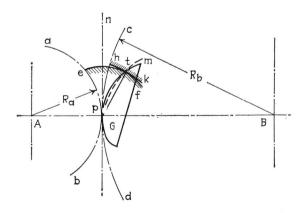

Fig. 11-4

Templates of many shapes may be used, but only two are commonly employed in modern gear practice. These are the circle for cycloidal teeth, and the logarithmic spiral for involute teeth. The involute, however, is more readily generated from a base circle as explained in Art. 11-5.

11-4   Cycloidal Gears.   The cycloidal form was one of the first regular profiles used for gear teeth. The difficulties encountered in producing accurate profiles have gradually forced this system into obsolescence. Cycloidal teeth are seldom used in gears at the present time, although in the composite system (see Art. 11-8) a portion of the tooth contour is cycloidal.

The cycloidal tooth curves are generated by rolling a circular template, or generating circle, on the outside and the inside of the pitch circle. In Fig. 11-5 the tracing point p traces the epicycloid pb, or face curve, when the circular template G_a is rolled on the pitch circle B, and traces the hypocycloid pa, or flank curve, when rolled on the inside of the pitch circle A. In an *interchangeable set*, all gears should operate properly with every other gear of the set. This condition is obtained by using the same size generating circle for the faces and flanks of each diametral pitch.

11-5   Involute Gears.   The involute curve is the basis of nearly all tooth profiles now in general use. The tooth profile is the involute of a *base circle*. In Fig. 11-6, a cord with a tracing point t is shown wrapped around two disks with centers at A and B. When A is rotated, the tracing point describes the involute cta on the disk A and the involute btd on the disk B. The tracing point is always on the tangent to the circles A and B (called base circles), and the tangent is always the normal to the tooth curves at their point of contact; hence p is the pitch point and Ap and Bp are the radii of the pitch circles. Pressure transmitted between the tooth surfaces will always act along the common normal; hence the pressure

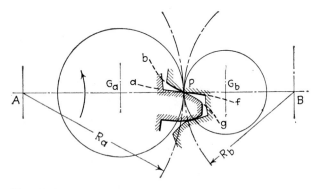

Fig. 11-5

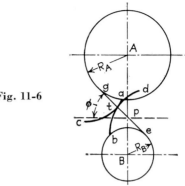

Fig. 11-6

angle ϕ is constant. The angular-velocity ratio is given by the equation

$$\frac{\omega_A}{\omega_B} = \frac{Bp}{Ap} = \frac{R_B}{R_A} = \text{constant} \qquad (11\text{-}1)$$

Hence, changing the center distance ApB does not alter the velocity ratio or destroy the proper tooth action of the involutes. The pitch diameters and the pressure angle change when the center distance is altered. The possibility of altering the center distance without destroying the correct tooth action is an important property of the involute gear.

In practice, the base circles and center distances are chosen so that particular values of the pressure angle are obtained, generally $14\frac{1}{2}$ or 20 deg and occasionally $17\frac{1}{2}$, $22\frac{1}{2}$, and 25 deg.

11-6   Interference in Involute Gears.   It is evident that the involute cannot extend inward beyond the base circle from which it is generated; hence, if the pinion revolves counterclockwise, the first point of contact between the tooth curves will be at e (Fig. 11-7), and the last point of contact will be at g, where the line of action, or pressure line, is tangent to the base circles. Any part of the tooth face of the pinion that extends beyond a circle drawn through g is useless; in fact it will interfere with the radial portion of the larger gear unless the flank is undercut. This interference is shown at i. The interference limits the permissible adden-

Fig. 11-7

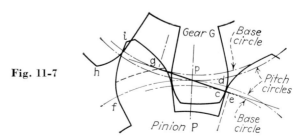

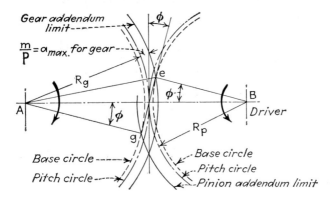

Fig. 11-8

dum length, and it is evident that, as the diameter of the pinion is decreased, the permissible addendum of the larger gear becomes smaller.

Let the addendum height be expressed in terms of the diametral pitch, making the addendum equal to m/P. m is a constant which when divided by the diametral pitch P gives the addendum, e.g., 1.0 or 0.8 as given in Table 11-1. From Fig. 11.8, the maximum outside radius of the gear A is

$$Ae = R_g + \frac{m}{P} = \sqrt{(Ag)^2 + (ge)^2}$$

and

$$R_g + \frac{m}{P} = \sqrt{R_g^2 \cos^2 \phi + (R_g + R_p)^2 \sin^2 \phi} \tag{11-2}$$

but

$$R_p = \frac{N_p}{2P} \quad \text{and} \quad R_g = \frac{N_g}{2P}$$

where N_p and N_g are the number of teeth in the pinion and gear, respectively.

By substituting these values in Eq. (11-2), expanding, and rearranging, it becomes

$$N_p^2 + 2N_pN_g = \frac{4m(N_g + m)}{\sin^2 \phi} \tag{11-3}$$

For a rack and pinion, $N_g = \infty$ and Eq. (11-3) becomes

$$N_p = \frac{2m}{\sin^2 \phi} \tag{11-4}$$

Equations (11-3) and (11-4) may be used to determine the smallest true involute pinion that will operate properly with a gear having N_g teeth, when the pressure angle and addendum ratio are known. Conversely, when N_p, the number of teeth on the pinion, is known, the number of teeth on the largest gear with which it will operate without interference may be determined.

Example. Determine the smallest pinion that will operate with a 60-tooth gear when both have full-depth $14\frac{1}{2}$-deg teeth. The addendum of full-depth teeth is $1/P$; hence m is 1. Then

$$N_p^2 + 2 \times 60 N_p = \frac{4(60 + 1)}{0.2504^2}$$

from which

$$N_p = 26.6, \text{ say 27 teeth}$$

11-7 Line of Action. The arc of action must be equal to or greater than the circular pitch, which for involute gears will be true when the line of action, or path of contact, is equal to or greater than $p \cos \phi$. In Fig. 11-8, the maximum line of action is

$$ge = (R_g + R_p) \sin \phi \gtreqless p \cos \phi = p_b \tag{11-5}$$

from which

$$p \lesseqgtr (R_g + R_p) \tan \phi$$

and

$$N_g + N_p \gtreqless \frac{2\pi}{\tan \phi} \tag{11-6}$$

This equation gives the number of teeth required on mating gears to obtain continuous action when the addenda of both gears are made as large as is possible without interference. For smooth power transmission, the line of action should be greater than the minimum. A value of at least 1.4 times the circular pitch is desirable.

The theoretical length of the line of action of any pair of true involute gears is given by the equation

$$L_a = \left[\sqrt{\left(R_p + \frac{m_p}{P}\right)^2 - R_p^2 \cos^2 \phi} \right.$$
$$\left. + \sqrt{\left(R_g + \frac{m_g}{P}\right)^2 - R_g^2 \cos^2 \phi} - (R_p + R_g) \sin \phi \right] \tag{11-7}$$

There will be interference when either radical in this equation is greater than $(R_p + R_g) \sin \phi$. In this case, the maximum line of action is determined by substituting $(R_p + R_g) \sin \phi$ for the larger radical.

11-8 Standard Involute Gears.* It is desirable that all gears of any system will operate with all other gears of the same system. The accepted

---

* Tooth Proportions for Coarse-pitch Involute Spur Gears, AGMA Standard, 201.02, Mar., 1958.

Table 11-1   **Gear-tooth proportions**

| Item | Symbol | Full-depth | | Helical, herringbone, and 20-deg stub |
		14½ deg, 22½ deg, cycloidal, and composite	20 deg, 25 deg	
Addendum...............	a	$1/P$	$1/P$	$0.8/P$
Dedendum, minimum......	b	$1.157/P$	$1.25/P$	$1/P$
Working depth...........	h_k	$2/P$	$2/P$	$1.6/P$
Whole depth, minimum....	h_t	$2.157/P$	$2.25/P$	$1.8/P$
Pitch diameter...........	D	N/P	N/P	N/P
Outside diameter.........	D_o	$(N+2)/P$	$(N+2)/P$	$(N+1.6)/P$
Tooth thickness on the pitch line, basic........	t	$1.5708/P$	$1.5708/P$	$1.5708/P$
Clearance, minimum......	c	$0.157/P$	$0.25/P$	$0.2/P$
Fillet radius, minimum....	r_f	$0.209/P$	$0.3/P$	$0.3/P$

P = diametral pitch; N = number of teeth.

Diametral pitch is used up to 1 diametral pitch inclusive. Circular pitch is used for 3-in. circular pitch and over. Diametral pitch varies by increments of $\frac{1}{4}$ from 1 to 2; $\frac{1}{2}$ from 2 to 5; 1 from 5 to 12; and 2 from 12 to 50.

In these systems the smallest pinion that will operate without interference with all gears, including a rack, has the following numbers of teeth: 14½-deg full-depth, 32 teeth; 20-deg full-depth, 18 teeth; 25-deg full-depth, 12 teeth; 14½-deg composite, 12 teeth; 20-deg stub, 14 teeth. These values for full-depth gears may be decreased when the addenda are modified according to Eq. (11-8).

standard tooth form has an involute profile. See Table 11-1 for standard proportions.

Fourteen-and-one-half-degree Composite System. When using small pinions with a pressure angle of 14½ deg, an addendum height of $1/P$, and a dedendum height of $1.157/P$, interference may be prevented by undercutting the flanks, which weakens the teeth, or by modifying the tooth form at the tip. The latter method is preferable and is used in the composite system. The portion of the tooth profile near the pitch line is a true 14½-deg involute, and the outer part of the face and the inner part of the flank are cycloidal in form. Standard proportions of the basic rack for this system are shown in Fig. 11-9.

Fourteen-and-one-half-degree Full-depth System. True involute profiles are obtained by the gear-generating process, using a straight-sided rack as the basic form. Small pinions have undercut flanks and do not have an arc of contact long enough to ensure continuous driving action. In the 14½-deg full-depth system, the smallest pair of equal gears with full-depth teeth that will operate together have 14 teeth

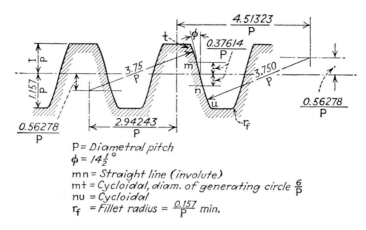

$P = Diametral\ pitch$
$\phi = 14\frac{1}{2}°$
$mn = Straight\ line\ (involute)$
$mt = Cycloidal,\ diam.\ of\ generating\ circle\ \frac{6}{P}$
$nu = Cycloidal$
$r_f = Fillet\ radius = \frac{0.157}{P}\ min.$

Fig. 11-9   Basic rack for the $14\frac{1}{2}$-deg composite tooth system.

each, and a 32-tooth pinion is the smallest that will operate properly without interference with all gears, including a rack. To obtain involute action with pinions having less than 32 teeth, it is sometimes desirable to increase the outside diameter of the pinion and to decrease the outside diameter of the gear by the same amount, while keeping the whole depth the same. The amount of increase and decrease is given by the formula

$$\text{Diameter increment} = [a - r_f(1 - \sin\ \phi)] - \frac{N_p \sin^2\ \phi}{2P} \qquad (11\text{-}8)$$

for $14\frac{1}{2}$-deg pinions having from 8 to 31 full-depth teeth. For 20-deg pinions having 8 to 17 full-depth teeth, the same correction should be made.

Twenty-degree and Twenty-five-degree Full-depth System. These systems use 20-deg and 25-deg pressure angles, which allow fewer teeth on the pinion without undercutting and greater load-carrying capacity than the $14\frac{1}{2}$-deg pressure angle.

Fine-pitch System.* This system is similar to the 20-deg full-depth system with a pressure angle of 20 deg but has a slight increase in whole depth to allow for the greater proportional clearance. Diametral pitches of 20 and finer are used in this system.

Stub-tooth Systems. Since interference in full-depth gears prevents the use of small pinions without modification, the stub, or shortened tooth, is often used. There are two accepted proportions for stub-tooth gears, the ASA or AGMA standard, and the Fellows stub. In both sys-

* 20-deg Involute Fine-pitch System, AGMA Standard, 207.04, June, 1956.

tems the pressure angle is 20 deg. In the Fellows system, the tooth is designated by two diametral-pitch numbers, thus, 5–7, 6–8, etc. The first figure is the diametral pitch used in computing the pitch diameter, circular pitch, fillet radius, and the number of teeth. The second figure is the diametral pitch used to compute the tooth height. The addendum height is $1/P$ and the dedendum is $1.25/P$. The fillet radius for Fellows is equal to one-fourth divided by the first figure of the diametral pitch.

11-9   Special Tooth Systems.   It is not always necessary or desirable that the gears be made interchangeable, and several systems have been developed to suit the particular needs of the industry employing them. In these systems, the addenda, dedenda, pressure angle, and center distances are chosen for each installation to give the best operating conditions and the strongest teeth. Space does not permit a complete discussion of these special systems. In brief, however, the long and short addendum systems as developed by Maag for spur gears and by Gleason for bevel gears afford less interference, provide stronger pinions, lend to better lubrication and less wear, and are quieter, but they sacrifice interchangeability.

11-10   Internal Gears.   The internal gear has its teeth cut on the inside of the rim and is used when the direction of rotation of both gears must be the same, and when short-center distances are required. The profile of the internal gear tooth is the same as that of the tooth space of a spur gear of the same pitch diameter. The addendum of the internal gear is inside the pitch circle and, to prevent interference, is generally limited in length so that the tip of the tooth passes through the interference point. If full-depth teeth are used, there may be fouling or interference between the tip of the pinion and the tip of the internal gear teeth, as shown in Fig. 11-10. To prevent this action, the gear should have at least 12 more teeth than the pinion when $14\frac{1}{2}$-deg full-depth

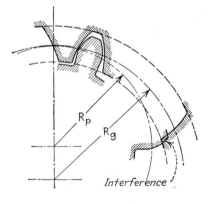

Fig. 11-10

teeth are used. With 20-deg stub-tooth gears, the gear should have at least seven more teeth than the pinion.

11-11 The Gear-design Problem. The determination of the proper gears to use in a particular application is a complex problem because of the many factors involved. First, the gears must operate together without tooth interference, with a proper length of contact, and without undue noise. The solution of this problem requires knowledge of gear geometry. Second, the gear teeth must have the ability to transmit the applied load without failure and with a certain margin of safety. They must resist not only the load resulting from the power transmitted but also the increases in load due to impact and shock caused by inaccuracy of tooth contour. The teeth must also be able to resist tooth deflection and accelerations, and stress concentration at the root of the teeth. The total resulting load is commonly referred to as the dynamic load. Third, the wearing qualities of the teeth must be considered. This is known as the wear load.

Up to about 1932, gear design was based upon the static strength of the tooth, assuming the tooth to be a cantilever beam acted upon by the moment resulting from the load obtained from the power source. The design stress was based upon the ultimate strength of the material with a factor of safety of about 3. In order to take into account the effects of tooth fabrication and additional loads due to impact, the design stress was further modified by a velocity factor. Since 1932, as a result of the research of Earle Buckingham,* gear design has been based upon the dynamic load, the endurance limit of the material, and the wear load. As a result, two general methods are now used: Buckingham and AGMA.†

---

* Earle Buckingham, Chairman of the Committee, Proposed Recommended Practice of the AGMA for Computing the Allowable Tooth Loads on Metal Spur Gears, report presented at the annual meeting of the AGMA, Cleveland, May 12, 1932.

Earle Buckingham, "Analytical Mechanics of Gears," McGraw-Hill Book Company, New York, 1949.

† Strength of Spur Gear Teeth, AGMA Standard 220.01, December, 1946.

Strength of Helical and Herringbone Gear Teeth, AGMA Standard 221.01, January, 1948.

Strength of Spur, Helical, Herringbone and Bevel Gear Teeth, AGMA Tentative Standard 225.01, October, 1959.

Surface Durability of Spur Gears, AGMA Standard 210.01, December, 1946.

Surface Durability of Helical and Herringbone Gears, AGMA Standard 211.01, April, 1944.

Surface Durability of Straight Bevel and Spiral Bevel Gears, AGMA Standard 212.01, April, 1944.

Surface Durability of Cylindrical-wormgearing, AGMA Standard 213.02, September, 1952.

Surface Durability of Double-enveloping-wormgearing, AGMA Standard 214.02, February, 1954.

The former uses Buckingham's general equations. The latter is based upon the findings of Buckingham and others but gives strength and durability formulas for obtaining directly the power transmitted by various types of gears. The AGMA has simplified the design by using constants which are obtained from graphs and tables. In both methods, the basis for design is the beam strength equation as developed by Wilfred Lewis. Since some gears may be satisfactorily designed by using the Lewis method and since the Lewis method will give tentative dimensions which can be later modified by the Buckingham method, both methods will be discussed. Space does not permit a complete discussion of the intricacies of gear design. Those interested are referred to the references in this article as well as to Darle W. Dudley, "Gear Handbook," McGraw-Hill Book Company, Inc., New York, 1962.

11-12   The Strength of Spur Gears. Each tooth may be considered to be a cantilever beam loaded as shown in Fig. 11-11. In the case of cast teeth not rigidly and accurately mounted, the load may be concentrated at an outer corner of the tooth; but with machine-cut teeth well mounted, the load may be considered to be distributed across the active face of the tooth.

The force F_n acting between the tooth surfaces is normal to the surface and in Fig. 11-11 is shown in the position producing the highest stress, i.e., at the beginning or ending of contact. The normal force may be replaced by components F_t and F_r, acting perpendicular and parallel to the center line of the tooth. The radial component produces compression stress in the tooth, which will be disregarded. The tangential component causes bending stresses, which are used as the basis of the design of the tooth for strength. The maximum bending stress may be located and computed as follows: Through the point a in Fig. 11-11 draw a parabola (shown in dash lines) tangent to the tooth curves at c and d. This parabola represents the outline of a beam of uniform stress, and

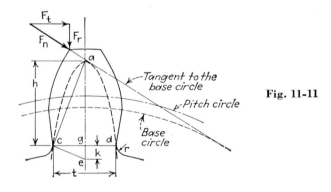

Fig. 11-11

therefore the maximum stress in the actual tooth will be at the points of tangency c or d. This stress is

$$s = \frac{Mc}{I} = \frac{6F_t h}{ft^2}$$

from which

$$F_t = sf\frac{t^2}{6h} \tag{11-9}$$

Both t and h are dependent upon the size of the tooth and its profile. Let

$$y = \frac{t^2}{6hp}$$

and substitute the value in Eq. (11-9)

$$F_t = sfyp = sfy\frac{\pi}{P} = \frac{sfY}{P} \tag{11-10}$$

where $Y = \pi y$
$\qquad F_t = $ force transmitted, lb
$\qquad s = $ stress, psi
$\qquad f = $ face width, in.
y and $Y = $ Lewis' form factor
p and $P = $ circular and diametral pitch, respectively

This equation was first developed by Wilfred Lewis and is used, with some modifications, for the determination of the strengths of all types of gears.

For convenience in gear computations, values of the form factor for various tooth systems are given in Table 11-2. To determine the form factor for any tooth profile not given in the tables draw the tooth outline to a large scale, locate the points where the inscribed parabola is tangent to the tooth outline, and scale the distances t and h. The points of tangency may be located without drawing in the parabola. In Fig. 11-11, triangles cge and agc are similar.

$$\frac{t}{2k} = \frac{2h}{t}$$

and

$$t^2 = 4hk$$

Then

$$y = \frac{t^2}{6hp} = \frac{4hk}{6hp} = \frac{2k}{3p} \tag{11-11}$$

Equation (11-10) indicates that the load which the gear tooth will carry is minimum when y is minimum, a condition that is found when c is the point of tangency of the parabola and the tooth outline. To deter-

Table 11-2 Form factors for the Lewis equation

Number of teeth	20-deg involute — Cycloidal, 14½-deg involute, composite, and generated y	Y	Full depth y	Y	AGMA stub y	Y	Fellows 20-deg stub teeth — 4-5 y	Y	5-7 y	Y	6-8 y	Y	7-9 y	Y	8-10 y	Y	9-11 y	Y	10-12 y	Y
10	0.056	0.176	0.064	0.201	0.083	0.261														
11	0.061	0.192	0.072	0.226	0.092	0.289														
12	0.067	0.210	0.078	0.245	0.099	0.311	0.096	0.302	0.111	0.348	0.102	0.320	0.100	0.314	0.096	0.302	0.100	0.314	0.093	0.292
13	0.071	0.223	0.083	0.261	0.103	0.324	0.101	0.318	0.115	0.361	0.107	0.336	0.106	0.332	0.101	0.317	0.104	0.327	0.098	0.308
14	0.075	0.236	0.088	0.276	0.108	0.340	0.105	0.330	0.119	0.374	0.112	0.352	0.111	0.348	0.106	0.332	0.108	0.339	0.102	0.320
15	0.078	0.245	0.092	0.289	0.111	0.349	0.108	0.339	0.123	0.386	0.115	0.364	0.115	0.361	0.110	0.346	0.111	0.348	0.105	0.330
16	0.081	0.255	0.094	0.295	0.115	0.361	0.111	0.348	0.126	0.396	0.119	0.374	0.118	0.370	0.113	0.355	0.114	0.354	0.109	0.340
17	0.084	0.264	0.096	0.302	0.117	0.368	0.114	0.358	0.129	0.405	0.122	0.383	0.121	0.380	0.116	0.364	0.116	0.366	0.111	0.349
18	0.086	0.270	0.098	0.308	0.120	0.377	0.117	0.368	0.131	0.411	0.124	0.390	0.124	0.390	0.119	0.374	0.119	0.374	0.114	0.358
19	0.088	0.277	0.100	0.314	0.123	0.387	0.119	0.374	0.133	0.414	0.127	0.398	0.127	0.398	0.122	0.383	0.121	0.380	0.116	0.364
20	0.090	0.283	0.102	0.320	0.125	0.393	0.121	0.380	0.135	0.425	0.129	0.405	0.129	0.405	0.124	0.390	0.123	0.386	0.118	0.371
21	0.092	0.289	0.104	0.327	0.127	0.399	0.123	0.386	0.137	0.431	0.131	0.411	0.131	0.411	0.126	0.396	0.125	0.392	0.120	0.377
23	0.094	0.296	0.106	0.333	0.130	0.408	0.126	0.396	0.141	0.441	0.134	0.422	0.135	0.422	0.129	0.407	0.128	0.402	0.123	0.387
25	0.097	0.305	0.108	0.339	0.133	0.417	0.129	0.405	0.143	0.449	0.137	0.432	0.138	0.432	0.133	0.417	0.130	0.409	0.126	0.396
27	0.100	0.311	0.111	0.349	0.136	0.427	0.132	0.414	0.146	0.458	0.140	0.440	0.140	0.440	0.135	0.425	0.133	0.417	0.129	0.405
30	0.102	0.320	0.114	0.358	0.139	0.436	0.135	0.425	0.149	0.468	0.143	0.449	0.144	0.452	0.138	0.433	0.136	0.427	0.132	0.415
34	0.104	0.326	0.118	0.371	0.142	0.446	0.139	0.438	0.152	0.478	0.147	0.460	0.148	0.465	0.142	0.447	0.139	0.436	0.136	0.426
38	0.107	0.335	0.122	0.383	0.145	0.455	0.141	0.443	0.155	0.487	0.150	0.471	0.150	0.471	0.145	0.455	0.141	0.443	0.139	0.437
43	0.110	0.345	0.126	0.396	0.147	0.465	0.144	0.452	0.158	0.496	0.153	0.481	0.153	0.481	0.148	0.465	0.144	0.452	0.141	0.443
50	0.112	0.352	0.130	0.408	0.151	0.474	0.147	0.461	0.161	0.506	0.156	0.490	0.156	0.490	0.151	0.471	0.147	0.461	0.144	0.452
60	0.114	0.358	0.134	0.421	0.154	0.484	0.150	0.471	0.164	0.515	0.159	0.500	0.159	0.500	0.154	0.483	0.150	0.471	0.148	0.469
75	0.116	0.364	0.138	0.434	0.158	0.496	0.154	0.484	0.167	0.525	0.162	0.509	0.162	0.509	0.157	0.493	0.153	0.480	0.151	0.474
100	0.118	0.371	0.142	0.446	0.161	0.506	0.158	0.496	0.171	0.536	0.166	0.521	0.166	0.521	0.160	0.503	0.156	0.490	0.154	0.484
150	0.120	0.376	0.146	0.459	0.165	0.518	0.162	0.509	0.174	0.546	0.170	0.534	0.169	0.531	0.164	0.515	0.160	0.503	0.158	0.496
300	0.122	0.383	0.150	0.471	0.170	0.535	0.167	0.525	0.179	0.562	0.174	0.548	0.172	0.542	0.168	0.527	0.165	0.518	0.163	0.512
Rack	0.124	0.390	0.154	0.484	0.175	0.550	0.173	0.543	0.184	0.578	0.179	0.562	0.176	0.553	0.172	0.540	0.170	0.534	0.168	0.528

mine the minimum value of k without drawing the parabola, select any point c near the narrowest part of the tooth, draw ac and then draw ce perpendicular to ac, draw cg perpendicular to the tooth center line, and scale eg. Repeat for several points close to c. A few trials will determine the minimum value of eg, or k.

11-13 Working Stress in Gear Teeth. The permissible working stress s in the Lewis equation depends upon the material, the heat treatment, the accuracy of the machine work, and the pitch-line velocity. Safe working stresses for common gear materials operating at very low velocities are usually assumed to be one-third the ultimate strength. Representative values are given in Table 11-3.

Slight inaccuracies in the tooth profile and tooth spacing, the fact that the teeth are not absolutely rigid, variations in the applied load, and repetitions of the loading cause impact and fatigue stresses that become

Table 11-3 **Safe beam stress or static stress of materials for gears** (values of s_w for use in the modified Lewis equations)

Material	Safe stress s_w	Ultimate strength s_{tu}	Yield stress s_{ty}
Cast iron, ordinary.	8,000	24,000	
Cast iron, good grade.	10,000	30,000	
Semisteel.	12,000	36,000	
Cast steel.	20,000	65,000	36,000
Forged carbon steel:			
SAE 1020 casehardened.	18,000	55,000	30,000
SAE			
1030 not treated.	20,000	60,000	33,000
1035 not treated.	23,000	70,000	38,000
1040 not treated.	25,000	80,000	45,000
1045 not treated.	30,000	90,000	50,000
1045 hardened.	30,000	95,000	60,000
1050 hardened.	35,000	100,000	60,000
Alloy steels:			
Ni, SAE 2320, casehardened.	50,000	100,000	80,000
Cr-Ni, SAE 3245, heat-treated.	65,000	120,000	100,000
Cr-Van, SAE 6145, heat-treated.	67,500	130,000	110,000
Manganese bronze, SAE 43.	20,000	60,000	30,000
Gear bronze, SAE 62.	10,000	30,000	15,000
Phosphor bronze, SAE 65.	12,000	36,000	20,000
Aluminum bronze, SAE 68.	15,000	65,000	25,000
Rawhide.	6,000		
Fabroil.	6,000		
Bakelite.	6,000	18,000 bending	
Micarta.	6,000	18,000 bending	

more severe as the pitch-line velocity increases. To allow for these additional stresses, it is customary to introduce a velocity factor into the Lewis equation. When V is the pitch-line velocity in feet per minute,

$$F_t = \frac{s_w f Y}{P} \frac{600}{600 + V} \qquad (11\text{-}12)$$

for ordinary industrial gears operating at velocities up to 2,000 fpm;

$$F_t = \frac{s_w f Y}{P} \frac{1,200}{1,200 + V} \qquad (11\text{-}13)$$

for accurately cut gears operating at velocities up to 4,000 fpm; and

$$F_t = \frac{s_w f Y}{P} \frac{78}{78 + \sqrt{V}} \qquad (11\text{-}14)$$

for precision gears cut with a high degree of accuracy and operating at velocities of 4,000 fpm and over.

The tangential force F_t at the pitch line may be obtained from the horsepower equation

$$F_t = \frac{33,000 \text{ hp}}{V} \qquad (11\text{-}15)$$

These equations are for gears operating under steady-load conditions. When operating more than 10 hr/day and when subjected to shock, the permissible tangential load should be modified according to the factors in Table 11-4.

For any set of operating conditions, there are many combinations of pitch, face width, and number of teeth that will satisfy the Lewis equation. However, well-proportioned gears should have a face width of from $8/P$ to $12.5/P$, or approximately $10/P$. Space requirements may

Table 11-4   Service factors

Type of load	Type of service		
	8–10 hr per day	24 hr per day	Intermittent, 3 hr per day
Steady..............	1.00	0.80	1.25
Light shock..........	0.80	0.65	1.00
Medium shock........	0.65	0.55	0.80
Heavy shock.........	0.55	0.50	0.65

Note: These factors are for completely enclosed gears well lubricated with the correct grade of oil. For nonenclosed gears, grease lubricated, use 65 per cent of the tabulated values.

require narrower teeth with a coarser pitch in installations such as automobile change gears. Turboreduction gears have much wider faces.

Example. A compressor running at 300 rpm is driven by a 20-hp, 1,200-rpm motor through a pair of $14\frac{1}{2}$-deg full-height gears. The center distance is 15 in., the motor pinion is to be forged steel (SAE 1045), and the driven gear is to be cast steel. Assume medium shock conditions. Determine the diametral pitch, the face width, and the number of teeth on each gear.

Solution. The pitch diameters will be 6 and 24 in., respectively. The pitch-line velocity will be

$$V = \frac{\pi \times 6 \times 1,200}{12} = 1,885 \text{ fpm}$$

The service factor from Table 11-4 is 0.65; hence the design tangential force at the pitch line will be

$$F_t = \frac{20 \times 33,000}{0.65 \times 1,885} = 538 \text{ lb}$$

If both gears were to be made of the same material, only the weaker pinion would have to be considered. In this example, the pinion may be considered in order to approximate the dimensions, after which any adjustments required by the gear can be made. For trial purposes assume a face width equal to $10/P$ and a value of Y equal to 0.30. The permissible stress at low speeds is 30,000 psi for the pinion. Substituting in the Lewis equation

$$538 = \frac{30,000 \times 10 \times 0.30}{P^2} \left(\frac{600}{600 + 1,885}\right)$$

from which

$$P^2 = 40.3$$

and

$$P = 6.35$$

This value suggests the use of a standard diametral pitch of 6, with 36 teeth on the pinion and 144 teeth on the gear. Values of Y are 0.330 and 0.374, respectively. Then the face width required is

$$f = \frac{538 \times 6}{30,000 \times 0.33} \left(\frac{2,485}{600}\right) = 1.35 \text{ in.} \qquad \text{for the pinion}$$

and

$$f = \frac{538 \times 6}{20,000 \times 0.374} \left(\frac{2,485}{600}\right) = 1.78 \text{ in.} \qquad \text{for the gear}$$

A face width of $1\frac{3}{4}$ in. is probably satisfactory, which is between $8/P$ to $12.5/P = 1.33$ to 2.1 in.

Note that the weaker gear is the one with the smaller product of Y and s_w.

11-14 Dynamic Loads on Gear Teeth. In Earle Buckingham's "Analytical Mechanics of Gears,"* a new equation was proposed, based

* McGraw-Hill Book Company, New York, 1949.

upon the determination of the effective mass acting at the pitch line of gears, acceleration load, separation of profiles and impact loads, for the determination of the dynamic load. However, the method presented here will be that which he introduced in 1932, because at present it is more widely used.

Small machining errors and the deflection of the teeth under load cause periods of acceleration, inertia forces, and impact loads on the teeth with an effect similar to that of a variable load superimposed on a steady load (see Art. 2-17). The total maximum instantaneous load on the tooth, or dynamic load, is

$$F_d = F_t + F_i = F_t + \frac{0.05V(Cf + F_t)}{0.05V + \sqrt{Cf + F_t}} \tag{11-16}$$

where F_d = total equivalent load applied at pitch line, lb
F_t = tangential load required for power transmission, lb
F_i = increment load (variable load), lb
C = a factor depending upon machining errors

To determine the proper value of C, the maximum errors for the various classes of gear cutting must be known. These errors are given in Table 11-5. The class of gear cutting required depends upon the speed of operation, and noise is a good measure of the accuracy required. The maximum errors that will permit reasonably quiet operation at different pitch-line velocities are given in Fig. 11-12. For extremely quiet operation, even these errors must be reduced. Knowing the class of

Table 11-5 **Maximum error in action between gears***

Diametral pitch	Class 1, industrial	Class 2, accurate	Class 3, precision
1	0.0048	0.0024	0.0012
2	0.0040	0.0020	0.0010
3	0.0032	0.0016	0.0008
4	0.0026	0.0013	0.0007
5	0.0022	0.0011	0.0006
6	0.0020	0.0010	0.0005
and finer			

* Courtesy AGMA.
Class 1, industrial gears cut with formed cutters.
Class 2, gears cut with great care.
Class 3, very accurate cut and ground gears.

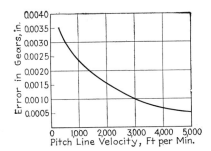

Fig. 11-12 Maximum permissible error in gears at various speeds.

gear cutting required, we can select the probable error from Table 11-5, and C from Table 11-6.

11-15 Stress Concentration. Where the tooth joins the bottom land, there is a concentration of stresses which are influenced by the fillet radius. This stress concentration causes the actual stress to be from 1.2 to 2 times the stress as obtained by substituting F_d in the Lewis equation. The exact value of the stress-concentration factor depends upon the material, the thickness of tooth at root, the load position on the tooth, the fillet radius, and the pressure angle. According to tests by Dolan

Table 11-6 **Values of the dynamic factor** C^*

Materials	Tooth form	Errors in gears					
		0.0005	0.001	0.002	0.003	0.004	0.005
Cast iron and cast iron	$14\frac{1}{2}$-deg	400	800	1,600	2,400	3,200	4,000
Cast iron and steel	$14\frac{1}{2}$-deg	550	1,100	2,200	3,300	4,400	5,500
Steel and steel	$14\frac{1}{2}$-deg	800	1,600	3,200	4,800	6,400	8,000
Cast iron and cast iron	20-deg, full depth	415	830	1,660	2,490	3,320	4,150
Cast iron and steel	20-deg, full depth	570	1,140	2,280	3,420	4,560	5,700
Steel and steel	20-deg, full depth	830	1,660	3,320	4,980	6,640	8,300
Cast iron and cast iron	20-deg, stub	430	860	1,720	2,580	3,440	4,300
Cast iron and steel........	20-deg, stub	590	1,180	2,360	3,540	4,720	5,900
Steel and steel	20-deg, stub	860	1,720	3,440	5,160	6,880	8,600

* Courtesy AGMA.
Semisteel and bronze have approximately the same values of C as cast iron.

and Broghamer* the stress-concentration factors are

$$K = 0.22 + \left(\frac{t}{r}\right)^{0.2} \left(\frac{t}{h}\right)^{0.4} \qquad \text{for } 14\tfrac{1}{2}\text{-deg gears} \qquad (11\text{-}17)$$

$$K = 0.18 + \left(\frac{t}{r}\right)^{0.15} \left(\frac{t}{h}\right)^{0.45} \qquad \text{for } 20\text{-deg gears} \qquad (11\text{-}18)$$

where the notation is as shown in Fig. 11-11 or

t = thickness of tooth at "theoretical weakest section," in.
r = radius of fillet, in.
h = height of load position above "theoretical weakest section," in.

This stress-concentration factor should be introduced into the Lewis equation (11-10) and the resulting dynamic stress will be

$$s_d = \frac{K F_d P}{f Y} \qquad (11\text{-}19)$$

The dynamic stress should be less than the endurance limit and give a reasonable margin of safety, depending upon the type of service. Dudley† points out that "many gear designers have found that gears do not have to be designed to carry as heavy a dynamic load as given by the Buckingham method. The author's work on high-speed aircraft gears indicates that in certain aircraft gas-turbine applications the dynamic load is really not more than about 135 per cent of the transmitted load. With the Buckingham method those applications are calculated to have dynamic loads in the order of 135 to 175 per cent. Even though the Buckingham method may not always give an answer that will agree with test data it is still the best method available at present." Table 11-7 gives values of the endurance limit for some common gear materials.

11-16   Design of Spur Gears for Wear.   The wear is dependent upon the materials used, the curvature of the tooth surfaces, the finish, the lubrication, and the amount of sliding action on the tooth surfaces. Pitting is a form of wear that occurs chiefly near the pitch line and is probably caused by fatigue failure of the material just under the surface, where the pressure between two curved surfaces produces the maximum shear stress, as pointed out in Art. 3-18. When the material below the surface develops a fatigue crack, the sliding action of the tooth surfaces will pull the surface material out, forming a pit. Wear may take the form of scratches or scores caused by pitted material or dirt carried in the lubricant. Failure of the lubrication may cause the surfaces to overheat

---

* T. J. Dolan and E. L. Broghamer, A Photoelastic Study of Stresses in Gear Tooth Profiles, *Univ. Illinois Eng. Exp. Sta. Bull.* 335, 1942.
  † Darle W. Dudley, "Practical Gear Design," p. 46, McGraw-Hill Book Company, New York, 1954.

Table 11-7   **Fatigue limits of gear materials**

Material	Brinell hardness number	Flexural endurance limit s_{ef}	Surface endurance limit s_{ew}
Gray cast iron..........................	160	12,000	90,000
Semisteel.............................	200	18,000	90,000
Phosphor bronze.......................	100	24,000	90,000
Steel................................	150	36,000	50,000
	200	50,000	70,000
For steel:	240	60,000	86,000
s_{ef} = 250 × Brinell number	250	62,500	90,000
For 400 Brinell number and above, use	280	70,000	102,000
s_{ef} = 100,000	300	75,000	110,000
s_{ew} = 400 × Brinell number − 10,000	320	80,000	118,000
	350	85,000	130,000
	360	90,000	134,000
	400	100,000	150,000
	450	100,000	170,000
	500	100,000	190,000
	550	100,000	210,000
	600	100,000	230,000

and seize. Abrasion and seizing can be practically eliminated if sufficient lubricant of the correct grade is supplied. Wear may occur in soft materials by the sliding action of the teeth pushing the metal toward the pitch line, where the sliding action becomes pure rolling and then reverses in direction. This action produces a hump near the pitch line and causes excessive noise.

Since pitting is due to fatigue failure, and abrasion and piling occur with soft material, it is evident that the load limit for wear is determined by the surface endurance limit of the material, the curvature of the surfaces, and the relative hardness of the surfaces. When mating gears are of different materials, the harder will mechanically work-harden the softer, raising its endurance limit, which for steels seems to increase in direct proportion to the Brinell hardness. The pinion should always be the harder to allow for work-hardening of the gear, to preserve the involute profile, to allow for greater abrasive wear on the pinion, and to decrease the possibility of seizing.

According to Buckingham, the load limit for wear is expressed by the equation

$$F_w = \frac{D_p f s_{ew}^2 \sin \phi}{1.4}\left(\frac{2N_g}{N_p + N_g}\right)\left(\frac{1}{E_p} + \frac{1}{E_g}\right) \tag{11-20}$$

where s_{ew} is the surface endurance limit from Table 11-7, N is the number of teeth, and E is the modulus of elasticity, the subscripts g and p referring to the gear and pinion, respectively. The value of F_w should not be less than the permissible value of the dynamic load F_d, from Eq. (11-16).

Example. Determine whether or not the gears in the example in Art. 11-13 are satisfactory as far as the dynamic and wear loads are concerned.

Solution. For a pitch-line velocity of 1,885 fpm, the permissible error e from Fig. 11-12 is 0.0017 in. For a 6-pitch gear, from Table 11-5 a Class 2 gear would have an e of 0.0010 in. and a Class 1 gear would have an e of 0.0020 in. Since the permissible error is 0.0017 in., a Class 2 gear must be used with a permissible error e of 0.001 in. From Table 11-6, the value of C is 1,600 for $14\frac{1}{2}$-deg involute steel and steel gears when e is equal to 0.001 in. The force transmitted is

$$F_t = \frac{20 \times 33,000}{1,885} = 350 \text{ lb}$$

Then from Eq. (11-16)

$$F_d = 350 + \frac{0.05 \times 1,885(1,600 \times 1.75 + 350)}{0.05 \times 1,885 + \sqrt{1,600 \times 1.75 + 350}} = 2,325 \text{ lb}$$

From a layout of the tooth profile it is determined that $h = 0.31$ in., $t = 0.18$ in., and $r = 0.035$ in. Substituting in Eq. (11-17)

$$K = 0.22 + \left(\frac{0.18}{0.035}\right)^{0.2} \left(\frac{0.18}{0.31}\right)^{0.4} = 1.34$$

The dynamic stress in the pinion from Eq. (11-19) is

$$s_d = \frac{1.34 \times 2,325 \times 6}{1.75 \times 0.33} = 32,360 \text{ psi}$$

From Table 11-7, s_{ef} is 36,000 psi for steel with a Brinell number of 150 and is satisfactory.

The dynamic stress in the gear from Eq. (11-19) is

$$s_d = \frac{1.34 \times 2,325 \times 6}{1.75 \times 0.374} = 28,550 \text{ psi}$$

From Table 2-5, s_{ef} is 30,000 psi for medium cast steel and is satisfactory. Check the gears for wear, using Eq. (11-20)

$$F_w = \frac{6 \times 1.75(50,000)^2 \times 0.25}{1.4} \left(\frac{2 \times 144}{36 + 144}\right) \left(\frac{1}{30,000,000} + \frac{1}{30,000,000}\right)$$
$$= 500 \text{ lb}$$

which is less than $F_d = 2,325$ lb.

Since F_w should not be less than F_d, it would be necessary to heat-treat the gears to an average Brinell number of 300, which gives a surface endurance limit of 110,000 psi. Then

$$F_w = \frac{6 \times 1.75(110,000)^2 \times 0.25}{1.4} \left(\frac{2 \times 144}{36 + 144}\right) \left(\frac{1}{30,000,000} + \frac{1}{30,000,000}\right)$$
$$= 2,420 \text{ lb}$$

which is greater than $F_d = 2,325$ lb and is satisfactory.

11-17 Gears with Cast Teeth. The teeth of gears may be cast. Improvements in the art of casting have led to gears with cast teeth which are in some cases more accurate than poorly cut teeth. Cast-teeth gears may be designed by the modified Lewis equation

$$F_t = s_w f y p \frac{600}{600 + V} \tag{11-21}$$

The working stress s_w may be obtained from Table 11-3 for cut teeth and the y factor from Table 11-2. The face width f is usually equal to $2.5p$ to $3p$. Standard circular pitches for cast teeth vary by $\frac{1}{8}$-in. increments from $\frac{1}{2}$ to $1\frac{1}{2}$ in., by $\frac{1}{4}$-in. increments from $1\frac{1}{2}$ to 3 in., and by $\frac{1}{2}$-in. increments from 3 to 4 in.

11-18 Nonmetallic Spur Gears. Gears made of rawhide, laminated fabric, and phenolic-resin materials, such as Bakelite and Micarta, are frequently used to reduce noise. Rawhide and laminated materials are not rigid and should be reinforced at both ends by metal flanges. To avoid charring by the heat of friction, rawhide gears should not be operated at pitch-line velocities greater than 2,500 fpm.

The permissible tangential force on these gears is

$$F_t = \frac{s_w f Y}{P} \left(\frac{150}{200 + V} + 0.25 \right) \tag{11-22}$$

11-19 Proportions of Gears. To complete the gear design, the first step is to determine the shaft size. If the pinion teeth are cut integral with the shaft, the root diameter should be slightly larger than the required shaft diameter. When a solid pinion is keyed to the shaft, the minimum pitch diameter is approximately

$$D_{\min} = 2 \times \text{bore} + \frac{0.25}{P} \tag{11-23}$$

The Nuttall Works of the Westinghouse Electric Company recommend that the minimum thickness of metal between the keyway and the root circle shall be

$$t_{\min} = \frac{1}{P} \sqrt{\frac{N}{5}} \tag{11-24}$$

where N is the number of teeth in the gear.

The outside diameter of the hubs of larger gears should be 1.8 times the bore for steel, 2 times the bore for cast iron, and 1.65 times the bore for forged steel or light service. The hub length should be at least $1\frac{1}{4}$ times the bore for light service and never less than the face width of the gear.

Small gears may be built with a web joining the rim to the hub. The web thickness should be from $1.6/P$ to $1.9/P$. The larger gears are provided with arms: four arms for split gears under 40 in. in diameter;

six arms for gears up to 120 in. in diameter; eight arms for larger gears. These arms are assumed to be cantilever beams loaded at the pitch line, with the load equally distributed to all arms. The design load is the stalling load, or load that will develop the maximum stress in the teeth at zero velocity. Hence

$$F_0 = \frac{s_w f Y}{P} \tag{11-25}$$

and

$$\frac{I}{c} = \frac{F_0 D}{2 n_a s_w} \tag{11-26}$$

where I/c = section modulus of arm section, in.^{3}
F_0 = stalling load, lb
n_a = number of arms
s_w = permissible working stress in tension, psi

Arms of various cross sections are used as shown in Fig. 11-13, and the elliptical arm is generally used except on very large and wide gears. Assuming the usual elliptical arm with the major axis twice the minor axis, the major axis at the outside of the hub will be

$$h = 4 \sqrt[3]{\frac{I}{\pi c}} = \sqrt[3]{20.4 \frac{I}{c}} \tag{11-27}$$

The arms are usually tapered toward the rim about $\frac{3}{4}$ in./ft.

The minimum thickness of the rim below the root circle is generally taken to be equal to the tooth thickness at the pitch line. This rule should be used only as a rough check since it does not allow for the differences in support furnished by different numbers of arms. According to the Nuttall Works, this thickness should be

$$t_r = \frac{1}{P} \sqrt[3]{\frac{N}{2 n_a}} \tag{11-28}$$

Other dimensions may be made as indicated in Fig. 11-13. In the final design of the gear, certain dimensions must be modified to prevent sudden changes in the thickness of adjoining parts. Unless large fillets are provided and sudden changes in section are avoided, there is danger of weakening, or even of breakage, due to shrinkage stresses during the cooling of the casting.

11-20   Gear Mountings and Bearings.   Gear mountings and bearings should be placed so that the bending deformation will be reduced to a minimum. It is often advisable to place a center bearing between the two sets of teeth on right- and left-handed helical gears. The deflection caused by bending or twist should not exceed 0.001 in. when measured

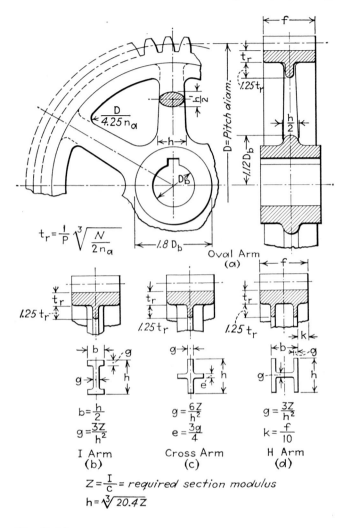

$$t_r = \frac{1}{P} \sqrt[3]{\frac{N}{2n_a}}$$

Oval Arm
(a)

$$b = \frac{h}{2}$$
$$g = \frac{3Z}{h^2}$$

I Arm
(b)

$$g = \frac{6Z}{h^2}$$
$$e = \frac{3g}{4}$$

Cross Arm
(c)

$$g = \frac{3Z}{h^2}$$
$$k = \frac{f}{10}$$

H Arm
(d)

$$Z = \frac{I}{c} = required\ section\ modulus$$
$$h = \sqrt[3]{20.4Z}$$

Fig. 11-13

over the length of the pinion. All computations for deformation are based on the pitch diameter. The load transmitted to the bearings of plain spur gears is the normal pressure between the tooth surfaces.

11-21   Efficiency of Gears. It is generally assumed that the friction loss in gear teeth depends on the tooth profile, pitch-line velocity, surface finish, and lubrication. However, when there is sufficient lubrication to prevent overheating and scoring, the friction appears to be practically independent of the velocity. The finish of the tooth surface is the most important factor in the efficiency of gears. Spur, helical, herringbone, and bevel gears cut in accordance with good commercial practice have an efficiency of 98 per cent or more. When lubrication is poor the efficiency may drop as low as 95 per cent. The efficiency of worm gears is 90 per cent or lower (see Fig. 12-11). Hypoid gears have an efficiency less than bevel gears but are capable of transmitting more power in less space. The power loss in the supporting bearings must be considered in addition to the loss in the teeth themselves.

11-22 Lubrication. To obtain the maximum life, the gears must be given a generous supply of the proper lubricant. The lubricant must maintain an oil film between the teeth and must also carry away the heat of friction, especially from the pinion, which, having more contacts per minute, tends to heat faster than the larger gear. The lubricant must be thin enough to penetrate the space between the teeth and heavy enough that the pressure will not break the oil film. Oil should be kept clean, since grit and metal dust carried in suspension in the oil will cause abrasive action on the tooth surfaces. With proper lubrication and correct alignment of the bearings, a properly designed and manufactured pair of gears will have an indefinite life.

12

Helical, Worm, and Bevel Gears

12-1 Helical Gears. In *helical gears*, the teeth are cut in the form of a helix about the axis of rotation. *Parallel helical gears* operate on parallel axes and have teeth of opposite hand. *Crossed-helical gears* operate on crossed axes and may have teeth of the same or of opposite hand. A *helical rack* has straight teeth which are oblique to the direction of motion. When spur gears begin to engage, the contact theoretically extends across the entire tooth on a line parallel to the axis of rotation. It has already been pointed out that this sudden application of load produces high impact stresses and excessive noise at high speeds. When helical gears begin to mesh, contact occurs only at the point of the leading edge of the tooth, gradually extending along a diagonal line across the tooth as the gears rotate. The gradual engagement and load application reduce the noise and the dynamic stress so that helical gears may be operated at higher speeds and can sustain greater tangential loads than straight-tooth spur gears of the same size. Pitch-line speeds of 4,000 to 7,000 fpm are common with automobile and turbine gears, and speeds of 12,000 fpm have been successfully used.

By referring to Fig. 12-1, it can be seen that the relation between the

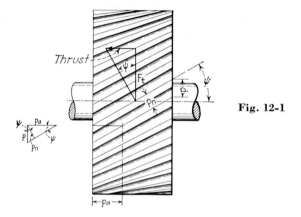

Fig. 12-1

normal circular pitch p_n (circular pitch in the plane normal to the teeth) and the *transverse circular pitch* p (circular pitch in the plane of rotation or the plane perpendicular to the gear axis and referred to as the *circular pitch*) is

$$p_n = p \cos \psi \tag{12-1}$$

where ψ is the helix angle, in degrees.

The relation of the *axial pitch* p_a to p is

$$p_a = \frac{p}{\tan \psi} \tag{12-2}$$

By designating the *normal diametral pitch* by P_n and knowing that $P_n p_n = \pi$, it can be seen that

$$P_n = \frac{P}{\cos \psi} \tag{12-3}$$

where P is the transverse diametral pitch, commonly called the *pitch*.

The relation between the pressure angles is

$$\tan \phi_n = \tan \phi \cos \psi \tag{12-4}$$

where ϕ_n is the normal pressure angle, in degrees, and ϕ is the transverse pressure angle (also called pressure angle), in degrees.

The end thrust, or axial load F_a is

$$F_a = F_t \tan \psi \tag{12-5}$$

12-2 Herringbone Gears.   The end thrust of two helical gears having opposite helices is counterbalanced if they are mounted on the same shaft, and if both sets of teeth are cut on a single gear blank, a *double-helical* or *herringbone* gear results. With the older methods of gear cut-

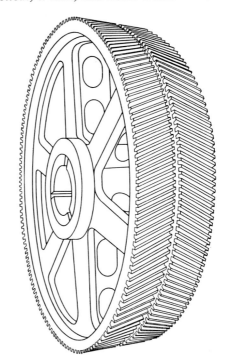

Fig. 12-2 Staggered-tooth herring-bone gear, Wuest type.

ting, a groove must be left at the center to provide clearance for the cutters, as shown in Fig. 12-2. The development of the Sykes gear-shaper permits the use of continuous teeth, as shown in Fig. 12-3.

12-3 Proportions of Helical and Herringbone Gears. Standard proportions for helical gears have not been developed, with the exception of a fine-pitch system.* Most manufacturers have developed proportions

* 20-deg Involute Fine-pitch System for Spur and Helical Gears, AGMA standard 207.04, June, 1956.

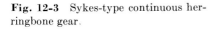

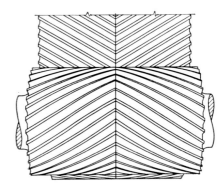

Fig. 12-3 Sykes-type continuous herringbone gear.

which are advantageous for their own tooling. Usually a 20-deg $2/P$ working depth, with a whole depth varying from $2.25/P$ to $2.35/P$, is used. Helix angles of 15 and 23 deg are preferred for single-helical gears and 30 and 45 deg for double-helical gears. Dudley* recommends two sets of basic proportions and suggests that the proportions may usually be modified so as to use the available tools. One set is based on the normal diametral pitch so that all helix angles in the series can be cut with the same hob, provided that the normal diametral pitch of the hob and gear is the same. The other set is based on the transverse diametral pitch and is recommended where quietness at high speed is important. This set requires different hobs for each helix angle even when the diametral pitch is the same. The first set gives helix angles from 0 to 30 deg, a working depth of $2/P$ and a whole depth of $2.25/P$. The other set uses four different helix angles: 15 deg with a working depth of $2/P$ and a whole depth of $2.35/P$; 23 deg with a working depth of $1.84/P$ and whole depth of $2.2/P$; 30 deg with a working depth of $1.74/P$ and whole depth of $2.05/P$; and 45 deg with a working depth of $1.42/P$ and whole depth of $1.7/P$.

The cutter and Y factor for helical gears should be based upon the virtual† or formative† number of teeth, N_v, obtained from the following equation

$$N_v = \frac{N}{\cos^3 \psi} \tag{12-6}$$

where N is the actual number of teeth.

12-4 Strength of Helical Gears. The strength of helical gears is determined by the equation

$$F_t = \frac{s_w f Y}{P} \frac{78}{78 + \sqrt{V}} \tag{12-7}$$

The form factor Y, based on actual number of teeth, may be taken from Table 11-2 when a standard profile in the normal plane is used. For any other profile, the method outlined in Art. 11-12 must be used. The other symbols have the same meaning as used in Chap. 11. In order that contact may be maintained across the entire active face of the gear, the minimum face width must be equal to the axial pitch just to give spur-gear action. In order to get the benefit of helical-gear action the face width should be at least twice the axial pitch. If the face width is too large, it is difficult to secure the accuracy to cause the whole face to carry the load uniformly. For average practice the face width should be

---

* Darle W. Dudley, "Practical Gear Design," pp. 95–97, McGraw-Hill Book Company, New York, 1954.

† Also called equivalent.

at least

$$f = \frac{2p}{\tan \psi} \qquad (12\text{-}8)$$

12-5 Dynamic Loads on Helical Gears. The method outlined in Art. 11-14 may be used with helical gears when Eq. (11-17) is modified as follows

$$F_d = F_t + \frac{0.05V(Cf \cos^2 \psi + F_t) \cos \psi}{0.05V + (Cf \cos^2 \psi + F_t)^{1/2}} \qquad (12\text{-}9)$$

12-6 Wear on Helical Gears. Ordinary industrial gears may be designed for wear by applying the service factors from Table 11-4 in Eq. (12-7). High-speed reduction gears for turbines are commonly designed to carry pitch-line loads of 100 lb per inch of face per inch of diameter, and loads as high as 320 lb have been successfully carried.

The better grades of helical gears should be checked for wear by the method outlined in Art. 11-16. For this purpose, Eq. (11-21) is modified as follows:

$$F_w = \frac{D_p f s_{ew}^2 \sin \phi}{1.4 \cos^2 \psi} \frac{2N_g}{N_p + N_g} \left(\frac{1}{E_p} + \frac{1}{E_g} \right) \qquad (12\text{-}10)$$

In an article, Determining Capacity of Helical and Herringbone Gearing,* W. P. Schmitter presented a very complete analysis with recommended equation for use in gear design. The following equation developed by Schmitter is the basis for the AGMA Standard Rating for Surface Durability of Helical and Herringbone Gears.†

The horsepower capacity is expressed by the equation

$$\text{hp} = \frac{C_m \times C_q \times C_c \times C_r \times C_v \times C_i \times D_p^2 fn}{126{,}000} \qquad (12\text{-}11)$$

where $C_m = (s_s/958)^2$ = material factor
  s_s = allowable working shear stress, equal to the endurance limit s_{es} or to $s_y/\sqrt{FS}$ depending on which gives the smaller value, psi
  FS = factor of safety, assumed to be $2\frac{1}{3}$
  $C_q = \dfrac{0.9L_a}{p \cos \phi \cos \psi}$ = contact-length factor
  L_a = length of the line of action [see Eq. (11-7)], in.
  $C_c = \sin \phi \cos \phi$ = curvature factor
  $C_r = \dfrac{N_g}{N_p + N_g}$ = ratio factor

---

* *Machine Design*, June and July, 1934.
† AGMA Standard 211.01, April, 1944.

$$C_v = \frac{78}{78 + \sqrt{V}} = \text{velocity factor}$$

$C_i = 0.64$ for gears up to 2 in. face width
   $= 0.667 - 0.0135f$ for gears from 2 to 18 in. face width
   $= 0.425$ for gears over 18 in. face width
$D_p = $ pitch diameter of pinion, in.
$f = $ face width, in.
$n = $ rpm of pinion

The factor C_i is called the inbuilt factor, and is used to provide for the increasing errors as the gear size increases. These errors increase partly because of machining difficulties and partly because the mountings for the larger gears cannot be made so accurately or so rigid as those of small gears.

When the gear proportions are not accurately known and for the more general applications, the product of C_c and C_q may be assumed to be 0.4.

This equation gives the rating for uniform load conditions assuming 8 hr/day service. For other conditions of service apply the factors from Table 11-4.

12-7   Crossed-helical Gears.   Crossed-helical gears shown in Fig. 12-4 are used to transmit power between shafts which are not parallel and do not intersect. These gears are essentially nonenveloping worm gears in that the gear blanks are cylindrical in shape. These gears are used to drive feed mechanisms on machine tools, camshafts, oil pumps on small internal-combustion engines, and similar units that require

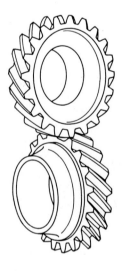

Fig. 12-4   Identical helical gears used on shafts at right angles.

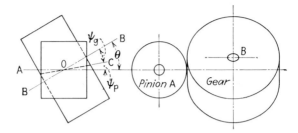

Fig. 12-5

only small amounts of power. This type of gearing should not be used to transmit heavy power loads since contact occurs only where the common normal to the axes cuts the tooth surface.

Gears of the same hand are usually paired; however, opposite hand gears may be used together. By reference to Fig. 12-5 it is seen that

$$\theta = \psi_p + \psi_g \qquad \text{for helices of the same hand} \qquad (12\text{-}12)$$

and

$$\theta = \psi_p - \psi_g \qquad \text{for helices of the opposite hand}$$

The angular-velocity ratio is not inversely proportional to the pitch diameters, as in spur gears, but is given by the equation

$$\frac{n_p}{n_g} = \frac{D_g \cos \psi_g}{D_p \cos \psi_p} \qquad (12\text{-}13)$$

The helix angle of each gear of a pair does not need to be the same. The normal pitches must be the same, but the pitches measured in the plane of rotation may or may not be the same. A special case, shown in Fig. 12-4, occurs when the axes are at right angles and the velocity ratio is unity. In this case the gears will be identical and the helix angles will be 45 deg.

12-8 Worm Gears. The maximum gear ratio advisable with helical gearing is about 10:1. For larger ratios, a gear train or double reduction, or a worm and worm gear, may be used. Worm and gear sets with ratios from 10:1 up to 100:1 are regularly employed, and ratios as high as 500:1 have been used. The worm and worm gear is a special case of helical gearing with nonparallel axes, the axes usually being at right angles.

The worm may be cut with a single or multiple thread as shown in Fig. 12-6. The *velocity ratio* does not depend on the diameters but is found by dividing the number of teeth on the gear by the number of threads on the worm, a 40-tooth gear and a double-thread worm having a velocity ratio of 20. The *lead* of the worm is the distance from any point

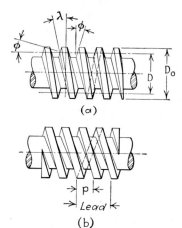

(a)

Fig. 12-6 (*a*) Single-thread worm. (*b*) Double-thread worm.

(b)

on one thread to the corresponding point on the next turn of the same thread, measured parallel to the axis. It is also the distance that a thread advances for one complete revolution of the worm. The *lead angle* λ is the angle between the tangent to the pitch helix and the plane of rotation. This angle is the complement of the helix angle as used with helical gears.

The *lineal pitch p*, equal to the circular pitch of the mating worm gear, is the distance from any point on one thread to the corresponding point on the adjacent thread, measured parallel to the worm axis. Note that, for a single-thread worm, the lineal pitch is the same as the lead; on a double-thread worm, the pitch is one-half the lead, etc.

The *pitch diameter* of the worm gear is the diameter measured on the central plane. The *throat diameter* is the outside diameter of the worm gear measured on the central plane.

12-9   Proportions of Worms.   The standard thread form is a straight-sided tooth, with a pressure angle of $14\frac{1}{2}$ deg for single and double threads, and 20 deg for triple- and quadruple-threaded worms, the angle in all

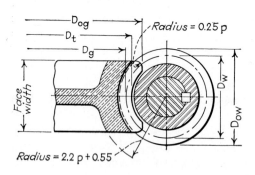

Fig. 12-7 AGMA standard worm form.

Table 12-1 **Proportions of AGMA standard industrial worms and gears**

Item	Symbol	Single and double		Triple and quadruple	
		Worm	Gear	Worm	Gear
Normal pressure angle...	ϕ_n	14½ deg	14½ deg	20 deg	20 deg
Linear pitch...	p	St'd	St'd	St'd	St'd
Pitch diameter...	D_w, D_g	$2.4p + 1.1$	$0.3183pN_g$	$2.4p + 1.1$	$0.3183pN_g$
Addendum...	a	$0.318p$	....	$0.286p$	
Whole depth...	h_t	$0.686p$	....	$0.623p$	
Outside diameter...	D_o	$3.036p + 1.1$	$D_t + 0.4775p$	$2.972p + 1.1$	$D_t + 0.3183p$
Throat diameter...	D_t	....	$D_g + 0.636p$	....	$D_g + 0.572p$
Normal tooth thickness..	t_n	$0.5p \cos \lambda$	....	$0.5p \cos \lambda$	
Face length...	......	$p\left(4.5 + \dfrac{N}{50}\right)$	....	$p\left(4.5 + \dfrac{N}{50}\right)$	
Face width...	......	....	$2.38p + 0.25$	....	$2.15p + 0.20$
Top round...	......	$0.05p$	....	$0.05p$	
Hub diameter...	......	$1.664p + 1$	$1.875 \times$ Bore	$1.726p + 1$	$1.875 \times$ Bore
Hub extensions...	......	p	$0.25 \times$ Bore	p	$0.25 \times$ Bore
Bore, maximum...	......	$p + 0.625$	....	$p + 0.625$	

N = number of threads on the worm. N_g = number of teeth on the gear.

cases being measured perpendicular to the pitch helix. The worm threads should be formed with a straight-sided milling cutter whose diameter is not less than the outside diameter of the worm or greater than $1\frac{1}{4}$ times the outside diameter. Standard linear pitches are $\frac{1}{4}$, $\frac{5}{16}$, $\frac{3}{8}$, $\frac{1}{2}$, $\frac{5}{8}$, $\frac{3}{4}$, 1, $1\frac{1}{4}$, $1\frac{1}{2}$, $1\frac{3}{4}$, and 2 in. Other dimensions are given in Fig. 12-7 and Table 12-1. The AGMA standard is intended for general industrial worms and does not cover worms of very large or small pitch, worms of more than four threads, worms of gear ratios over 100:1, or worms cut directly on the shaft or where the use justifies greater refinements in the design.

In the design of worms not covered by this standard, the following recommendations of the Brown and Sharpe Manufacturing Company* are useful. Since the velocity ratio does not depend on the worm diameter, the diameter is limited only by the shaft size required to transmit the power without excessive deformation. This company recommends that the minimum pitch diameter be

$$D_w = 2.35p + 0.4 \text{ in.} \qquad \text{for worms cut on the shaft} \qquad (12\text{-}14)$$

and

$$D_w = 2.40p + 1.1 \text{ in.} \qquad \text{for worms bored to fit over the shaft} \qquad (12\text{-}15)$$

The following pressure angles are recommended; with lead angles up to 15 deg, $14\frac{1}{2}$ deg; with lead angles up to 25 deg, 20 deg; with lead

* "Treatise on Gearing," Brown and Sharpe Manufacturing Company, Providence, R.I.

(a)   (b)   (c)

Fig. 12-8 (*a*) Straight-face. (*b*) Hobbed straight-face. (*c*) Concave face.

angles up to 35 deg, 25 deg; and with lead angles up to 45 deg, 30 deg. For worms mating with gears having 24 teeth or more, the 20-deg pressure angle is recommended. Pressure angles of 30 deg are common in automotive gears and in industrial reduction units. The larger pressure angles are used with the larger lead angles because of the difficulty of machining the high-lead threads. The threads are usually formed by a revolving milling cutter or a hob; and if the lead is large and the pressure angle small, the interference of the cutter will undercut the flanks of the worm.

12-10 Proportions of Worm Gears. Worm gears are made in three general types shown in Fig. 12-8. The straight-faced gear is simply a helical gear; and since it has only point contact with the worm thread, it is used for only very light loads. The hobbed straight-faced gear is cut with a hob, after which the outer surface is turned. This form is used for light loads and indexing wheels. The concave-faced gear is cut with a hob of the same pitch diameter as the mating worm so that the teeth fit the contour of the worm and present a larger contact area. The concave face is the accepted standard form and is used for all heavy services and general industrial uses.

The permissible width of gear face is limited by the fact that hobcut gears tend to become pointed near the ends of wide gears. For this reason, the sides of the gear are usually beveled as shown in Fig. 12-9. According to O. F. Shepard,* the extreme value of the face angle is

* *Trans. AGMA*, vol. 11, p. 201.

$$\tan \frac{\beta}{2} = \frac{\tan \phi_n}{\tan \lambda} \tag{12-16}$$

where β = face angle
ϕ_n = normal pressure angle
λ = lead angle

The face angle ranges from 60 to 90 deg, 60 deg being the usual angle. The proportions of worm gears to mate with AGMA standard industrial worms are given in Table 12-1.

12-11 Design of Worm and Gear for Strength. Since the teeth of the worm gear are always weaker than the worm threads, the strength may be determined by applying the Lewis equation to the gear. Because of the sliding action between the worm and gear teeth, the dynamic forces are not so severe as in the regular forms of gearing. Including the velocity factor, the Lewis equation for worm gears is

$$F_t = s_w p f y \frac{1,200}{1,200 + V_g} \tag{12-17}$$

where F_t is the tangential pitch-line load on the gear, in pounds.

The other symbols have the same meaning as before. The value of the form factor taken from Table 11-2 for full-height spur teeth is not strictly correct, but will give results on the safe side. Table 12-2 gives values of s_w.

The strength of the worm gear is usually not the determining factor in the design of a worm drive, but the strength should be checked before the design is finally approved.

12-12 Design of Worm and Gear for Wear. The wear and the heating are usually the determining factors in the successful application of a worm and gear. Although many methods of design have been proposed, none is generally accepted, and the methods proposed by various authorities do not agree.

Lubrication is important in the determination of the wearing qualities, since a thin or low-viscosity lubricant will be squeezed out, and

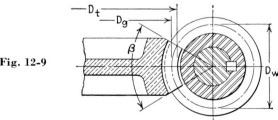

Fig. 12-9

Table 12-2   **Safe beam stress or static stress of materials for worm gears for use in Eq. (12-17)**

Material	Safe stress s_w	Ultimate strength s_{tu}	Yield stress s_{ty}
Cast iron, good grade......................	10,000	30,000	
Semisteel................................	12,000	36,000	
Cast steel...............................	20,000	60,000	27,000
Manganese bronze, SAE 43.................	20,000	70,000	25,000
Bronze, SAE 63, leaded gun metal...........	8,000	30,000	12,000
Phosphor bronze, SAE 65....................	15,000	35,000	20,000

Bronze SAE 63 is preferred with unhardened worms, and SAE 65 for chilling to harden for use with worms of great accuracy and hardness.

Different authorities give allowable stresses ranging from 65 to 150 per cent of the tabulated values. The tabulated values represent average practice.

metal-to-metal contact will result in high friction and overheating. If it is assumed that the proper grade of lubricant is used, the capacity of the worm and gear may be determined from the formula

$$F_w = A \cos \lambda \, \frac{600}{600 + V_w} \frac{s_c}{C_s} \qquad (12\text{-}18)$$

where F_w = permissible tangential load on gear, lb
A = projected area of tooth, in.^{2}
s_c = permissible working compressive stress from Table 12-3, psi
V_w = pitch-line velocity of worm, fpm
C_s = a service factor from Table 12-4

The projected area of the tooth is

$$A = h_w \frac{D_w}{2} \frac{\beta}{57.3} \qquad (12\text{-}19)$$

where h_w is the working depth of the worm, in inches, and β is the face angle, in degrees.

12-13   Materials for Worms and Gears.   Experience indicates that the gear should be made of softer material than the worm, and that the worm should be hard and carefully finished. Worms are universally made of hardened steel, and the gear is generally made of bronze. Cast-iron gears with soft-steel worms are used for light loads and installations where the service required is unimportant. Representative materials are given in Table 12-2. Also refer to AGMA Standards 243.51, 243.61, and 243.71, January, 1963.

Table 12-3 **Permissible surface pressures** s_c **for use in the wear formula (12-18)**

Material		Number of teeth in the gear							
Worm	Gear	10	20	30	40	50	60	70	80 and over
0.20C steel, untreated	Cast iron	75	225	425	750	900	1,080	1,250	1,350
0.40C steel, untreated	Bronze, SAE 63, sand cast	112	340	625	1,075	1,350	1,625	1,900	2,000
0.40C steel, heat-treated, ground	Bronze, SAE 63, sand cast	170	510	940	1,600	2,000	2,425	2,850	3,000
0.10C alloy steel, carburized, hardened, ground	Bronze, SAE 65, sand cast	225	675	1,250	2,150	2,700	3,250	3,800	4,000
	Bronze, SAE 65, chill cast	310	930	1,725	2,950	3,700	4,500	5,250	5,500
	Nickel bronze, sand cast	375	1,125	2,500	3,600	4,500	5,450	6,350	6,700
	Nickel bronze, chill cast	450	1,350	2,980	4,300	5,400	6,500	7,600	8,000

Values tabulated are for $14\frac{1}{2}$-deg pressure angles.
Multiply by 1.05 for 20-deg pressure angle.
Multiply by 1.10 for 30-deg pressure angle.

Table 12-4 **Service factors for worm gears for use in Eq. (12-18)**

Type of Service	Factor C_s
Intermittent with light shock	1.0–1.5
Continuous with medium shock: line shafts, crushers, etc	1.5–2.0
Continuous with heavy shock: reciprocating pumps, paper and rubber mills, etc	2.0–2.5
Continuous with frequent and very heavy shocks: main-line drives, steel mills, etc	2.5–3.0

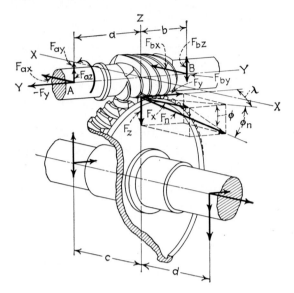

Fig. 12-10

12-14 Bearing Forces. A typical worm and gear is shown in Fig. 12-10. In this figure, let the pressure between the teeth and normal to the tooth surface be F_n. The sliding action between the surfaces acts chiefly along the tangent to the pitch helix. The friction force is fF_n, f being the coefficient of friction. Considering the forces acting on the gear, we have

$$F_x = F_n \cos \phi_n \sin \lambda + fF_n \cos \lambda = F_n(\cos \phi_n \sin \lambda + f \cos \lambda)$$
$$F_y = F_n \cos \phi_n \cos \lambda - fF_n \sin \lambda = F_n(\cos \phi_n \cos \lambda - f \sin \lambda)$$
$$= \frac{F_x(\cos \phi_n \cos \lambda - f \sin \lambda)}{\cos \phi_n \sin \lambda + f \cos \lambda} \tag{12-20}$$
$$F_z = F_n \sin \phi_n = \frac{F_x \sin \phi_n}{\cos \phi_n \sin \lambda + f \cos \lambda}$$

where ϕ_n is the pressure angle in the normal plane and $\tan \phi_n = \tan \phi \cos \lambda$, and f is the coefficient of friction (so used in Arts. 12-14 to 12-16). The force F_x is the tangential turning force on the worm and the end thrust on the gear. The force F_y is the tangential turning force on the gear and the end thrust on the worm. The force F_z is a separating force tending to force the worm and gear apart.

Consider the worm only, with bearings centered at A and B. The force F_z produces the forces F_{az} and F_{bz} acting on the bearings parallel to the Z axis. The magnitudes of these bearing forces are

$$F_{az} = F_z \frac{b}{a + b} \tag{12-21}$$

and

$$F_{bz} = F_z \frac{a}{a + b} \tag{12-22}$$

Similarly, the force F_x produces the forces F_{ax} and F_{bx} acting on the bearings and parallel to the X axis.

The force F_y produces the forces F_{ay} and F_{by} equal but opposite in direction and parallel to the Y axis. Then

$$F_{ay} = F_{by} = F_y \frac{D_w}{2(a + b)} \tag{12-23}$$

The total force on each bearing is the vector sum of the three component forces just determined. The bearing forces on the gear are found in the same manner.

12-15 Efficiency of the Worm and Gear. The efficiency is the ratio of the work output of the gear to the work input of the worm. The work done by the worm per minute is

$$W_i = F_x V_w$$

and the work output of the gear per minute is

$$W_0 = F_y V_g = F_y V_w \tan \lambda$$

Hence the efficiency is

$$
\begin{aligned}
\text{Eff} &= \frac{F_y \tan \lambda}{F_x} = \frac{\tan \lambda (\cos \phi_n \cos \lambda - f \sin \lambda)}{\cos \phi_n \sin \lambda + f \cos \lambda} \\
&= \frac{\tan \lambda (\cos \phi_n - f \tan \lambda)}{\cos \phi_n \tan \lambda + f}
\end{aligned} \tag{12-24}
$$

A critical study of this equation brings out some interesting things that must be considered in high-efficiency drives. For a given pressure angle, the efficiency depends upon the lead angle and the coefficient of friction. If the equation is differentiated and equated to zero, it is found that the efficiency is maximum when

$$\tan \lambda = \sqrt{1 + f^2} - f \tag{12-25}$$

Different values of the coefficient of friction result in the efficiency curves shown in Fig. 12-11. The curve of maximum efficiency obtained with each coefficient of friction is plotted in the same figure. Note that for any coefficient of friction, the efficiency curve rises rapidly with

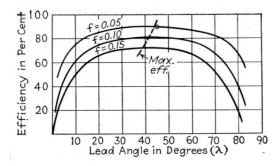

Fig. 12-11 Relation between lead, coefficient of friction, and efficiency of worm drives.

increasing lead angle, remains fairly constant over a wide range, and then drops off rapidly at high values of the lead angle. Small lead angles generally indicate inefficient worm drives; hence single-thread worms are seldom used except where high gear ratios are required, or where the worm is to be self-locking. A self-locking worm is one in which the gear cannot drive the worm. Although this condition depends upon a function of the lead angle, the pressure angle and the coefficient of friction, it is approximately correct to assume the worm to be self-locking if the tangent of the lead angle is less than the coefficient of friction.

The coefficient of friction varies with the lubricant used, the finish of the tooth surfaces, and the sliding velocity. The sliding velocity is

$$V_r = \frac{V_w}{\cos \lambda} = \frac{\pi D_w n_w}{12 \cos \lambda} = \frac{0.262 D_w n_w}{\cos \lambda} \qquad (12\text{-}26)$$

where D_w = worm pitch diameter, in.
n_w = speed of the worm, rpm
V_r = sliding velocity, fpm

Table 12-5   Coefficient of friction for worm gears

V_r	f	V_r	f	V_r	f
0	0.200	200	0.037	3,000	0.071
20	0.099	400	0.033	4,000	0.082
40	0.076	500	0.036	5,000	0.092
60	0.064	750	0.038	6,000	0.100
80	0.055	1,000	0.042	7,000	0.109
100	0.049	1,500	0.051	8,000	0.116
150	0.041	2,000	0.058	10,000	0.130

These coefficients and the corresponding efficiencies do not include the losses in the bearings. With proper alignment and good bearings, the bearing loss may be neglected.

Values for the coefficient of friction are not too definite. Many formulas, based upon the work of Wilfred Lewis and others, have been suggested. Buckingham* has provided a table of recommended values. Table 12-5 gives fairly reasonable values.

12-16 Heat Dissipation from Worm Drives. Considerable trouble may be experienced if the housing does not have sufficient heat-dissipating capacity. All the heat generated by friction must be dissipated through the oil to the housing and thence to the atmosphere. Most oils lose their lubricating properties at temperatures around 200°F, and the operating temperature of the worm should be limited to 180°F. The heat to be dissipated is

$$H = \frac{F_n f V_w}{778 \cos \lambda} \qquad \text{Btu/min} \tag{12-27}$$

where F_n is the force normal to the tooth surface, in pounds.

The dissipating capacity of the bearing depends on its construction. The ordinary housing will dissipate about 1.8 Btu/(°F)(hr) per square foot of surface. If the housing is exposed to good air currents, its dissipating capacity will be increased. In some installations it is necessary to circulate the oil through coolers, to provide cooling coils inside the housing, or to provide radiating fins on the housing.

* Earle Buckingham, "Analytical Mechanics of Gears," p. 414, McGraw-Hill Book Company, New York, 1949.

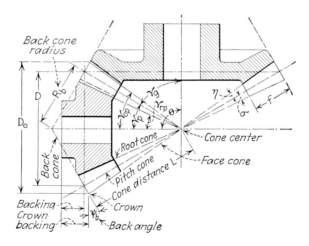

Fig. 12-12 Bevel gear nomenclature. σ = gear addendum angle; η = gear dedendum angle; γ_g = gear pitch angle; γ_p = pinion pitch angle; γ_{op} = pinion face angle; γ_{rp} = pinion root or cutting angle; D = pinion pitch diameter; D_o = pinion outside diameter.

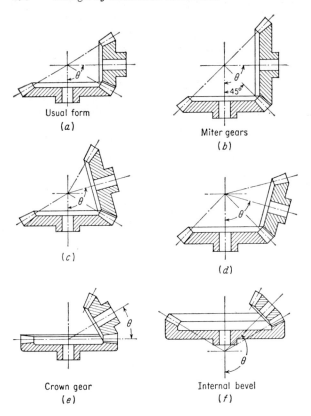

Fig. 12-13   Bevel-gear arrangements.

12-17 Bevel Gears. All gears previously discussed have had teeth cut or cast on cylindrical pitch surfaces. When the shafts intersect, the pitch surfaces are conical, and the gears are called bevel gears. The shafts may intersect at any angle. Bevel gears may have external or internal contact, and may have straight, curved, or spiral teeth. Unless otherwise stated, a pair of bevel gears is assumed to be straight toothed and to have the axes intersect at right angles. Figure 12-12 shows some of the terminology used in bevel gears. Figure 12-13 shows usual gear arrangements. The teeth of bevel gears are usually formed by the generating process but may be milled or cast if the application does not require the accuracy of generated teeth.

The tooth profile at the large end of the gear should theoretically be laid out on a sphere. The profile laid out on the surface of the back cone differs only sightly from the spherical profile, and the cone can be developed into a plane surface on which to study the tooth form and the tooth action. The profile of the tooth at the large end is therefore the same as the profile of a spur gear laid out on a pitch radius equal to the back-cone radius. The number of teeth on this imaginary spur gear is called the *virtual* or *equivalent number of teeth* and is found by the equation (see Fig. 12-12)

$$N_v = \frac{N}{\cos \gamma} \tag{12-28}$$

where N_v = virtual number of teeth
N = actual number of teeth
γ = pitch cone angle

$$\cos \gamma_p = \frac{R_g}{L} \quad \text{and} \quad \cos \gamma_g = \frac{R_p}{L} \tag{12-29}$$

where R_g and R_p are the radius of gear and pinion respectively, in inches, and L is the cone distance, in inches.

The virtual number of teeth, not the actual number of teeth, must be used in selecting the proper cutters and in all computations for the strength of bevel gears (Gleason system excepted).

12-18 Proportions of Bevel Gears. Industrial gears were formerly made with standard full-height $14\frac{1}{2}$-deg involute teeth with the same proportions as given in Table 11-1 for spur gears. These had undercut teeth on pinions of low tooth numbers. To prevent this undercutting, the use of long- and short-addendum teeth was introduced for pinions having less than 32 teeth with the $14\frac{1}{2}$-deg pressure angle, and less than 18 teeth with the 20-deg pressure angle. The total working depth with this system is taken as $2/P$, and the addendum of the pinion is made 0.7 of the working depth. Proportions for this system are given in Table 12-6.

The Gleason Company developed a system, applicable to generated straight-, spiral-tooth, and Zerol bevel gears, which was later adopted as the AGMA's recommended practice. In this system the pressure angle for straight-tooth bevel gears is 20 deg. However, a pressure angle of

Table 12-6 **Proportions of long- and short-addendum bevel gears**

Item	Pinion (driver)	Gear (driven)
Addendum......................	$1.4/P$	$0.6/P$
Dedendum......................	$0.7571/P$	$1.5571/P$
Tooth thickness on the pitch line,		
$14\frac{1}{2}$-deg......................	$1.778/P$	$1.364/P$
20-deg......................	$1.862/P$	$1.280/P$

$14\frac{1}{2}$ deg may be used without undercutting with selected gear ratios. The spiral bevel-gear system uses $14\frac{1}{2}$-, 16-, and 20-deg pressure angles, depending upon the gear ratios. The addenda are chosen so that the sliding action during approach is slightly less than that during recess, thus obtaining smoother action and quieter operation. Bevel gears are not usually interchangeable.

The recommended proportions for straight bevel gears are given in Table 12-7; and for spiral bevel gears the reader is referred to the "Gleason Spiral Bevel Gear System," Gleason Works, Rochester, N.Y., 1952.

12-19   Strength of Bevel Gears.   Since the size of the tooth and the force per unit of face length vary across the face of the tooth, the Lewis equation must be modified for use with bevel gears.

Table 12-7   **Proportions for straight-tooth bevel gears***

$$\text{Working depth} = \frac{2}{P} \quad \text{Total depth} = \frac{2.188}{P} + 0.002$$

$$\text{Addendum of gear} = \frac{\text{addendum from table}}{P}$$

$$\text{Addendum of pinion} = \frac{2}{P} - \text{addendum of gear}$$

$$\text{Dedendum of gear}\dagger = \frac{2.188}{P} - \text{addendum of gear}$$

$$\text{Dedendum of pinion}\dagger = \frac{2.188}{P} - \text{addendum of pinion}$$

Gear ratios		Add., in.	Gear ratios		Add., in.	Gear ratios		Add., in.	Gear ratios		Add., in.
From	To		From	To		From	To		From	To	
1.00	1.00	1.000	1.15	1.17	0.880	1.42	1.45	0.760	2.06	2.16	0.640
1.00	1.02	0.990	1.17	1.19	0.870	1.45	1.48	0.750	2.16	2.27	0.630
1.02	1.03	0.980	1.19	1.21	0.860	1.48	1.52	0.740	2.27	2.41	0.620
1.03	1.04	0.970	1.21	1.23	0.850	1.52	1.56	0.730	2.41	2.58	0.610
1.04	1.05	0.960	1.23	1.25	0.840	1.56	1.60	0.720	2.58	2.78	0.600
1.05	1.06	0.950	1.25	1.27	0.830	1.60	1.65	0.710	2.78	3.05	0.590
1.06	1.08	0.940	1.27	1.29	0.820	1.65	1.70	0.700	3.05	3.41	0.580
1.08	1.09	0.930	1.29	1.31	0.810	1.70	1.76	0.690	3.41	3.94	0.570
1.09	1.11	0.920	1.31	1.33	0.800	1.76	1.82	0.680	3.94	4.82	0.560
1.11	1.12	0.910	1.33	1.36	0.790	1.82	1.89	0.670	4.82	6.81	0.550
1.12	1.14	0.900	1.36	1.39	0.780	1.89	1.97	0.660	6.81	∞	0.540
1.14	1.15	0.890	1.39	1.42	0.770	1.97	2.06	0.650			

In case of choice use the larger.

* "Gleason 20-deg Straight Bevel Gear System," Gleason Works, Rochester, N.Y., 1954.

† Actual dedendum 0.002 in. larger.

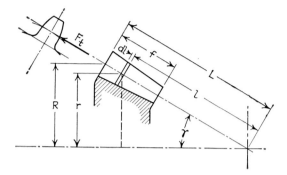

Fig. 12-14

In Fig. 12-14, consider a very short length of tooth dl over which the force may be considered as uniform in intensity. The force acting on this portion of the tooth is dF, and the Lewis equation becomes

$$dF = s p_l y \, dl$$

where p_l is the circular pitch at distance l from apex of pitch cone and the turning moment about the axis produced by this force is

$$r \, dF = r s p_l y \, dl$$

The tooth thickness, the circular pitch, and the radius r at any point are proportional to the distance from the apex of the pitch cone. Hence

$$r \, dF = \frac{R l s}{L} \frac{p l}{L} y \, dl = R s p y \left(\frac{l}{L}\right)^2 dl$$

By integration of this equation, the total turning moment is

$$T = \frac{R s p y}{L^2} \int_{L-f}^{L} l^2 \, dl = R s p f y \left(1 - \frac{f}{L} + \frac{f^2}{3L^2}\right)$$

The force that, applied at the pitch line (large end of the gear), will produce this torque is

$$F_t = s f p y \left(1 - \frac{f}{L} + \frac{f^2}{3L^2}\right) \tag{12-30}$$

The width of the tooth face in bevel gears is limited to one-third the length of the cone distance; i.e., the maximum value of f is $L/3$. Hence the last term in the parenthesis will never be greater than $\frac{1}{27}$ and can be disregarded without appreciable error. By elimination of this term, the usual form of the Lewis equation for bevel gears is obtained.

$$F_t = sfpy \frac{L-f}{L} = \frac{sfY}{P} \frac{L-f}{L} \tag{12-31}$$

where p or P = pitch at large end
F_t = equivalent tangential force at large end, lb
f = face width, in.
y or Y = form factor for the virtual number of teeth

The cone distance L in inches is

$$L = \sqrt{R_g^2 + R_p^2} \tag{12-32}$$

where R_g is the pitch radius of the gear, in inches, and R_p is the pitch radius of the pinion, in inches.

The stress to be used in this equation is

$$s = s_w \frac{600}{600 + V} \qquad \text{for full-depth teeth finished with formed cutters} \tag{12-33}$$

and

$$s = s_w \frac{78}{78 + \sqrt{V}} \qquad \text{for the generated system} \tag{12-34}$$

The velocity to be used is that at the large end. Form factors may be obtained from Table 11-2 and s_w from Table 11-3 for spur gears.

The dynamic and wear loads may be checked by Eqs. (11-16), (12-31), and (11-20), using the virtual numbers of teeth for N_p and N_g, the pitch-line velocity at the large diameter, and F_t as the equivalent tangential force at this velocity.

Well-proportioned bevel gears have a face width from $6/P$ to $10/P$ but never exceeding $L/3$.

12-20   AGMA Procedure for Bevel-gear Design.   An AGMA Standard* gives the horsepower rating, for peak load, of straight and spiral bevel gears by the equation

$$hp = \frac{snD_pfY(L - 0.5f)}{126,000PL} \frac{78}{78 + \sqrt{V}} \tag{12-35}$$

where s = 250 times Brinell hardness number for gears hardened and not hardened after cutting
= 300 times Brinell hardness number for casehardened gears
f = face width, in.
n = speed of pinion, rpm
D_p = pitch diameter of pinion, in.
Y = form factor from Table 12-8
L = pitch cone distance, in.
V = pitch-line velocity at large end, fpm

* AGMA Standard 222.01-1944.

Table 12-8   Form factors Y for bevel gears [Eq. (12-35)]

Actual number of teeth in pinion	Gear ratios														
	1.00 to 1.25	1.25 to 1.50	1.50 to 1.75	1.75 to 2.00	2.00 to 2.25	2.25 to 2.50	2.50 to 2.75	2.75 to 3.00	3.00 to 3.25	3.25 to 3.50	3.50 to 3.75	3.75 to 4.00	4.00 to 4.50	4.50 to 5.00	5.00 to ∞
Straight-tooth bevel gears															
10	0.231	0.260	0.280	0.294	0.305	0.315	0.324	0.332	0.340	0.347	0.353	0.358	0.365	0.371	0.377
11	0.268	0.264	0.273	0.286	0.296	0.303	0.309	0.315	0.320	0.324	0.328	0.332	0.336	0.340	0.342
12	0.248	0.265	0.281	0.295	0.308	0.318	0.328	0.335	0.341	0.345	0.348	0.351	0.353	0.355	0.256
13	0.264	0.278	0.291	0.280	0.278	0.286	0.291	0.295	0.298	0.299	0.301	0.303	0.305	0.307	0.310
14	0.242	0.254	0.263	0.272	0.281	0.288	0.294	0.299	0.304	0.307	0.310	0.313	0.316	0.318	0.319
15	0.248	0.258	0.266	0.274	0.283	0.290	0.296	0.301	0.305	0.308	0.312	0.315	0.318	0.319	0.320
16	0.252	0.261	0.269	0.277	0.285	0.292	0.298	0.304	0.308	0.312	0.314	0.317	0.319	0.321	0.323
17–18	0.257	0.265	0.273	0.281	0.288	0.295	0.302	0.307	0.311	0.315	0.318	0.320	0.322	0.325	0.326
19–21	0.265	0.272	0.279	0.286	0.294	0.300	0.307	0.312	0.317	0.320	0.324	0.326	0.328	0.330	0.382
22–25	0.274	0.281	0.288	0.295	0.301	0.307	0.314	0.319	0.324	0.327	0.331	0.332	0.335	0.337	0.338
26–30	0.284	0.291	0.297	0.304	0.310	0.317	0.322	0.327	0.332	0.336	0.339	0.342	0.344	0.346	0.347
Spiral-tooth bevel gears															
11	0.316	0.335	0.343	0.325	0.327	0.333	0.338	0.344	0.350	0.356	0.361	0.367	0.375	0.384	0.390
12	0.298	0.318	0.333	0.343	0.351	0.357	0.363	0.368	0.372	0.377	0.379	0.381	0.384	0.386	0.388
13	0.302	0.320	0.334	0.343	0.351	0.358	0.365	0.371	0.376	0.381	0.384	0.386	0.388	0.391	0.393
14	0.306	0.322	0.334	0.345	0.354	0.362	0.369	0.374	0.378	0.382	0.386	0.389	0.391	0.393	0.395
15	0.314	0.330	0.342	0.352	0.360	0.368	0.374	0.380	0.385	0.389	0.392	0.394	0.397	0.399	0.402
16	0.322	0.335	0.347	0.358	0.367	0.374	0.381	0.386	0.390	0.394	0.397	0.400	0.402	0.404	0.406
17–18	0.329	0.343	0.354	0.364	0.373	0.382	0.389	0.394	0.398	0.400	0.403	0.406	0.407	0.409	0.410
19–21	0.339	0.351	0.362	0.373	0.382	0.389	0.396	0.401	0.405	0.407	0.410	0.411	0.412	0.414	0.415
22–25	0.351	0.363	0.373	0.382	0.391	0.398	0.403	0.407	0.410	0.412	0.413	0.414	0.415	0.417	0.418
26–30	0.364	0.374	0.384	0.393	0.399	0.404	0.407	0.410	0.412	0.414	0.415	0.416	0.417	0.418	0.419

Courtesy AGMA.

An AGMA Standard* gives the horsepower rating for durability by the equations

$$\text{hp} = 0.8 C_m C_B f \qquad \text{for straight bevel gears} \qquad (12\text{-}36)$$

and

$$\text{hp} = C_m C_B f \qquad \text{for spiral bevel gears} \qquad (12\text{-}37)$$

where C_m = materials factor from Table 12-9.

The AGMA Standard provides curves for obtaining C_b. The approximate value of C_b may be obtained from the formula

$$C_B = \frac{\sqrt{D_p}\, D_p n}{233}\left(1.4 - \frac{V}{4400}\right)$$

where n is the speed of the pinion, in rpm.

* AGMA Standard 212.01-1944

Table 12-9   **Material factor for Eqs. (12-36) and (12-37)**

Gear			Pinion			C_m
Material	Brinell	Rockwell C	Material	Brinell	Rockwell C	
I	160–200	...	II	210–245	...	0.30
II	210–245	...	II	245–280	...	0.35
II	245–280	...	II	285–325	...	0.40
II	270–310	...	II	335–360	...	0.45
II	285–325	...	II	335–360	...	0.50
II	210–245	...	III	....	50*	0.40
II	245–280	...	III	....	50*	0.45
II	285–325	...	III	....	50*	0.55
II	210–245	...	IV	....	55*	0.40
II	245–280	...	IV	....	55*	0.50
II	285–325	...	IV	....	55*	0.60
III	....	50*	III	....	50*	0.80
III	....	50*	IV	....	55*	0.90
IV	....	55*	IV	....	55*	1.00

I. Annealed steel   III. Oil or surface-hardened steel
II. Heat-treated steel   IV. Casehardened steel
* Minimum values.

12-21 Hypoid Gears.*

These gears are approximations of hyperboloidal gears, i.e., gears whose pitch surfaces are hyperboloids of revolution. As shown in Fig. 12-15, they resemble spiral bevel gears, but the axes do not intersect. The advantages of this type of gear are somewhat smoother action and the possibility of extending the shafts past each other so that bearings can be used on both sides of the gear and the pinion.

* A. L. Stewart and E. Wildhaber, The Design and Manufacture of Hypoid Gears, *S.A.E. Journal*, vol. 18, June, 1926.

A. H. Candee, Large Spiral and Hypoid Gears, *Trans. ASME*, vol. 51, 1928.

"Gleason Method for Designing Hypoid Gear Blanks," Gleason Works, Rochester, N.Y., 1960.

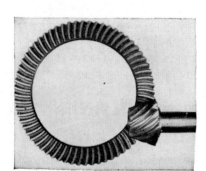

Fig. 12-15 Hypoid gear and pinion. (*Courtesy Gleason Works.*)

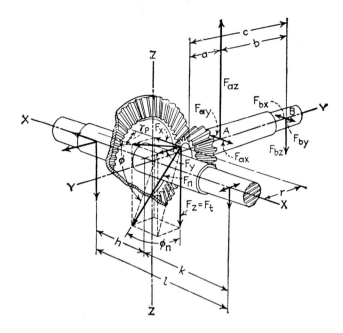

Fig. 12-16

12-22 Zerol Bevel Gears. These gears are spiral bevel gears with a zero spiral angle but with curved teeth, and they are cut on the same machines as spiral bevels. They have the advantages of localized tooth contact and the possibility of ground tooth surfaces. They have the same tooth action and end thrust as straight bevel gears and may be used in the same mountings. The basic pressure angle is 20 deg, but the $22\frac{1}{2}$-deg pressure angle is used with small numbers of teeth to prevent undercutting. The long and short addenda are used. The reader is referred to Gleason Zerol Bevel Gear System (1956) for design information.

12-23 Gleason Design Procedure. The Gleason Works of Rochester, N.Y., have had years of experience in the design and manufacture of straight and spiral-teeth bevel gears, Zerol and Hypoid gears. Through their effort proportioning and design of these gears have been standardized. This work has been generally accepted by the gear industry and in time has been adopted by the AGMA. Formulas, charts, and tables have been provided to simplify the design procedure.*

* "Gleason 20° Straight Bevel Gear Systems" (1959), "Gleason Spiral Bevel Gear Systems" (1959), "Gleason Bending Stresses in Bevel Gear Teeth" (1960), "Gleason Surface Durability Pitting Formulas for Bevel Gear Teeth" (1960), and "Formulas for Bevel Gear Teeth" (1960), Gleason Gear Works, Rochester, N.Y.

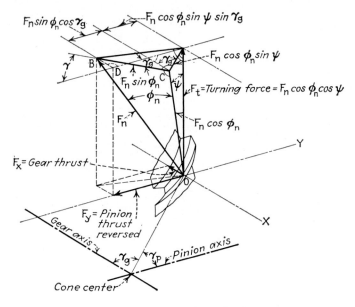

Fig. 12-17   Forces acting on a spiral bevel gear.

12-24   Bearing Loads on Bevel Gears.   Bevel gears may be mounted so that they overhang their supports. This condition, together with the thrust loads, makes the bearing loads much more severe than in spur gearing. In Fig. 12-16 a pair of straight-tooth bevel gears is shown with the loads exerted by the pinion on the gear tooth indicated. The force

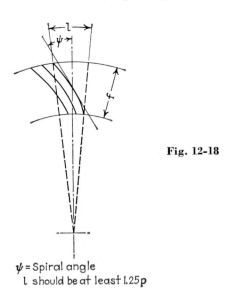

Fig. 12-18

$\psi =$ Spiral angle
l should be at least $1.25p$

F_n is the actual force between the teeth. The forces acting on the pinion tooth are the reverse of those shown in the figure. An examination of the figure shows that

$$F_z = F_t = F_n \cos \phi_n \qquad\qquad\qquad (12\text{-}38)$$
$$F_x = F_n \sin \phi_n \cos \gamma_p = F_t \tan \phi_n \cos \gamma_p$$

and

$$F_y = F_n \sin \phi_n \sin \gamma_p = F_t \tan \phi_n \sin \gamma_p$$

where ϕ_n is the pressure angle in a plane normal to the cone element.

The force F_t is the turning force tangential to the pitch cone. The force F_x is the end thrust on the gear and a radial force on the pinion. The force F_y is a radial force on the gear and the end thrust on the pinion. The actual forces on the bearings are indicated in the figure and are found in the same manner as described for worm gears in Art. 12-14.

Note that the forces are assumed to be concentrated at the center of pressure on the tooth surface, which is practically at the center of the face. In the Lewis equation, the value of F_t was the equivalent force concentrated at the large end of the gear. Hence the values in the Lewis equation must be multiplied by the factor $D/(D - f \sin \gamma)$ to obtain the value of F_t in the bearing force equations.

The end thrust is much more serious in spiral bevel gearing than in

Table 12-10   Axial thrust on spiral bevel gears

Driving pinion			Axial thrust
Spiral	Rotation viewed toward the cone center		(+ indicates that the gear, or pinion, is forced away from the cone center)
RH	Clockwise		$\left[+ \dfrac{\tan \phi_n \sin \gamma}{\cos \psi} - \tan \psi \cos \gamma \right] F_t$ for the driver
or			
LH	Counter-clockwise		$\left[+ \dfrac{\tan \phi_n \sin \gamma}{\cos \psi} + \tan \psi \cos \gamma \right] F_t$ for the driven
RH	Counter-clockwise		$\left[+ \dfrac{\tan \phi_n \sin \gamma}{\cos \psi} + \tan \psi \cos \gamma \right] F_t$ for the driver
or			
LH	Clockwise		$\left[+ \dfrac{\tan \phi_n \sin \gamma}{\cos \psi} - \tan \psi \cos \gamma \right] F_t$ for the driven

ψ = spiral angle of pinion.
γ = pitch angle of pinion.
ϕ_n = tooth-pressure angle in plane normal to the cone element.
F_t = turning force, lb.

straight bevel gearing, and the thrust reverses in direction when the rotation of the gears reverses. Hence spiral bevel gears must be mounted with thrust bearings capable of carrying thrust in both directions. In addition to the thrust of a straight bevel tooth there is an axial force due to the spiral form of the tooth. The total end thrust on the pinion, from Fig. 12-17, is

$$F_y = \frac{F_t \tan \phi_n \sin \gamma}{\cos \psi} \pm F_t \tan \psi \cos \gamma \qquad (12\text{-}39)$$

the plus or minus sign depending on the direction of the spiral and the direction of rotation of the gear. Thrust or axial force is assumed to be plus when it forces the gear, or pinion, away from the cone center. The spiral angle ψ is measured as shown in Fig. 12-18.

For convenience, the end thrusts are given in Table 12-10 for the different combinations of spiral and rotation. In determining the rotation, look at the gear or pinion toward the cone center.

13

Springs

13-1 Springs. In most machine members, the deformation must be kept low, that is, the member must be kept stiff and rigid. In a spring, the reverse effect is desired, and the deformation must be relatively large, the spring being a machine member built to have a high degree of resilience.

Springs are used as cushions to absorb shock, as in machine supports, on automobile frames, and in airplane landing gear; as a source of power since they store up energy that is later delivered as driving power, as in clocks, trigger mechanisms, etc.; and to maintain contact between machine members by exerting a direct force, as in clutches, brakes, valve springs, and cam followers. Springs are also used as load-measuring devices, as in spring balances, power dynamometers, and in instruments such as gages, meters, and engine indicators.

13-2 Stress in Coil Springs of Round Wire. The coil, or helical, spring consists of a wire or rod wound into a helix and is primarily intended for axial direct compression or tension loads.

The action of the force F in Fig. 13-1 tends to rotate the wire, thereby causing torsional stresses in the wire. Also, bending, direct com-

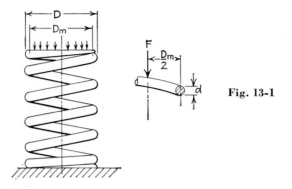

Fig. 13-1

pression, and direct shearing stresses, which are neglected in the conventional spring equations, are set up in the wire. Considering only the torsion and assuming a round wire, the stress is

$$s_s = \frac{Tc}{J} = \frac{FD_m}{2}\frac{16}{\pi d^3} = \frac{8FD_m}{\pi d^3} \qquad (13\text{-}1)$$

where F = axial load, lb

D_m = mean coil diameter, in.

d = wire diameter, in.

In this equation, the bending, direct shear, compression, and wire curvature are neglected. An analysis by A. M. Wahl* that considers these stresses indicates that the stress at the inner surface of the coiled wire may reach values 60 per cent higher than those given by Eq. (13-1), when the spring index is low. The *spring index* is the ratio of the mean coil diameter to the wire diameter. According to Wahl, the maximum stress in the spring wire is

$$s_s = K\frac{8FD_m}{\pi d^3} \qquad (13\text{-}2)$$

where

$$K = \frac{4C-1}{4C-4} + \frac{0.615}{C}$$

and C is the spring index D_m/d. To simplify the computations, the value of K may be taken from the curve in Fig. 13-2.

K is the product of two factors, K_c and K_s. The stress across the diameter of the wire in the coil is greatest at the inner surface. K_c is a stress-concentration factor caused by the curvature of the coil. K_s is a factor to care for the transverse shear effect. Then

$$K = K_c K_s$$

* A. M. Wahl, "Mechanical Springs," McGraw-Hill Book Company, Inc., New York, 1963.

where

$$K_s = 1 + \frac{0.5}{C}$$

When a spring is made of ductile material the curvature factor K_c would be unity because of the yielding of the material, and the stress for a static load would be

$$s_s = K_s \frac{8FD_m}{\pi d^3} \tag{13-3}$$

13-3 Stress in Coil Springs of Noncircular Wire. Coil springs made of square or rectangular wire are frequently used, since a stronger spring can be built into the same space required for a spring of round wire. However, spring manufacturers do not recommend noncircular wire springs unless necessary because the wire is not uniform in cross section and is produced in smaller quantities than round wire, which fact does not result in improved fabrication and availability. The wire will become trapezoidal in forming the coil. The stress in a square-wire spring, based on Saint Venant's torsion theory for noncircular bars, is

$$s_s = K \frac{FD_m}{0.416b^3} = \frac{2.4FD_m}{b^3} K \tag{13-4}$$

For rectangular wires with b/t between 1 and 2.5, indices greater than 3, and with the long dimension parallel to the spring axis, the stress is

$$s_s = \frac{KFD_m(3b + 1.8t)}{2b^2t^2} \tag{13-5}$$

where t is the short dimension of the rectangular cross section of the wire, in inches, and b is the long dimension of the rectangular cross section of the wire, in inches.

For rectangular wires with the short dimension parallel to the spring axis, Eq. (13-5) will give values of stresses as much as 20 per cent too high

Fig. 13-2 Stress factors for coil springs.

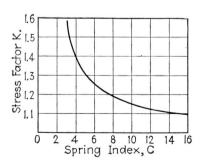

for the smaller spring indices and fairly close approximations of stresses for the larger indices.

The spring index for coil springs of noncircular wire may be closely approximated by using the ratio of the mean coil diameter to the dimension of the wire perpendicular to the axis.

13-4   Deflection of Coil Springs.   The axial deflection of the spring is found from the angular twist of the coiled wire. Then

$$\theta = \frac{TL}{JG} = \frac{FD_m}{2}\frac{\pi D_m n}{\cos \psi}\frac{32}{\pi d^4}\frac{1}{G} = \frac{16FD_m^2 n}{d^4 G \cos \psi} \tag{13-6}$$

where θ = angular twist, rads
ψ = helix angle
n = number of effective coils
G = modulus of rigidity, psi

In most springs, the helix angle is small, and $\cos \psi$ is practically unity. The movement of the point of load application is practically $\theta D_m/2$. Hence the deflection of a spring of round wire is

$$y = \frac{\theta D_m}{2} = \frac{8FD_m^3 n}{d^4 G} = \frac{8FC^3 n}{Gd} \tag{13-7}$$

where y is the axial deflection, in inches, and the other symbols have the same meaning as before.

The deflection of a coil spring of square wire is

$$y = \frac{5.575FD_m^3 n}{b^4 G} \tag{13-8}$$

The deflection of a coil spring with rectangular wire is

$$y = \frac{2.45FD_m^3 n}{Gt^3(b - 0.56t)} \tag{13-9}$$

13-5   Conical Coil Springs.   Certain installations require a spring with increasing stiffness as the load increases, i.e., a decreasing rate of deflection per unit load. This can be accomplished by winding the wire in a conical form so that the larger coils, which have the greater deflection rate, will successively drop out of action by seating on the next smaller coil. In order to save space, some springs are wound so that the coils telescope into each other.

The maximum stress will generally be in the coil of largest diameter, but since the spring index decreases at the small end, the stress should be checked in the coil of least diameter. The stress equations already developed hold for conical springs.

The deflection of a conical spring of round wire is

$$y = \frac{2nF(D_1 + D_2)(D_1^2 + D_2^2)}{d^4G} \tag{13-10}$$

and for a conical spring of flat wire with the long dimension parallel to the axis

$$y = \frac{0.71nF(b^2 + t^2)(D_1 + D_2)(D_1^2 + D_2^2)}{b^3t^3G} \tag{13-11}$$

where D_1 and D_2 are the mean diameters of the smallest and largest coils, respectively.

13-6 Design of Compression and Tension Springs. By changing the mean diameter, the wire diameter, and the number of coils, any number of springs may be obtained to support a given load with a given deflection. However, there are certain limitations imposed by the use to which the spring is to be put. The usual procedure is to assume a mean diameter and a safe working stress, after which the wire diameter is found by substitution in the proper stress equation. The number of effective coils is then found from the deflection equation. Several trials are usually required before a suitable combination is obtained. The stress factor K depends on the spring index that must be assumed in the first trial solution for the wire diameter. For general industrial uses the spring index should be 8 to 10; for valve and clutch springs 5 is common; and 3 is a minimum value to be used only in extreme cases. Because of slight variations in the modulus of rigidity, variations in wire diameter, and other manufacturing tolerances, the deflection equations do not give extremely accurate values, and if extreme accuracy is required, the manufacturer should be consulted.

Compression springs should not be compressed solid when subjected to the maximum load. A clearance between the effective coils should be provided so as to prevent wear, or to prevent a foreign substance lodging between the coils and causing a fatigue crack failure. The *free length* is the solid length plus the clearance plus the maximum deflection. The *solid length* depends upon the number of effective coils and the type of end provided. Different types of ends are illustrated in Fig. 13-3, together with equations to determine the free length. Note that the *effective number of coils* n may be less than the actual number in the spring.

A helical compression spring that is too long compared to its mean diameter may buckle at comparatively low axial loads since such a spring is a very flexible column. The *critical axial load* that will cause buckling is indicated by the formula*

---

* A. M. Wahl, When Helical Springs Buckle, *Machine Design*, May, 1943.

$$F_{cr} = K_S K_L L_O \qquad (13\text{-}12)$$

where F_{cr} = axial load to produce buckling, lb
$\quad K_S$ = spring constant, or load per inch of axial deflection
$\quad K_L$ = factor depending on the ratio L_O/D_m, from Fig. 13-4
$\quad L_O$ = free, or open, length of spring, in.
$\quad D_m$ = mean diameter of coil, in.

In the selection of the value of K_L from Fig. 13-4, a hinged-end spring may be considered as one supported on pivots at both ends, and a built-in end as one in which a squared and ground-end spring is compressed between two rigid and parallel flat plates.

Tension springs are usually wound with the coils closed and under an initial tension; i.e., it is necessary to apply from 20 to 30 per cent of the maximum load before the coils begin to separate. When loops or end hooks are provided, the small radius where the hook joins the first coil is a region of high local stress. To provide for this local stress and for the greater possibility of excess deformation, the design stress for tension springs should not exceed 70 per cent of that used with compression springs. Figure 13-5 shows examples of extension spring ends.

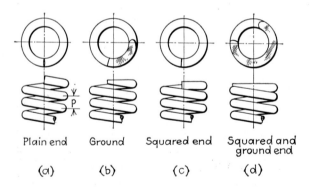

	Plain end	Ground	Squared end	Squared and ground end
	(a)	(b)	(c)	(d)

Fig. 13-3 Compression springs: n is the effective number of coils; p is the pitch, in inches; and d is the diameter of the wire, in inches.

	Actual no. coils	Solid length	Free length
(a)	n	$(n+1)d$	$np + d$
(b)	n	nd	np
(c)	$n+2$	$(n+3)d$	$np + 3d$
(d)	$n+2$	$(n+2)d$	$np + 2d$

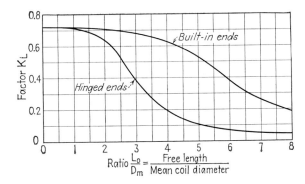

Fig. 13-4 Buckling factor for helical compression springs.

13-7 Eccentric Loads on Springs. The requirements of the design may be such that the load line does not coincide with the axis of the coils. This not only reduces the safe load for the spring but may affect the stiffness. When the load is offset a distance a from the spring axis, the safe load on the spring should be reduced by multiplying the safe load with axial loading by the factor $D_m/(2a + D_m)$.

13-8 Materials for Coiled Springs. Tables 13-1 to 13-4 provide the properties of some of the materials used in springs.* Washburn and

* Space limitation prevents a full discussion of materials. The reader is referred to "Handbook of Mechanical Spring Design," pp. 10–13, Associated Spring Corp., Bristol, Conn., 1963, and H. C. R. Carlson, Selection and Application of Spring Materials, *Mechanical Eng.*, pp. 331–334, May, 1956.

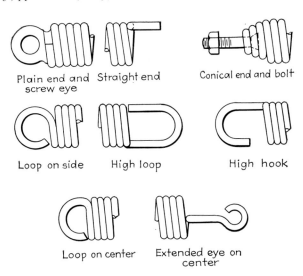

Fig. 13-5 Extension spring ends.

Moen (W&M) gauge is used for the diameter of ferrous wire, and American wire gauge (AWG) or Brown and Sharpe (B&S) gauge is used for nonferrous wire. To prevent error the diameter of the wire should also be given in inches. The majority of coil springs are made of oil-tempered carbon-steel wire containing 0.60 to 0.70 per cent carbon and 0.60 to 1.00 per cent manganese. Music wire is used for high-grade springs using wire less than $\frac{1}{8}$ in. in diameter. Annealed wire containing 0.85 to 0.95 per cent carbon and 0.30 to 0.40 per cent manganese is used in the larger sizes, which are coiled hot and are hardened and drawn after coiling. Nickel steel, chromium steel, stainless steel, brass, phosphor bronze, Monel metal, and other metals that can be hard-drawn are used in special cases to increase fatigue resistance, corrosion resistance, and temperature resistance. Shot-peening increases the endurance range of steel springs considerably and is recommended for ferrous helical springs for service involving high stresses and fatigue. The stress range for high-

Table 13-1   Strength of oil-tempered steel wire

	C = 0.60%	Ph = 0.04% max
	Mn = 0.80%	S = 0.04% max
	Si = 0.15%	
	E = 30,000,000 psi	G = 11,600,000 psi

W&M gauge, steel wire gauge	Diam., in. d	Ultimate strength, minimum, in tension s_{tu}	Yield* stress, torsion s_{sy}	Diam., in. d	Ultimate strength, minimum, in tension s_{tu}	Yield* stress, torsion s_{sy}
34	0.0104	300,000	135,000	$\frac{3}{32}$	234,000	105,000
24	0.0230	287,000	129,000	$\frac{1}{8}$	220,000	99,000
20	0.0348	274,000	123,000	$\frac{5}{32}$	209,000	94,000
18	0.0475	262,000	118,000	$\frac{3}{16}$	200,000	90,000
16	0.0625	251,000	113,000	$\frac{7}{32}$	193,000	87,000
14	0.080	240,000	108,000	$\frac{1}{4}$	189,000	85,000
12	0.1055	228,000	102,000	$\frac{9}{32}$	182,000	82,000
10	0.135	216,000	97,000	$\frac{5}{16}$	178,000	80,000
8	0.162	207,000	93,000	$\frac{11}{32}$	176,000	79,000
6	0.192	198,000	89,000	$\frac{3}{8}$	171,000	77,000
4	0.2253	191,000	86,000	$\frac{13}{32}$	169,000	76,000
2	0.2625	186,000	84,000	$\frac{7}{16}$	167,000	75,000
0	0.3065	180,000	81,000	$\frac{15}{32}$	165,000	74,000
00	0.331	176,000	79,000	$\frac{1}{2}$	162,000	73,000

Higher strengths may be obtained, but the ductility will be decreased.

* Stresses in the table are the maximum stresses that a compression spring will sustain without permanent set and should be used only when the maximum deflection is mechanically limited and the load is virtually steady. The stress in springs subject to fluctuations should be decreased as indicated in the notes below Table 13-2.

Table 13-2 **Properties of common spring materials**

Material	Type of spring	Type of stress	Maximum working stress s_w based on $d = \frac{1}{8}$ in.	Modulus of elasticity	Stress at maximum temperature
Spring steel, oil-tempered, SAE 1360	Compression Extension Torsion	Torsion Torsion Flexure	100,000 70,000 120,000	$E = 28,500,000$ $G = 11,600,000$	
Spring steel, hard-drawn, SAE 1360	Compression Extension Torsion	Torsion Torsion Flexure	70,000 50,000 85,000	$E = 30,000,000$ $G = 11,600,000$	
Music wire, SAE 1095	Compression Extension Torsion	Torsion Torsion Flexure	100,000 70,000 120,000	$E = 30,000,000$ $G = 11,600,000$	
Silico-manganese alloy steel, SAE 9260	Compression Extension Torsion	Torsion Torsion Flexure	100,000 70,000 120,000	$E = 30,000,000$ $G = 11,600,000$	
Chrome-vanadium, tempered, SAE 6150	Compression Extension Torsion	Torsion Torsion Flexure	100,000 70,000 120,000	$E = 30,000,000$ $G = 11,600,000$	60,000 at 450°F
Chrome-vanadium, hard-drawn, SAE 6150	Compression Extension Torsion	Torsion Torsion Flexure	70,000 50,000 85,000	$E = 28,000,000$ $G = 10,800,000$	
*Stainless steel, tempered C, 0.35 %; Cr, 18 %; Ni, 8 %	Compression Extension Torsion	Torsion Torsion Flexure	100,000 70,000 120,000	$E = 30,000,000$ $G = 11,600,000$	60,000 at 550°F
Stainless steel, hard-drawn, Cr, 18 %; Ni, 8 %	Compression Extension Torsion	Torsion Torsion Flexure	70,000 50,000 85,000	$E = 28,000,000$ $G = 10,800,000$	42,500 at 450°F
Phosphor bronze, SAE 81	Compression Extension Torsion	Torsion Torsion Flexure	45,000 31,500 54,000	$E = 16,000,000$ $G = 6,000,000$	
Brass, spring, SAE 80	Compression Extension Torsion	Torsion Torsion Flexure	40,000 30,000 48,000	$E = 12,000,000$ $G = 5,000,000$	
Monel metal, Cu, 28 %; Mn, 2 %; Ni, 67 %; Fe, 3 %	Compression Extension Torsion	Torsion Torsion Flexure	45,000 31,500 54,000	$E = 23,000,000$ $G = 9,250,000$	40,000 at 350°F
Everdur, Cu, 95 %; Mn, 1 %; Si, 4 %	Compression Extension Torsion	Torsion Torsion Flexure	45,000 31,500 54,000	$E = 16,000,000$ $G = 6,000,000$	

Stresses tabulated are the maximum stresses without permanent set and should only be used where the spring load is constant with no fluctuation, and when the maximum spring deformation is mechanically limited.

Springs subject to infrequent fluctuations of variable amount from zero to the maximum or from an intermediate to the maximum should have maximum stresses 90 per cent of the tabulated values.

Springs subject to rapid fluctuations of variable amount from zero to the maximum or from an intermediate to the maximum should have maximum stresses 80 per cent of the tabulated values.

Springs subject to rapidly repeated and regular fluctuations from zero to the maximum or from an intermediate to the maximum should have maximum stresses 70 per cent of the tabulated values.

* Chromium spring steels should not be used where subjected to low temperatures, since at 0°F, and below, these steels may become brittle under stress.

nickel alloy springs cannot be increased significantly by shot-peening. Shot-peening of extension or close-wound torsion springs does not produce the beneficial effects it does for compression springs.

The strength of drawn steel wire increases with the reduction in size, and the hardness increases with a corresponding loss in ductility. The minimum tensile strength of steel spring wire is given very closely by the equation

$$s_u = \frac{C_1}{\sqrt[4]{d}} - \frac{C_2}{d} \tag{13-13}$$

where $C_1 = 177,500$ and $C_2 = 2,050$ for music wire
$C_1 = 138,000$ and $C_2 = 1,600$ for oil-tempered 0.60% C steel wire
$C_1 = 124,000$ and $C_2 = 1,450$ for hard-drawn and stainless-steel wire

The yield stress increases from 75 to 85 per cent of the ultimate strength as the diameter decreases. Since coil springs are in torsion, the design stress must be based on the torsional shear strength. Tests indicate that the yield stress in torsion is approximately 45 per cent of the ultimate strength in tension for oil-tempered wire, and 35 per cent for hard-drawn wire.

Safe working stresses should not exceed 80 per cent of the yield stress in torsion, and if the spring is continuously subjected to rapid load fluctuations, the working stress must be reduced from 25 to 50 per cent. A method of estimating suitable working stresses is developed in Art. 2-15.

When springs are to be used in high temperatures, the modulus of rigidity should be taken from Table 13-3. This table is only approximate since the modulus of rigidity varies with temperature, wire size, analysis, and heat treatment.

13-9 Critical Frequency of Coiled Springs. Any spring having periodical applications of load may be subject to surging. Thus, at certain critical speeds, the valves of internal-combustion engines may surge or vibrate after the valve has closed, permitting the valve to flutter on its seat and interfere with the proper operation of the engine. Since the surges require time to travel from coil to coil and return, a periodic force

Table 13-3   Variation of the modulus of rigidity with temperature

Temperature, °F	Modulus of rigidity, G		
	Spring steel	Stainless steel	Brass
−200	11,700,000		
100	11,600,000	11,600,000	5,000,000
200	11,500,000	11,600,000	4,900,000
400	11,200,000	11,450,000	4,300,000
600	10,800,000	11,200,000	
800	10,200,000	10,800,000	
1000	9,500,000	10,200,000	

may cause surges to be superimposed upon each other; and if this occurs, high stresses that have been known to cause spring breakage are induced.

The speeds at which surging will occur correspond to the natural frequency of the spring, which depends upon the wire diameter, the coil diameter, the number of effective coils, and the elastic properties of the spring wire. The natural frequency is expressed by the equation

$$f = \frac{d}{2\pi D_m^2 n}\left(\frac{6Gg}{w}\right)^{1/2} \qquad (13\text{-}14)$$

where f = frequency, cycles/sec
$\quad w$ = weight of the wire, lb/in.³
$\quad g$ = acceleration due to gravity = 32.2 ft/sec²

Experimental results* indicate that the actual frequency of the spring is from 10 to 15 per cent below that given by the equation. According to W. M. Griffith,† the critical frequency of the spring should be at least 20 times the frequency of application of a periodic load, in order to avoid resonance with all harmonic frequencies up to the twentieth order.

13-10 Torsion Springs. Torsion springs are used as cushions on flexible drives transmitting rotary motion or torque and as sources of power for driving such mechanisms as clocks. The springs may be coiled in a helix or in a spiral. See Figs. 13-6 and 13-7.

In the helical torsion spring (Fig. 13-6), the wire is subjected to flexural stresses. The wire forms a curved beam, the stresses in which may be found by Eq. (3-69). Starting with the curved beam equations and making certain simplifications, A. M. Wahl shows that the stress in a

---

* C. H. Kent, Don't Overlook Surge in Designing Springs, *Machine Design,* October, 1935.

† W. M. Griffith, Engineering Standards for the Design of Springs, *Prod. Eng.,* July, 1933.

Fig. 13-6 Helical torsion spring.

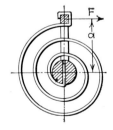

Fig. 13-7 Spiral torsion spring.

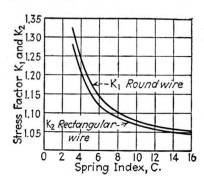

Fig. 13-8 Stress factors for torsion springs.

torsion spring of round wire is

$$s_f = \frac{32M}{\pi d^3} \frac{4C^2 - C - 1}{4C(C-1)} = K_1 \frac{32M}{\pi d^3} = K_1 \frac{32Fa}{\pi d^3} \tag{13-15}$$

where M = bending moment, in.-lb
 K_1 = stress factor from Fig. 13-8
 F = load, lb
 a = moment arm of load or distance from load line to spring axis, in.

For rectangular wire, the stress is

$$s_f = \frac{6M}{bt^2} \frac{3C^2 - C - 0.8}{3C(C-1)} = K_2 \frac{6M}{bi^2} = K_2 \frac{6Fa}{bt^2} \tag{13-16}$$

where b is the wire dimension parallel to the axis, in inches, and t is the dimension perpendicular to the axis, in inches.

The deflections of helical torsion springs are given by the equations

$$y = \frac{64FD_m a^2 n}{Ed^4} \quad \text{and} \quad \theta = \frac{3,665FD_m an}{Ed^4} \quad \text{for round wire} \tag{13-17}$$

and

$$y = \frac{12\pi FD_m a^2 n}{Ebt^3} \quad \text{and} \quad \theta = \frac{2,160FD_m an}{Ebt^3} \quad \text{for rectangular wire} \tag{13-18}$$

where y = movement of point of load application, i.e., at a distance a from the spring axis, in.
 θ = angular movement, deg
 a = moment arm, in.

The stress in spiral torsion springs (Fig. 13-7) is given by Eqs. (13-15) and (13-16). The deflections are given by

$$y = \frac{64FLa^2}{\pi Ed^4} \quad \text{and} \quad \theta = \frac{1,168FLa}{Ed^4} \quad \text{for round wire} \tag{13-19}$$

and

$$y = \frac{12FLa^2}{Ebt^3} \quad \text{and} \quad \theta = \frac{687.6FLa}{Ebt^3} \quad \text{for rectangular wire} \quad (13\text{-}20)$$

where L is the length of wire in the spiral, in inches.

For all practical purposes, the length may be taken as $\pi n D_m$.

Deflection formulas for torsion springs give values slightly less than the actual deflections, since the equations make no allowance for the decrease in diameter as the coils wind up under load. All torsion springs should be installed so that the applied load will wind up the wire, reducing the diameter, and clearance must be provided when the spring operates around a mandrel. The springs should also be wound with a small clearance between adjacent coils to prevent sliding friction.

13-11 Protective Coatings. Before forming the springs, the wire may be coated with copper or tin and immediately drawn through a die to obtain a high burnish. The coating is thin and is not an efficient corrosion resistant. After forming, the springs may be electroplated with cadmium, chromium, nickel, brass, or copper. The acid cleaning before plating may cause embrittlement of the material unless carefully done. The formed springs may be dipped in or sprayed with japan, which has a good resistance to rusting. Music-wire springs are usually not protected, and the high-chromium or stainless steels need no protection. To prevent clashing of the coils and injury to the coatings, the clearance between the coils when the spring is fully compressed in service should not be less than 10 per cent of the wire diameter.

13-12 Leaf Springs. A single thin plate supported at one end and loaded at the other may be used as a spring. The stress in such a spring, shown in Fig. 13-9, is

$$s = \frac{Mc}{I} = \frac{6FL}{wt^2} \tag{13-21}$$

and the deflection is

$$y = \frac{1}{3}\frac{FL^3}{EI} = \frac{4FL^3}{wt^3E} = \frac{2sL^2}{3tE} \tag{13-22}$$

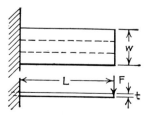

Fig. 13-9

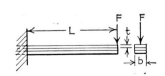

Fig. 13-10

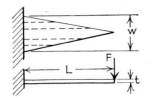

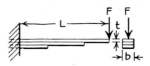

Fig. 13-11   Fig. 13-12

If the plate is cut into a series of n strips of width b, and these are placed as shown in Fig. 13-10, the above equations become

$$s = \frac{6FL}{nbt^2} \tag{13-23}$$

and

$$y = \frac{4FL^3}{nbt^3E} = \frac{2sL^2}{3Et} \tag{13-24}$$

which gives the stress and deflection in a leaf spring of uniform cross section.

The stress in such a spring is maximum at the support. If a triangular plate is used, as in Fig. 13-11, the stress will be uniform throughout. If this triangular plate is cut into strips and placed as in Fig. 13-12 to form a graduated leaf spring, then

$$s = \frac{6FL}{nbt^2} \tag{13-25}$$

and

$$y = \frac{6FL^3}{nbt^3E} \tag{13-26}$$

If bending stress alone is considered, the graduated spring may have zero width at the loaded end, but sufficient metal must be provided to support the shear. Hence it is necessary to have one or more leaves of uniform cross section extending to the end. (See Fig. 13-13.) Examination

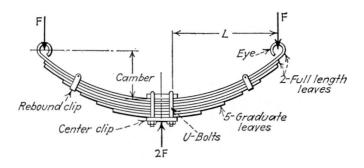

Fig. 13-13   Typical semielliptic leaf spring.

of the deflection equations indicates that, for the same deflection, the stress in the uniform-section leaves is 50 per cent greater than in the graduated leaves, assuming that each spring element deflects according to its own elastic curve. If the subscripts f and g are used to indicate the full-length (uniform section) and graduated leaves respectively, then

$$s_f = \frac{3s_g}{2}$$

and

$$\frac{6F_f L}{n_f b t^2} = \frac{3}{2}\frac{6F_g L}{n_g b t^2}$$

from which

$$F_g = \frac{2n_g F_f}{3n_f} = \frac{2n_g}{2n_g + 3n_f} F \tag{13-27}$$

and

$$F_f = \frac{3n_f}{2n_g + 3n_f} F \tag{13-28}$$

From Eq. (13-23)

$$s_f = \frac{6FL}{n_f b t^2}\frac{3n_f}{2n_g + 3n_f} = \frac{18FL}{bt^2(2n_g + 3n_f)} \tag{13-29}$$

which is the relation between the maximum stress and the load applied at the end of the spring when all leaves have the same thickness.

The deflection of the spring is

$$y = \frac{12FL^3}{bt^3 E(2n_g + 3n_f)} \tag{13-30}$$

13-13 Equalized Stress in Spring Leaves. In the spring just discussed, the stress in the full-length leaves was found to be 50 per cent greater than the stress in the graduated leaves. In order to utilize the material to the best advantage, all leaves should be stressed the same. This condition can be obtained if the full-length leaves are given a greater radius of curvature than that used in the graduated leaves, before the leaves are assembled to form the spring. As shown in Fig. 13-14, this will

Fig. 13-14

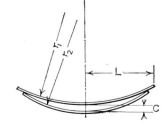

leave a gap or clearance space between the leaves. When the center spring clip is drawn up tight, the upper leaf will bend back and have an initial stress opposite to that produced by the normal loading, and the lower leaf will have an initial stress of the same nature as that produced by the normal loading. As the normal load is applied, the stress in the upper leaf is relieved and then increased in the normal way. The initial gap between the leaves can be adjusted so that under maximum-load conditions the stress in all leaves will be the same; or, if desired, the top leaves may have the lower stress. The latter case is desirable in automobile springs, since the top leaf must carry additional loads caused by the swaying of the car, twisting, and in some cases compression due to driving the car through the rear springs.

Consider the case of equal stress in all leaves at the maximum load. Then at maximum load the total deflection of the graduated leaves will exceed the deflection of the full-length leaves by an amount equal to the initial gap c. Hence

$$c = \frac{6F_g L^3}{n_g bt^3 E} - \frac{4F_f L^3}{n_f bt^3 E}$$

But, since the stresses are equal,

$$F_g = \frac{n_g}{n_f} F_f = \frac{n_g}{n} F$$

and

$$F_f = \frac{n_f}{n} F$$

When the proper substitutions are made,

$$c = \frac{6FL^3}{nbt^3 E} - \frac{4FL^3}{nbt^3 E} = \frac{2FL^3}{nbt^3 E} \tag{13-31}$$

The load on the clip bolts F_b required to close the gap is determined by the fact that the gap is equal to the sum of the initial deflections of the full and graduated leaves. Hence

$$c = \frac{2FL^3}{nbt^3 E} = \frac{4L^3}{n_f bt^3 E} \frac{F_b}{2} + \frac{6L^3}{n_g bt^3 E} \frac{F_b}{2}$$

from which

$$F_b = \frac{2n_g n_f F}{n(2n_g + 3n_f)} \tag{13-32}$$

The final stress in the spring leaves will be the stress in the full-length leaves due to the applied load less the initial stress Then

$$s = \frac{6L}{n_f bt^2} \left(F_f - \frac{F_b}{2} \right)$$

$$= \frac{6FL}{n_f bt^2} \left(\frac{3n_f}{2n_g + 3n_f} - \frac{n_f n_g}{n(2n_g + 3n_f)} \right)$$

from which

$$s = \frac{6FL}{nbt^2} \qquad (13\text{-}33)$$

The deflection caused by the applied load is the same as in the spring without initial stress.

13-14 Materials for Leaf Springs. Plain carbon steel of 0.90 to 1.00 per cent carbon, properly heat-treated, is commonly used. Chrome-vanadium and silico-manganese steels are used for the better-grade springs. The alloy steels do not have strengths greatly in excess of the carbon steels, but have greater toughness and a higher endurance limit and are better suited to springs subjected to rapidly fluctuating loads. The properties of common spring materials are given in Table 13-4. The selection of the proper working stress is complicated by the fact that the spring is essentially a shock-absorbing device. Hence, the working stress should be based on the endurance limit and, of course, should never exceed the yield stress. The endurance limit in reversed bending varies from 40 to 50 per cent of the ultimate strength in tension. In order to provide for surface defects in the rolled steel from which leaf springs are made, and for large variations in stress when in service, the springs should be designed for a working stress of about one-half the endurance limit when at the resting load. Where the load variation can be definitely predicted, suitable working stresses may be determined by the method outlined in Art. 2-15.

13-15 Design of Leaf Springs. The stress and deflection as determined by the usual spring equations are slightly altered by several items

Table 13-4 **Properties of leaf-spring materials**

Material	Yield stress s_{ty}	Flexural endurance limit s_{ef}
Spring steel:		
SAE 1095.........	175,000	100,000
Chrome-vanadium:		
SAE 6140.........	200,000	110,000
SAE 6150.........	200,000	110,000
Silico-manganese:		
SAE 9250.........	190,000	115,000
SAE 9260.........	200,000	120,000

Spring-steel plate is rolled to the Birmingham wire gauge (BWG). Thickness should also be given in inches.

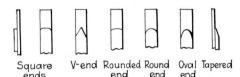

Square   V-end Rounded Round Oval Tapered
ends   end   end   end

Fig. 13-15   Types of spring-leaf ends.

that must be considered before the final design is accepted. In all important designs, a sample spring should be tested before being put into service.

In the development of the equations it was assumed that the leaves were uniform in thickness and that the ends of the graduated leaves were pointed. Heavy machinery, truck, and railway springs are usually designed in this way. Automobile and similar springs are usually rounded and thinned at the ends, and this condition slightly changes the deflection.

Friction between the leaves tends to reduce the deflection and make the springs stiffer. To reduce wear and to obtain uniform spring action, separator pads are often used between the ends of adjacent leaves.

The spring leaves are held together by a center bolt, a bolted clip, or a band shrunk on. The center bolt reduces the metal area at the section of maximum stress and also causes highly localized stresses near the hole. This is poor design and such springs usually break through the center hole. Shrunk bands are superior construction, but are used only on heavy springs. With spring clips, the bolts must be drawn up tightly or bending of the leaves will occur under the clip, increasing the effective spring length and hence the deflection and stress. The effective length of a spring with bolted center clips should be measured to a point about one-third the distance from the clip edge to the center.

As the load is applied to the spring, the curvature and the effective length change, thus altering the deflection rate. By properly choosing the initial camber, the spring may be made to soften or stiffen as the load is increased.

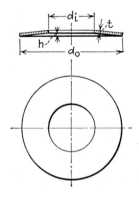

Fig. 13-16   Disk spring.

If the leaves are all given the same curvature before assembly, they will separate during the rebound, and dirt and grit may enter between them. This condition may be prevented by rebound clips, or the springs may be nipped, i.e., the shorter leaves are given a slightly greater initial curvature, thus maintaining contact at all times.

To protect the spring from excessive stress during the rebound, short rebound leaves may be placed above the master leaves. The rebound clips also aid in distributing the load to all the leaves during the rebound.

Some springs are made with the top leaf thinner than the others. Equation (13-24) shows that as t is decreased, the stress for the same deflection is decreased. Hence, when the top leaf is subjected to twisting or compression loads, it is desirable to use a thin top leaf to increase the margin of safety.

13-16 Disk Springs.* Disk or "Belleville" springs are used where space limitations require high capacity units. Each element consists of an annular disk, initially dished to a conical shape as shown in Fig. 13-16.

* For a more complete discussion see "Handbook of Mechanical Spring Design," pp. 70–73, Associated Spring Corporation, Bristol, Conn., 1963.

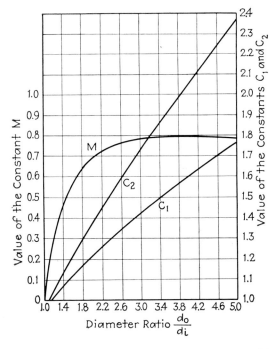

Fig. 13-17 Disk-spring constants to be used with Eqs. (13-34) to (13-36). (*Machine Design, June, 1936.*)

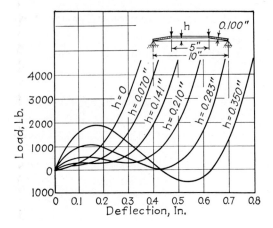

Fig. 13-18 Curves showing variation of load-deflection rate of disk spring identical except for the height h. (*Machine Design, June*, 1936.)

When the load is applied uniformly around the edge, the relation between the applied load and the axial deflection is given by the equation

$$F = \frac{4Ey}{(1 - \mu^2)Md_o^2}\left[(h - y)\left(h - \frac{y}{2}\right)t + t^3\right] \tag{13-34}$$

where y = axial deflection of each disk, in.
      t = thickness, in.
      d_o = outside diameter, in.
      μ = Poisson's ratio
      h = free height of truncated cone, in.
      M = a constant depending upon ratio of outside to inside diameter, see Fig. 13-17

The maximum stresses occurring at the edges are given by the equations

$$s = \frac{4Ey}{(1 - \mu^2)Md_o^2}\left[C_1\left(h - \frac{y}{2}\right) + C_2t\right] \qquad \text{at the inner edge} \quad (13\text{-}35)$$

and

$$s = \frac{4Ey}{(1 - \mu^2)Md_o^2}\left[C_1\left(h - \frac{y}{2}\right) - C_2t\right] \qquad \text{at the outer edge} \quad (13\text{-}36)$$

where C_1 and C_2 are constants depending upon the ratio of the outside to the inside diameter. The constants can be taken from Fig. 13-17.

The true stresses are unknown, but experience with this type of spring indicates that, under static conditions, the stress as given by these equations may be as high as 220,000 psi for steel having a yield stress of 120,000 psi. Fatigue tests indicate that the maximum stress in these equations may be as high as 180,000 psi. The fatigue life is greatly increased when the corners of the disk edges are rounded off.

Some properties of such springs are indicated in the curves of Fig. 13-18. With a flat spring ($h = 0$), the load increases rapidly as the deflection increases. When h is approximately $t \sqrt{2}$, there is a period of considerable increase in deflection with constant load. Such springs may be used to support bearings and other members where temperature changes must be accommodated without appreciable variation in load. When $h > t \sqrt{2}$, the load increases with the deflection for a short period, followed by a period during which the load actually decreases with increase in deflection.

The load capacity of disk springs may be increased by moving the point of load application outward from the inner edge. The load capacity varies approximately as the ratio of the distances from the outer edge to the point of load application. The deflection varies approximately as this distance ratio. Initially flat springs have a maximum flexibility when the ratio of outside to inside diameters is approximately 2. In general the ratio d_o/d_i should be between $1\frac{1}{2}$ and 5.

14

Couplings and Clutches

14-1 Couplings. A *coupling* is a mechanical device for uniting or connecting parts of a mechanical system. This chapter concerns itself with those couplings which are used on machine shafts for the purpose of transmitting torque. Couplings may be employed for a permanent or semipermanent connection between shafts, or for disconnection of machine components to permit one member to run while the other is stationary.

Commercial shafts are limited in length by manufacturing and shipping requirements so that it is necessary to join sections of long transmission shafts with couplings. Couplings are also required to connect the shaft of a driving machine to a separately built driven unit. Permanent couplings are referred to simply as couplings, while those which may be readily engaged to transmit power, or disengaged when desired, usually are called clutches.

14-2 Rigid Couplings. *Rigid couplings* are permanent couplings which by virtue of their construction have essentially no degree of angular, axial, or rotational flexibility; they must be used with collinear shafts.

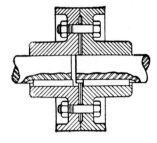

Fig. 14-1   Flange coupling.

The *flange coupling* shown in Fig. 14-1 is perhaps the most common coupling. It has the advantage of simplicity and low cost, but the connected shafts must be accurately aligned to prevent severe bending stresses and excessive wear in the bearings. The length of the hub is determined by the length of key required, and the hub diameter is approximately twice the bore. The thickness of the flange is determined by the permissible bearing pressure on the bolts. Although usually not critical, the shearing stress on the cylindrical area where the flange joins the hub should be checked.

When large flanges are objectionable, *compression couplings* similar to the coupling in Fig. 14-2 may be used. Torque is transmitted by keys between the shafts and the cones and by friction between the cones and the outer sleeve. The inner cones may be split so that they will grip the shaft when drawn together.

A *collar coupling*, shown in Fig. 14-3, consists of a cylindrical collar pressed over the ends of the two collinear shafts being connected, approximately one-half of the collar contacting each shaft. Usually one or more radial pins completely through each shaft and the collar, or setscrews, may be used to ensure that there is no undesired radial movement. A variation of the collar coupling results if a cylindrical part of one shaft fits concentrically into a female bored portion in the second shaft and pins, setscrews, or other suitable fasteners are used.

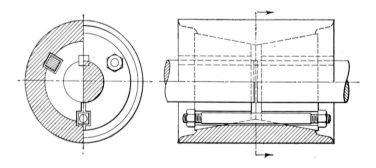

Fig. 14-2   Compression coupling.

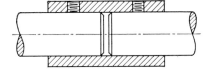

Fig. 14-3 Collar coupling with setscrews.

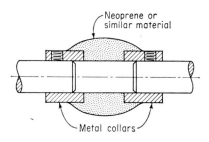

Fig. 14-4 Simple elastic material bonded coupling.

14-3 Flexible Couplings.   Upon occasion it is desired that couplings be able to accommodate reasonable amounts of axial angularity between shafts, a small amount of eccentricity between parallel shafts, or axial movement of shafts during use. *Flexible couplings* may be employed for any or all of these cases. In addition, torsionally flexible couplings may be employed to absorb some of the torque in the shaft, or to permit large amounts of torsional flexibility as in fluid couplings or torque converters.

Perhaps the simplest flexible coupling permitting small amounts of axial misalignment and/or torsional flexibility consists of a piece of

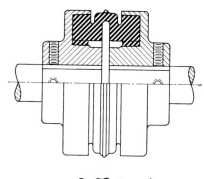

Fig. 14-5 Elastic material bonded coupling.

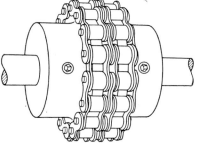

Fig. 14-6 Chain coupling.

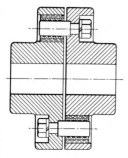

Fig. 14-7   Elastic material bushed coupling.

elastic material (such as neoprene, Teflon, or similar material) bonded to two separated collinear cylindrical members which in turn can be fastened onto the two shafts by means of setscrews or other fasteners. Such an *elastic-material bonded coupling* is shown in Fig. 14-4. A more complicated version of this coupling with several smaller pieces of elastic material separating the two halves of the coupling is depicted in Fig. 14-5.

The *chain couplings* of Fig. 14-6 consist essentially of two chain sprockets connected with a short continuous length of roller or silent chain. They find application with angular misalignment as well as with slight eccentricities of shafts.

A slight amount of torsional and angular flexibility is provided by use of the *elastic-material bushed coupling* of Fig. 14-7, with some elastic material cylindrically surrounding the flange bolts in the flange in one portion of the coupling.

Steel, leather, fabric, or plastic material bolted at alternate points to the two flanges constitutes the *flexible disk coupling* of Fig. 14-8. The flexible *toroidal spring coupling* of Fig. 14-9 and the flexible *gear-type coupling* of Fig. 14-10 are other examples of couplings respectively providing for angular and axial flexibility.

The *Oldham coupling* of Fig. 14-11 can be employed for connecting two parallel shafts with axial eccentricities from zero to a reasonable

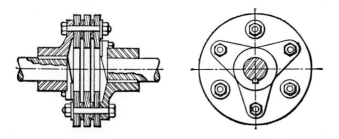

Fig. 14-8   Flexible disk coupling.

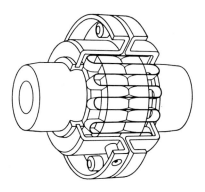

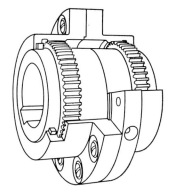

Fig. 14-9   Flexible toroidal spring coupling.

Fig. 14-10   Flexible gear-type coupling.

amount. Regardless of the amount of the eccentricity, the two shafts rotate with no rotational misalignment at all; hence this coupling finds application where the two parallel shafts have to turn through exactly the same angle per unit of time.

The single *Hooke's joint* of Fig. 14-12, commonly called a *universal joint*, frequently is used for shafts with angular misalignment. The driving shaft rotates at a constant speed and the driven shaft rotates at a continuously varying angular velocity, the variation increasing with angular misalignment of the shafts. If two universal joints and an intermediate shaft are used as indicated in Fig. 14-13, and if the angular misalignment α between each shaft and the intermediate shaft is equal, the driving and driven shaft will remain in exact angular alignment, and the intermediate shaft will rotate with a varying speed. This application of universal joints permits comparatively large angular misalignments between the input and output shafts. Any relative axial motion between the intermediate shaft and the other shafts must be taken care of, usually by use of splines

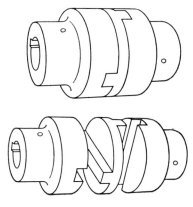

Fig. 14-11   Oldham coupling.

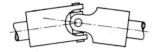

Fig. 14-12   Single Hooke's (universal) joint.

on the intermediate shaft and the mating universal joints. Such a system often is used in automotive propulsion systems.

Several *constant-velocity universal joints* have been developed which when used singly permit the output and input shafts to rotate at exactly the same angular velocity, regardless of whether the two shafts are collinear or angularly misaligned. The Bendix-Weiss and the Rzeppa joints are two examples of recent developments along this line. Motion in the *Bendix-Weiss joint* is transmitted through four steel balls which are constrained by the intersection of curved grooves or races; these joints accommodate axial motion without the necessity of a sliding-spline connection. The *Rzeppa joint* utilizes a cage to keep six steel balls in a homokinetic plane, thereby maintaining a constant angular speed ratio between output and input shafts. Important factors in proper determination of joint size required are the angular speed of rotation and the angle between the shafts. As an illustration, the capacity of a joint with a 35-deg shaft angle is approximately one-third its capacity when the shaft angle is 3 deg or less.

14-4   Hydraulic Couplings.   A *hydraulic coupling*, often called a *fluid coupling*, employs a fluid to provide angular flexibility between the input and output shafts. An oil or water is usually employed as the fluid, although in some applications air has been used. As depicted in Fig. 14-14, a fluid coupling consists of a primary rotor connected to the input shaft, a secondary rotor connected to the output shaft, a quantity of fluid, and some sort of housing with suitable seals to retain the fluid. *Slip* is defined as the difference in angular speeds of rotation of the input and output shafts; the output shaft always rotates more slowly than the input shaft. For 100 per cent slip, the fluid coupling acts as a clutch, permitting the output shaft to remain stalled at rest but with some torque applied to it.

The hydraulic coupling is a turbo-type transmission consisting of a rotating driving impeller, corresponding to a centrifugal pump, and a

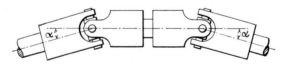

Fig. 14-13   Use of two universal joints and an intermediate shaft.

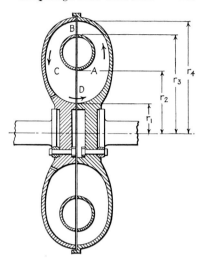

Fig. 14-14 Simple hydraulic coupling.

rotating driven runner, corresponding to the runner of a hydraulic turbine. Both members are enclosed in a housing, which usually rotates with the impeller and confines the fluid to the working circuit. The right-hand member shown in Fig. 14-14 is the driving impeller. Both members have radial guide vanes in the passages A and C, and when the impeller is rotated, liquid in the passage A is given a rotational velocity. Centrifugal force causes the liquid to flow outward between the radial vanes, and at B the axial component of the velocity carries the liquid across the gap into the runner where it moves inward through the passage C to D, where it again enters the impeller. At B the fluid has a rotational velocity v_1 and a kinetic energy of $\frac{1}{2}Mv_1^2$. At D the rotational velocity has decreased to v_2 and the kinetic energy to $\frac{1}{2}Mv_2^2$. Energy equal to $\frac{1}{2}M(v_1^2 - v_2^2)$ has been imparted to the runner, producing driving torque.

The torque imparted to the runner is always equal to the torque developed by the impeller; hence there is no torque multiplication. Since fluid circulation is dependent upon centrifugal force, the runner always must rotate at a lower speed than the impeller. The efficiency of power transmission is equal to 100 per cent less the slip in per cent. The slip between the impeller and the runner is due to friction and turbulence losses and often amounts to 1 or 2 per cent at full torque; when space limits the diameter, slip as high as 4 per cent may be used. The limit of torque transmission is reached when the slip increases to the point where liquid circulation becomes turbulent. By designing for a maximum torque, at 100 per cent slip, which is slightly less than the maximum torque capacity of the driving-power source, it will be impossible to stall the driving engine or motor. Load changes cause changes in the speed of the driven runner; hence these couplings are not suitable where precise speed control is required.

For a limited range of slip, corresponding to the usual conditions of usage, the torque capacity is given by

$$T = KSN^2W(r_o^2 - r_i^2) \tag{14-1}$$

where K = a coefficient, approximately 36×10^{-7}
      N = input impeller speed, rpm
      r_i = mean radius of inner passage, in. $= \dfrac{2}{3} \dfrac{r_2^3 - r_1^3}{r_2^2 - r_1^2}$
      r_o = mean radius of outer passage, in. $= \dfrac{2}{3} \dfrac{r_4^3 - r_3^3}{r_4^2 - r_3^2}$
      $r_1, r_2, r_3,$ and r_4, in., are shown in Fig. 14-14
      S = slip, per cent
      T = torque transmitted, in.-lb
      W = weight of fluid circulating in operating portion of coupling, lb

Thus it is seen that the torque transmitted by a fluid coupling varies directly as the square of the speed of the input impeller and directly as the slip. The power varies as the cube of the input impeller speed and directly as the slip, and depends upon the coupling diameter, speed of rotation, fluid density, amount of fluid circulated, and the slip. The power loss is due largely to friction of the fluid. The best general-purpose fluid has been found to be a light mineral oil with a viscosity of about 150 sec Saybolt at 100°F.

The amount of power transmitted may be regulated by varying the mass of fluid circulating in the coupling. Two methods are in general use. In one, calibrated nozzles permit the fluid to leak from the periphery of the coupling and collect in a sump from which an independent pump returns it to the driving impeller. A valve in the pump discharge regulates the liquid feed and, hence, the mass of fluid in the circulating passages. The second method omits the pump but uses a reservoir surrounding the coupling and rotating with the impeller, as in Fig. 14-15. Centrifugal force causes the fluid to hug the outside of the reservoir chamber where a scoop, adjustable in position, picks up the fluid and returns it to the impeller. By changing the position of the scoop, the amount of fluid in the circulating passages, and hence the power transmitted, can be changed at will. To disconnect the coupling, the fluid must be discharged from the circulating passages. This is done by opening valves in the outer portion of the coupling members.

To reduce the drag torque or creeping tendency of the driven member, a baffle is added at the point D on the driven runner, in Fig. 14-15. This baffle is a radial rib blocking off approximately the inner third of the circulating passage. When starting, the impeller rotates at the regular operating speed, imparting a high velocity to the fluid, which on entering the runner passages clings to the outer surface of the slow-moving runner.

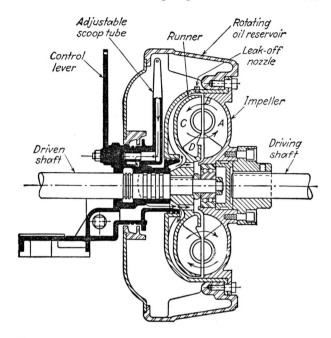

Fig. 14-15 Hydraulic coupling.

This stream is broken up by the baffle ring, reducing the drag. As the runner speed increases, the fluid circulates above the baffle, and the drag and slip reach their normal value.

14-5 Hydraulic Torque Converters. If a stationary vaned guide wheel or stator is inserted between the two elements of the fluid coupling, a *hydraulic torque converter* results. The vanes of the stationary stator are so shaped as to reduce the kinetic energy of the fluid leaving the driving impeller, causing a change of momentum of the fluid in the driven impeller and a resulting increase of torque.

The torque on the driven unit may be as much as five to six times that of the prime mover or more. One or more stationary blades with an equal number of blades on the runner may be used. A simple single-stage torque converter is shown in Fig. 14-16.

The power transmitted by a hydraulic torque converter is given by the equation

$$T = KN^2D^5 \tag{14-2}$$

where D = outer diameter of vanes, in.
  K = design coefficient, in.^{4}-rev^2/(lb)(min^2)
  N = driven shaft speed, rpm
  T = torque, in.-lb

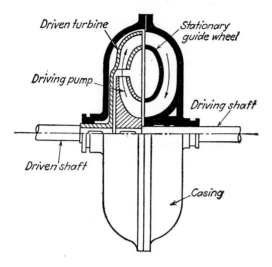

Fig. 14-16   Hydraulic torque converter.

The coefficient K varies with the design and must be determined for any given design and vane arrangement. It then can be used to determine the torque capacity of any other geometrically similar converter.

The power absorbed by the impeller and transmitted to the circulating fluid is proportional to the cube of the speed, and the impeller must be designed to absorb the driving engine or motor power at operating speed. The engine powering the converter then will theoretically operate at approximately constant speed regardless of the speed of the driven runner. Actually, the speed of an internal-combustion engine will drop 15 to 20 per cent from maximum speed of the runner to stalling speed. The runner operates from zero rpm, or stalling speed, to approximately two-thirds of the engine speed, at which speed the output torque is approximately equal to the engine torque.

Well-designed hydraulic torque converters have fairly constant efficiencies, 70 to 85 per cent, from one-fourth to full load. This flat efficiency permits economical operation at various torque-speed ratios.

In order to obtain greater torque multiplication than is possible from a single torque converter, upon occasion two or more hydraulic torque converters in series have been employed with success. Of course a set of gears would accomplish the same maximum torque multiplication with different torque characteristics. In general, for fluid couplings and hydraulic torque converters of equal diameter and essentially the same volume, the maximum torque output is higher for the hydraulic torque converter.

In both the fluid couplings and hydraulic torque converters, the power loss results in heating of the fluid. Adequate means must be provided to dissipate this heat. In small-torque-capacity units, the natural heat dissipation capacity often is adequate. In high-power units, such as those used on automobiles, auxiliary means of cooling must be provided. Frequently this is accomplished by circulating water through the fluid sump or by circulating the fluid through a radiator or some other type of heat exchanger. In these units, it is not uncommon for 25 to 35 per cent of the input power to be dissipated in the form of heat.

14-6 Jaw Clutches. The simplest positive clutch is the *square-jaw* clutch shown in Fig. 14-17. These clutches produce positive coupling between the two shafts, regardless of direction of rotation. Although they must be connected while the shafts are essentially stationary and unloaded, they are suitable for situations requiring simple and rapid disconnection. The proportions are more or less empirical, the jaw size being determined by the permissible compression on the material used. The *spiral-jaw clutch* (often called *bevel-jaw clutch*) of Fig. 14-18 is essentially the same as the square-jaw clutch with the exception that it is capable of transmitting torque in only one direction. If an attempt is made to transmit torque in the wrong direction, the two parts of the clutch disconnect because of the spiral jaws. There are applications where this is desirable, such as the disconnect of a starter motor from the engine of an automobile.

14-7 Friction Clutches. In contrast to the previously mentioned clutches which depend upon shear within metal, elastic material, or fluid to transmit torque, *friction clutches* depend upon frictional contact between two or more members for torque transmission. The major types of friction clutches are plate, cone, rim, dry-fluid, magnetic-particle, and eddy-current clutches.

In general, the torque transmitted by a friction clutch may be obtained by

$$T = nF_a fr_f \tag{14-3}$$

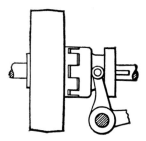

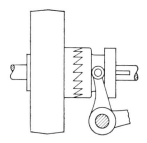

Fig. 14-17 Square-jaw clutch.

Fig. 14-18 Spiral-jaw clutch.

where F_a = total axial force between the surfaces in friction contact, lb
$\quad\;\; f$ = coefficient of friction at the friction surfaces (see Arts. 14-11 through 14-13)
$\quad\; n$ = number of pairs of identical mating friction surfaces (number of mating surfaces at which slippage can occur)
$\quad r_f$ = average friction radius, in.
$\quad\; T$ = total torque transmitted, in.-lb

14-8   Plate or Disk Clutches.   In Fig. 14-19 are shown two flanges, one keyed rigidly to the driving shaft, and the other fitted to the driven shaft by a feather key or spline so that it may be moved along the shaft. The amount of torque transmitted is dependent upon the axial pressure distribution, the dimensions of the friction surfaces, and the coefficient of friction. The pressure p per square inch of friction surface may or may not be uniform. The torque capacity of the clutch of Fig. 14-19 is obtained by

$$T = \int rfp\,da = f\int_0^{2\pi}\int_{r_i}^{r_o} rp(r\,d\theta\,dr) = f\int_0^{2\pi} d\theta \int_{r_i}^{r_o} pr^2\,dr$$

$$= 2\pi f \int_{r_i}^{r_o} pr^2\,dr \quad (14\text{-}4)$$

If the clutch were a multidisk clutch as indicated in Fig. 14-20 with n identical pairs of mating friction surfaces, each annular friction area having outside and inside radii of r_o and r_i, the resulting torque capacity would be n times that from Eq. (14-4).

14-9   Uniform Pressure Disk Clutches.   If the pressure distribution of the disk clutches of Figs. 14-19 and 14-20 is assumed to be a uniform p, then Eq. (14-4) becomes

$$T = 2\pi npf \int_{r_i}^{r_o} r^2\,dr = \frac{2\pi npf}{3}(r_o^3 - r_i^3) \tag{14-5}$$

Since the total axial force F_a is

$$F_a = \int p\,da = p\int_0^{2\pi}\int_{r_i}^{r_o} r\,d\theta\,dr = p\int_0^{2\pi} d\theta \int_{r_i}^{r_o} r\,dr$$

$$= \pi p(r_o^2 - r_i^2) \tag{14-6}$$

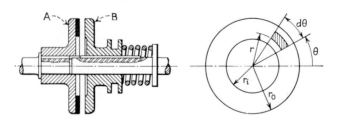

Fig. 14-19   Single-disk friction clutch.

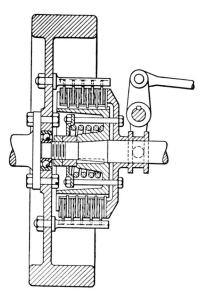

Fig. 14-20   Multiple-disk clutch for automotive synchromesh transmission.

Eq. (14-5) may be rewritten as

$$T = nfF_a r_f \tag{14-7}$$

where r_f is the mean friction radius based upon uniform pressure distribution as given by

$$r_f = \frac{2}{3} \frac{r_o^3 - r_i^3}{r_o^2 - r_i^2} \tag{14-8}$$

For new unworn clutches, the assumption of uniform pressure distribution is realistic. Usually this assumption is not valid after the clutch has been in use for a period of time.

14-10   "Uniform Wear" Clutches.   The condition of uniform pressure distribution assumed in the preceding section is essentially correct for new contact surfaces, but on account of relative motion between the surfaces during the engaging period, there is a certain amount of wear, especially when lubrication is absent, as is usually the case in clutches depending on friction for power transmission. In order that the surfaces will remain in contact, the wear in the axial direction must be the same for all values of r. But wear W is proportional to the work done by friction, which is in turn proportional to the product of the normal pressure and the velocity of rubbing. Hence

$$W = kpv = Kpr$$

from which

$$p = \frac{W}{Kr} = \frac{C}{r} \tag{14-9}$$

where C = a pressure constant, lb/in.
  K = a wear constant, in.^{2}/lb
  k = a wear constant, in.^{2}-sec/lb
  p = pressure between the friction surfaces at radius r, lb/in.^{2}
  v = relative slippage velocity between friction surfaces at radius r, in./sec
  W = axial wear of friction surface at radius r, in.

Substituting Eq. (14-9) into (14-4) yields

$$T = 2\pi nfC \int_{r_i}^{r_o} r \, dr = \pi nfC(r_o^2 - r_i^2) \tag{14-10}$$

But

$$F_a = \int p \, da = \int_0^{2\pi} \int_{r_i}^{r_o} pr \, d\theta \, dr = C \int_0^{2\pi} d\theta \int_{r_i}^{r_o} dr$$
$$= 2\pi C(r_o - r_i) \quad (14\text{-}11)$$

therefore

$$T = \frac{nfF_a(r_o + r_i)}{2} = nfF_a r_f \tag{14-12}$$

where

$$r_f = \frac{r_o + r_i}{2} \tag{14-13}$$

Equation (14-9) indicates that for "uniform wear," the pressure p varies from a maximum p_{max} at $r = r_i$ linearly to a minimum at $r = r_o$. Thus the pressure constant C may be defined by

$$C = p_{max}r_f \tag{14-14}$$

Substituting this into Eqs. (14-10) and (14-11) gives

$$T = \pi nfp_{max}r_f(r_o^2 - r_i^2) \tag{14-15}$$
$$F_a = 2\pi p_{max}r_f(r_o - r_i) \tag{14-16}$$

14-11 Friction Materials. Up until approximately 1930, the friction materials in principal usage were leather, wood, cork, or felt operating against metal, usually cast iron, steel, or brass. Although these organic compounds were reasonably durable with acceptable coefficients of friction, they are adversely affected by rather small temperature rises, moisture, and lubrication. In most of these organic friction materials, damage became noticeable with temperatures of approximately 175°F and above.

Cork and *felt* have high resilience and work well in clutches for light service, such as in home sewing machines and tension controls of tape recorders.

Metallic contact clutches of the past usually employed bronze or brass on steel, cast iron on cast iron, or steel on cast iron. These metallic materials are suitable for higher operating temperatures than the organics previously mentioned, but they have comparatively low coefficients of friction, low torques, and usually demand some form of suitable lubrication.

Asbestos-base and powder-metal friction materials have been developed to provide less wear, higher-temperature operation, higher coefficients of friction, and less lubrication and moisture effects than their predecessors.

Woven asbestos consists of asbestos fiber woven around copper, zinc, brass, or lead wire and impregnated with rubber, drying oils, asphalts, or other resins, baked, and then heavily compressed. Woven asbestos materials are rather durable and possess a reasonable amount of flexibility.

Molded asbestos material consists of short asbestos fiber bonded with materials similar to those employed for woven asbestos.

Molded semimetallics consist of asbestos and copper powder with a synthetic bonding resin. Usually they are bonded in $\frac{1}{64}$ to $\frac{1}{8}$ in. thicknesses directly onto a metal shoe or other backing. Because of the added heat transmission due to the presence of the copper powder, molded semi-metallic friction materials have higher power ratings than molded asbestos materials.

Powder metal friction materials consist of various combinations of powdered copper, zinc, iron, tin, and silicon along with nonmetallic powders such as alumina, silica, silicon carbide, and graphite. The resulting materials have a low wear rate and consequently can be used in rather small thicknesses, such as 0.006- to 0.010-in.-thick facings on steel, cast iron, copper, or aluminum backings.

14-12 Wet versus Dry or Lubricated Friction Clutches. Most friction materials employed in clutches can be successfully used dry, wet, or lubricated. The dry-operated materials have higher coefficients of friction. In general, a wet or lubricated clutch wears less and is substantially better suited for heavy duty and rapid cycling. A wet or lubricated clutch may employ a spray, a shallow dip into liquid, or complete immersion in the liquid, with a resulting lower coefficient of friction than for dry conditions. Usually the working pressure for wet or lubricated clutches can be higher than is possible for dry clutches, thereby counteracting the torque reduction due to the lowering of the coefficient of friction.

A dry clutch usually is simpler and more compact in construction,

Table 14-1 Typical coefficients of friction and maximum pressures for friction clutches*

Contact surfaces		Coefficient of friction $f\ddagger$		Maximum operating temperature, °F	Maximum unit pressure p_{max}, psi	Relative cost	Comment
Wearing	Opposing†	Wet	Dry				
Cast bronze	Cast iron or steel	0.05	....	300	80–120	Low	Subject to seizing
Cast iron	Cast iron	0.05	0.15–0.2	600	150–250	Very low	Good at low speeds
Cast iron	Steel	0.06	....	500	120–200	Very low	Fair at low speeds
Hard steel	Hard steel	0.05	....	500	100	Moderate	Subject to galling
Hard steel	Hard steel, chromium plated	0.03	....	500	200	High	Durable combination
Hard-drawn phosphor bronze	Hard steel, chromium plated	0.03	....	500	150	High	Good wearing qualities
Powder metal§	Cast iron or steel	0.05–0.1	0.1–0.4	1000	150	High	Good wearing qualities
Powder metal§	Hard steel, chromium plated	0.05–0.1	0.1–0.3	1000	300	Very high	High energy absorption
Wood	Cast iron or steel	0.16	0.2–0.35	300	60–90	Lowest	Unsuitable at high speed
Leather	Cast iron or steel	0.12–0.15	0.3–0.5	200	10–40	Very low	Subject to glazing
Cork	Cast iron or steel	0.15–0.25	0.3–0.5	200	8–14	Very low	Cork-insert type preferred
Felt	Cast iron or steel	0.18	0.22	280	5–10	Low	Resilient engagement
Vulcanized fiber or paper	Cast iron or steel	....	0.3–0.5	200	10–40	Very low	Low speeds, light-duty
Woven asbestos§	Cast iron or steel	0.1–0.2	0.3–0.6	350–500	50–100	Low	Prolonged slip service ratings given
Woven asbestos	Cast iron or steel	0.1–0.2	....	500	100–200	Low	This rating for short infrequent engagements
Woven asbestos§	Hard steel, chromium plated	0.1	0.2–0.5	....	1,200	Moderate	Used in aircraft
Molded asbestos§	Cast iron or steel	0.08–0.12	0.2–0.5	500	50–150	Very low	Wide field of applications
Impregnated asbestos	Cast iron or steel	0.12	0.32	500–750	150	Moderate	For demanding applications
Carbon graphite	Steel	0.05–0.1	0.25	700–1000	300	High	For critical requirements
Molded phenolic plastic, macerated cloth base	Cast iron or steel	0.1–0.15	0.25	300	100	Low	For light special service

* A. F. Gagne, Jr., Clutches, *Machine Design*, vol. 24, No. 8, p. 136, August, 1950.

† Steel, where specified, should have a carbon content of approximately 0.70 per cent. Surfaces should be ground true and smooth.

‡ Conservative values should be used to allow for possible glazing of clutch surfaces in service and for adverse operating conditions.

§ For a specific material within this group, the coefficient usually is maintained within plus or minus 5 per cent.

idles with less disengaged friction drag and resultant heating, and is less affected by changes in ambient temperature.

14-13 Coefficients of Friction and Operating Pressures. Since the torque capacity of friction clutches is a direct function of the coefficient of friction as well as the axial force, it is important that the designer be able to obtain suitable values for coefficients of friction and maximum unit pressures between the friction surfaces. Typical coefficients of friction, maximum operating pressures, and other information for friction materials under normal conditions of clutch operation are given in Table 14-1.

14-14 Cone Clutches. The equations for r_f and T developed for disk clutches apply to cone clutches with the exception that the pressure must be measured perpendicular to the cone surface. After the clutch is engaged and for steady operation, the axial force F_a required to produce the normal force F_n (see Fig. 14-21) is

$$F_a = F_n \sin \alpha \tag{14-17}$$

If engagement takes place when one member is rotating and the other is stationary, there appears to be a force resisting engagement and theoretically

$$F_a = F_n (\sin \alpha + f \cos \alpha) \tag{14-18}$$

Experimental results indicate that the $f \cos \alpha$ term is only 25 per cent effective; therefore

$$F_a = F_n \left(\sin \alpha + \frac{f \cos \alpha}{4} \right) \tag{14-19}$$

The clutch may be designed for free disengagement, in which case the tan α must be greater than the coefficient of friction, and a spring must be used to keep the cone surfaces in contact. If the clutch is not to disengage of its own accord, the tan α must be less than the coefficient of friction and a force must be exerted to disengage the clutch. Angle α typically varies from $7\frac{1}{2}$ to 30 deg, depending upon the type of disengagement desired and upon the clutch facing material. The most common

Fig. 14-21 Cone clutch.

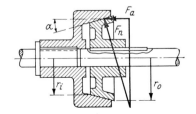

angle is about $12\frac{1}{2}$ deg. The proper angles may be obtained from the coefficients of friction given in Table 14-1.

The axial force necessary may be exerted by suitable springs or by other means. If exerted by springs, the springs must be designed to sustain the load impressed when the clutch is released, since the spring is then still further compressed, and the load increases in direct proportion to the compression of the spring. The increase in spring load is usually 15 to 20 per cent of the operating load.

The torque transmitted by the cone clutch of Fig. 14-21 is given by

$$T = \frac{F_a f r_f}{\sin \alpha} \tag{14-20}$$

14-15   Block Clutches.   In block clutches, torque is transmitted by friction between one or more blocks radially pressed against the cylindrical surface of a drum. Usually the blocks are outside the drum, but internal block or shoe clutches also exist. A single block or shoe of such a clutch is shown in Fig. 14-22. If the arc subtended by the block is not more than 60 deg, it often is assumed that the radial pressure is uniformly distributed over the contact area. Assuming the block moves radially, the torque transmitted by the single block of a clutch is

$$T = wf p_n r^2 \theta \tag{14-21}$$

where f = coefficient of friction

p_n = uniform normal pressure between the block and the drum, psi

r = radius of drum, in.

T = torque transmitted by the one block, in.-lb

w = width of block parallel to the axis of rotation of the drum, in.

θ = angle of contact between the drum and the block, rad

The total radial force F_r required to force the block against the drum is

$$F_r = 2p_n wr \sin \frac{\theta}{2} = p_n wL \tag{14-22}$$

where L is the chord length of the arc of contact of the block, in inches.

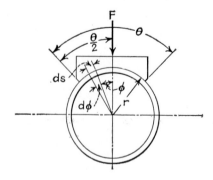

Fig. 14-22   Diagram of single block clutch.

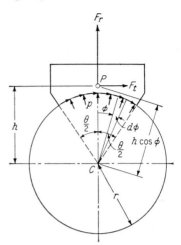

Fig. 14-23 Forces acting on block of block clutch.

Referring to Fig. 14-23, the net drum forces acting on the shoe can be transformed into the radial force F_r and the tangential force F_t acting through point p a distance h from C, the center of the drum.

Summing moments about point p of all forces acting on the block from the drum yields

$$\int_{-\frac{\theta}{2}}^{\frac{\theta}{2}} fpwr \, (h \cos \phi - r) \, d\phi = 0 \tag{14-23}$$

from which

$$h = \frac{r\theta}{2 \sin \dfrac{\theta}{2}} \tag{14-24}$$

Therefore

$$T = fF_r h = \frac{fF_r r\theta}{2 \sin \dfrac{\theta}{2}} \tag{14-25}$$

If the angle θ is sufficiently large that p_n cannot be considered constant, then p_n usually is assumed to be given by

$$p_n = p_{\max} \cos \phi \tag{14-26}$$

This yields

$$h = \frac{4r \sin \dfrac{\theta}{2}}{\theta + \sin \theta} \tag{14-27}$$

$$T = fF_r h = \frac{4fF_r r \sin \dfrac{\theta}{2}}{\theta + \sin \theta} \tag{14-28}$$

and

$$F_r = \int_{-\frac{\theta}{2}}^{\frac{\theta}{2}} p_n wr \cos \phi \, d\phi = p_{max} wr \int_{-\frac{\theta}{2}}^{\frac{\theta}{2}} \cos^2 \phi \, d\phi$$

$$= \frac{p_{max} wr}{2} (\theta + \sin \theta) \quad (14\text{-}29)$$

This gives

$$T = 2fp_{max}wr^2 \sin \frac{\theta}{2} \quad (14\text{-}30)$$

If a block clutch has more than one block contacting the drum, Eqs. (14-21) through (14-30) may be applied to each block and their effects added to obtain the total torque capacity of the clutch. If each block is identical in size and loading, the total torque capacity may be obtained by multiplying Eq. (14-21), (14-25), (14-28), or (14-30) by n, the number of identical blocks.

Since clutches normally are used on rotating members, the centrifugal forces of the masses of the blocks affect the pressure between the blocks and the drum. This should be taken into account in determining the value of p to be used in the previous equations. Unless the rotational speeds are comparatively high, the effects of centrifugal forces are often neglected.

14-16 Expanding-ring Clutches. Expanding-ring clutches consist of a drum or shell attached to the driving shaft and an expandable split ring or band placed inside this shell and connected to the driven shaft. The ring may be expanded by a cam or wedge so that it presses against the inside of the driving shell. The radial pressure is commonly assumed to be uniformly distributed over the entire contact surface, although this is only a rough approximation.

In Fig. 14-24, the force applied to the end of each of the two identical split rings to expand the ring is F_a. If the pressure p is assumed uniformly distributed over the contact surface, then the total frictional torque T is given by

$$T = 2 \int fpwr \, ds = 2 \int_0^\theta fpwr^2 \, d\phi = 2fpwr^2\theta \quad (14\text{-}31)$$

where θ is the arc of contact, in rads, between each ring and the drum.

If the ring is made in two sections as shown in Fig. 14-24, the applied force F_a required to separate the parts is found by taking moments about the pivot point O. The moment of the normal forces acting on the area ds, about the pivot point O, is

$$M_o = \int dM_o = \int_0^\theta pwr \, d\phi \left(\frac{L}{2} \sin \phi\right) = \frac{pwrL}{2} (1 - \cos \theta) \quad (14\text{-}32)$$

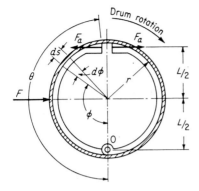

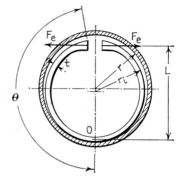

Fig. 14-24 Forces acting on the two split rings of an expanding ring clutch.

Fig. 14-25 Expanding forces on the one-piece ring of an expanding ring clutch.

For many clutches of this type, θ is nearly π rads with a corresponding $\cos \theta$ of approximately -1. Using this value for θ, Eq. (14-32) becomes

$$M_o = pwrL \qquad \text{for each half of the band} \tag{14-33}$$

The moment of F_a about O must be equal to M_o. Hence

$$F_a L = M_o = pwrL \tag{14-34}$$
$$F_a = pwr$$

Therefore

$$T = 2fF_a r\theta \tag{14-35}$$

If the ring is made in one piece, there is an additional force required to expand the inner ring elastically before contact is made with the inner surface of the shell. Referring to Fig. 14-25,

$$\frac{M}{EI} = \frac{1}{r_1} - \frac{1}{r} \tag{14-36}$$

where r_1 is the original unstressed radius of the ring, in inches, and M is the bending moment about O, in inch-pounds.

In this case, M is equal to $F_a L$, and I is $wt^3/12$, the moment of inertia of the cross-sectional area of the ring. Hence

$$F_e = \frac{Ewt^3}{12L}\left(\frac{1}{r_1} - \frac{1}{r}\right) = \frac{Ewt^3}{6L}\left(\frac{1}{d_1} - \frac{1}{d}\right) \tag{14-37}$$

where d_1 is the original unstressed diameter of the ring, in inches, and d is the inner diameter of the drum, in inches.

Usually d_1 is from $\frac{1}{64}$ to $\frac{1}{16}$ in. smaller than d.

The total force F required to expand the unsplit ring and to produce

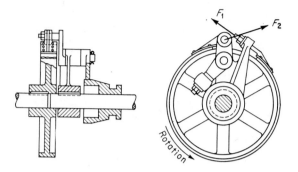

Fig. 14-26 Band clutch.

the necessary pressure between the contact surfaces is

$$F = F_a + F_e = pwr + \frac{Ewt^3}{6L}\left(\frac{1}{d_1} - \frac{1}{d}\right) \tag{14-38}$$

14-17 Band Clutches. *Band clutches* are especially suitable for mine hoists and other services where heavy loads are accompanied by severe shock. The clutch usually consists of a flexible steel band lined with wood or composition blocks, asbestos fabric, or molded semimetallics, one of which is fixed to either the driving or driven member; the other end is pulled around the circumference of a drum on the mating member. Figure 14-26 illustrates a clutch of this type.

F_1 and F_2 are the total tensions in the band. For the direction of impending drum rotation relative to the band shown in Fig. 14-26, $F_1 > F_2$. Thus

$$F_1 - F_2 = \frac{2T}{D} \tag{14-39}$$

where D = diameter of friction surface of the drum, in.
$\quad\quad F_1$ = the largest force in the band, lb
$\quad\quad F_2$ = the smallest force in the band, lb
$\quad\quad T$ = torque transmitted, in.-lb

Permitting F_c to be zero in Eq. (16-4), and letting θ be the angle of contact (in rads) of the band on the drum yields

$$\frac{F_1}{F_2} = e^{f\theta} \tag{14-40}$$

where f is the coefficient of friction.

The width of the clutch band may be determined by the permissible unit pressure p_{max} normal to the drum. The maximum normal pressure is at the high-tension end of the band and is

$$p_{max} = \frac{F_1}{wr} \tag{14-41}$$

14-18 Centrifugal Clutches. Whenever it is desired to engage a load after the driving member has attained a predetermined minimum rotational speed, one of the various types of *centrifugal clutches* can be used. Such clutches are particularly useful with internal-combustion engines which cannot be started under load, helicopter rotor drives, and the cutting-chain drives of engine-powered chain saws. Two major types of centrifugal clutches are the weight type and the mercury type. The weight type of Fig. 14-27 consists of three or more weights loosely attached to the input shaft, usually with a spring or springs applying a radial inward force on the weights. As the angular speed of the input shaft increases, the inertial forces of the rotating weights increase, causing the weights to move radially outward until they contact the internal cylindrical friction surface of the driven member, thereby transmitting torque to the driven drum. The torque transmitted is given by

$$T = 2nrf \left(\frac{w\rho N^2}{70,400} - S \right) \tag{14-42}$$

where f = coefficient of friction
N = rotational speed of driver, rpm
n = number of identical inertial weights
r = radius of friction surface of the drum, in.
S = radial spring force on each inertial weight, lb
T = torque transmitted, in.-lb
w = weight of each inertial weight, lb
ρ = radial distance from center of driving shaft to mass centroid of each inertial weight, in.

If a continuous garter spring is employed to hold the weights toward the driving member, then S is given by

$$S = T_s \cos \left(\frac{\pi}{2} - \frac{\pi}{n} \right) \tag{14-43}$$

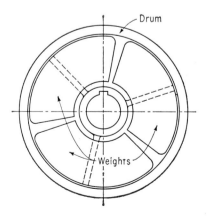

Fig. 14-27 Schematic of centrifugal weight clutch (weights move radially on pins attached to the shaft hub).

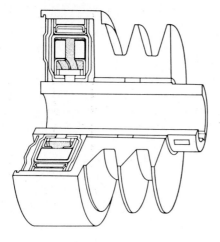

Fig. 14-28 Centrifugal clutch operated by mercury inside a flexible
enclosure.

where T_s is the initial tension in the garter spring, in pounds, and the
angle is in rads.

If it is desired to effect an engagement at a speed N_e, rpm, of the
driven member, S is obtained from

$$S = \frac{w\rho N_e^2}{70,400} \qquad \text{for radial springs} \qquad (14\text{-}44)$$

or

$$S = \frac{w\rho N_e^2}{70,400 \cos (\pi/2 - \pi/n)} \qquad \text{for garter springs} \qquad (14\text{-}45)$$

Figure 14-28 depicts a centrifugal clutch operated by mercury contained in a plastic gland. As the rotational speed of the driver increases,
the centrifugal force causes most of the mercury to travel radially outward, forcing friction shoes or other friction surfaces against the inside
of the driven drums and effecting a transfer of torque.

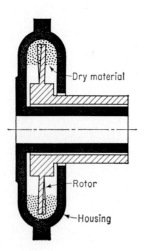

Fig. 14-29 Dry fluid coupling.

14-19   Dry-fluid Clutches.   The clutch of Fig. 14-29 is a *dry-fluid clutch* or *coupling*. It consists of a driving housing, a driven rotor plate, and a quantity of dry material, usually steel shot within the housing. As the driving housing begins to rotate, the steel shot travels radially to the inside periphery of the housing and causes a wedging action with the driven rotor, transmitting torque to the rotor. In general, the torque capacity is a function of the amount of steel shot in the housing.

14-20   Magnetic-particle Clutches.   The *magnetic-particle clutch* of Fig. 14-30 depends upon the effect of a magnetic field upon ferromagnetic particles which may be in an oil-slurry mixture or dry. The torque transmitted depends upon the shear resistance of the ferromagnetic particle layer, which is a function of the radial magnetic field between the field coil through the particles to the housing. The torque is an almost linear function of the current in the field coil. An alternative construction of the clutch consists of the excitation coil in the outer housing and a ferromagnetic central drum, an inversion of what is shown in Fig. 14-30.

14-21   Eddy-current Clutches.   The magnetic reaction produced by eddy currents induced in the surfaces of driven and driving rotors is the main feature of *eddy-current clutches.* When the two rotors move relative to each other in a magnetic field, eddy currents are induced in the two rotors, causing a torque reaction between the rotors which is a function of the amount of angular slip between the rotors and the current in the coil creating the magnetic field.

14-22   Slip Clutches.   Any clutch which will transmit torque up to a predetermined amount and will slip when this is exceeded is known as a *slip clutch*, or *overload release clutch.* Multiple disk clutches make excellent slip clutches with proper setting of the preloading in the springs producing the axial force on the friction disks. A slip clutch is used to protect a machine in case of jamming and for overload protection of motors and engines. Prolonged slippage causes overheating, hence a thermal trip

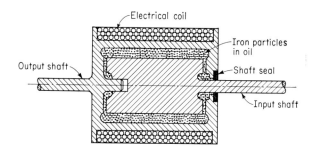

Fig. 14-30   Magnetic-particle clutch schematic diagram.

often is employed with overload clutches to shut off the driving member and/or to sound an alarm.

14-23   Trip Clutches.   When certain undesirable conditions exist, such as overspeed of a passenger elevator, a *trip clutch* can be employed to disconnect the driving member and/or apply braking effort. In its usual condition, a trip clutch applies no torque whatsoever to the driven member until it is tripped, often by a toggle device, at which time torque is applied to the driven member until manually reset.

14-24   Overrunning Clutches.   Clutches which are designed to transmit torque for one direction of rotation of the driver and then free-wheel or transmit essentially no torque when the direction of the driver rotation is reversed are known as *over-running, free-wheeling,* or *unidirectional clutches.* They may be divided into clutches which depend upon friction and a wedging action to transmit torque and those which depend upon shear and compression bearing of parts.

The *roller type overrunning clutch* of Fig. 14-31 may consist of spring-loaded balls or cylinders, and may be either external or internal in construction. For the direction of rotation shown in the figure, the balls or rollers will become wedged between the inner and outer members and torque will be transmitted. For opposite direction of rotation of the inner member, the balls or rollers will not wedge and essentially no torque

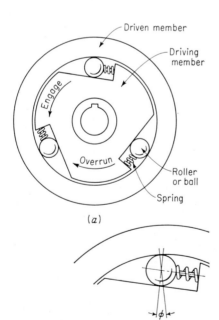

Fig. 14-31   (*a*) Roller-type overrunning clutch.   (*b*) Angle ϕ between ramp and radial line.

will be transmitted. The torque capacity in inch-pounds is

$$T = nFr \tan \frac{\phi}{2} \tag{14-46}$$

where F = radial force between one roller or ball and the race, lb
$\quad n$ = number of identical balls or rollers
$\quad r$ = radius of race against which the balls or rollers wedge, in.
$\quad \phi$ = the angle a perpendicular to the driving ramp makes with a radial line [see Fig. 14-31(b)]

The maximum allowable value of F is determined by the dimensions of the ball or roller and the contacting surfaces, the material modulus, and the allowable contact stress. For a straight ramp with cylindrical rollers, F becomes

$$F_{\max} = rLK_c \tag{14-47}$$

where K_c is a combined material-contact stress factor, psi, and L is the length of each roller, in. Typical values of K_c are 15,000 to 24,000 psi for carburized carbon-steel rollers and mating surfaces, 9,000 to 14,000 psi for medium-carbon-steel rollers and mating surfaces, and 1,000 to 5,500 psi for cast iron rollers with hardened steel mating surfaces.[*]

The *sprag clutch* of Fig. 14-32 transmits torque between two races by the wedging action of the sprags between the races. The sprags are so shaped that for one direction of relative rotation, each sprag tends to become tightly wedged between the two races, thereby permitting torque transmission. In the opposite direction of relative rotation the wedging action is ended and free-wheeling results. The torque capacity in inch-pounds is

$$T = \frac{nr_iLK_c \tan \theta}{1/r_i + 1/r_s} \tag{14-48}$$

where L = axial length of each sprag, in.
$\quad n$ = number of sprags
$\quad r_i$ = radius of inner race, in.
$\quad r_s$ = radius of sprag contact surface, in. [see Fig. 14-32(b)]
$\quad \theta$ = sprag gripping angle [usually between 2 and 7 deg, see Fig. 14-32(b)]

and all other symbols are as previously defined. The maximum torque capacity of a sprag clutch is comparatively high for any given physical size.

The *molded sprag clutch* of Fig. 14-33 depends upon elasticity of

[*] D. J. Myatt, "Machine Design," p. 51, McGraw-Hill Book Company, New York, 1962.

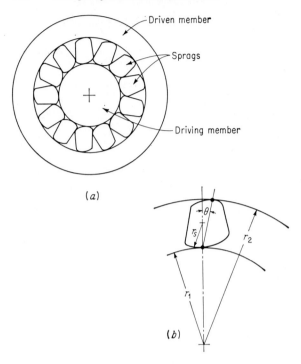

(a)

(b)

Fig. 14-32 (a) Sprag-type overrunning clutch. (b) Detail of one sprag.

the molded sprags for its unidirectional torque transmission. It should be used for light service.

The *slip-spring coupling clutch* or coupling of Fig. 14-34 works upon the principle that torsional rotation of a helical spring causes a change of diameter, an increase or decrease depending upon the direction of rotation. This clutch is widely used on the dial drive shaft of telephones.

The *pawl clutches* of Fig. 14-35 wedge the driving and driven members together tightly for unidirectional torque transmission, and free-wheeling action for opposite conditions of relative rotation of the two

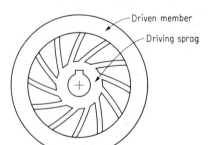

Fig. 14-33 Molded sprag overriding clutch (driver made of Teflon, nylon, or similar material).

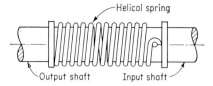

Fig. 14-34 Slip-spring overriding clutch.

members. These clutches find frequent application in ready-pull starters of outboard motors and lawn mower engines.

The *spring-loaded spiral-jaw clutch* of Fig. 14-36 permits torque transmission for one direction of driver rotation. Opposite driver rotation causes the spiral surfaces of the jaws to rachet or free-wheel with essentially no torque being applied to the driven member.

The *disengaging idler gear clutch* of Fig. 14-37 relies upon the tooth loads on the idler either to keep the idler and the other gears in contact or to cause the idler to disengage by the center of the idler moving along a slot. Once the idler is disengaged, no torque can be transmitted until the idler is manually moved back into the position engaging the driving and the driven gears.

14-25 Clutch Design Considerations. A clutch of good design must have ability to withstand and dissipate heat, adequate reserve torque capacity (other than a slip clutch), and long life. For the high speeds encountered in aircraft and automotive practice, the driven members should have low moments of inertia, and parts should be accurately balanced. The clutch should engage smoothly with a low operating force, have a positive release, and be easy to repair.

The coefficient of friction depends upon the facing material and may vary with the normal pressure, temperature, cleanliness, and wetness of the surfaces. Design values of the coefficient of friction may be taken from Table 14-1. When the contact surfaces are sliding over each other, as in starting, the coefficient is decreased. When starting, the inertia of

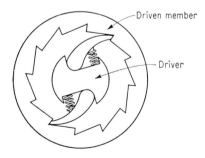

Fig. 14-35 Spring-loaded pawl overriding clutch.

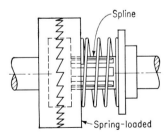

Fig. 14-36 Spring-loaded spiral-jaw clutch.

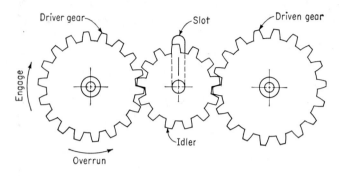

Fig. 14-37 Disengaging idler gear clutch.

the driven parts must be overcome; hence to provide sufficient starting capacity, some designers design clutches for overload capacities of 75 to 100 per cent. Often this is not desirable because slippage permits smooth and gradual engagement; of course clutches without overload capacity must be capable of transmitting the desired torque after slippage has stopped. If the load is variable or subject to shock, additional service factors often are used. Suggested service factors are given in Table 14-2.

Table 14-2   Typical service factors for clutches

Type of service	Service factor (not including starting factor)
Driving machine:	
Electric motor, steady load	1.0
Fluctuating load	1.5
Gas engine, single cylinder	1.5
Multiple cylinder	1.0
Diesel engine, high speed	1.5
Large, slow speed	2.0
Driven machine:	
Generator, steady load	1.0
Fluctuating load	1.5
Blower	1.0
Compressor, depending on the number of cylinders	2.0–2.5
Pumps, centrifugal	1.0
Single acting	2.0
Double acting	1.5
Line shaft	1.5
Woodworking machinery	1.75
Hoists, elevators, cranes, shovels	2.0
Hammer mills, ball mills, crushers	2.0
Brick machinery	3.0
Rock crushers	3.0

15

Brakes

15-1 Purpose of Brakes. Brakes depend upon friction between contacting surfaces or viscous or electromagnetic damping resistance for their absorption of energy. The heat-dissipating capacity of brakes often limits their capacity, although coefficient of friction between the braking surfaces, unit pressures between these surfaces, viscosity of hydrodynamic braking fluid, and strength of electromagnetic fields also may be important factors in the design of brakes.

The novice designer frequently sizes brakes upon their ability to apply a desired braking torque without giving much thought to how the brake will be able to dissipate the energy that it is required to absorb. Under these conditions, he may be surprised to discover that automotive brakes frequently are required to absorb power at a rate several times the horsepower capacity of the engine powering the automobile.

15-2 Types of Brakes. Among the varieties of *mechanical brakes* are band, block, shoe, disk, and spot brakes. The hydrodynamic brake is the major type of *fluid brake*, although additional sorts of fluid agitator brakes exist. Electric generators and eddy-current brakes are the more important types of *electrical brakes*.

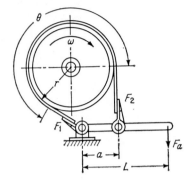

Fig. 15-1 Simple band brake.

Mechanical brakes are capable of stopping the motion of a machine member as well as retarding its motion; fluid and electrical brakes are incapable of bringing a machine member to rest, and hence are useful only in controlling the motion of an element through the absorption of energy. There are many applications where the latter is satisfactory, such as controlling the downhill travel of a loaded truck or the travel of a mine skip.

15-3   Band Brakes.   Perhaps the simplest of all brakes are the *band brakes*. They consist of a rope, belt, or flexible steel band (lined with friction material) acting against the external surface of a cylindrical drum. A simple band brake is shown in Fig. 15-1. From the derivations of Art. 16-6, the ratio of the tension forces at the band ends is expressed by the belt equation

$$\frac{F_1}{F_2} = e^{f\theta} \tag{15-1}$$

where F_1 = force in band on high-tension side, lb
$\quad\ F_2$ = force in band on low-tension side, lb
$\quad\ e$ = 2.7183
$\quad\ f$ = coefficient of friction
$\quad\ \theta$ = angle of contact, rad

The torque T developed by the braking action is

$$T = (F_1 - F_2)r \tag{15-2}$$

where r = radius of friction surface on the drum, in.
$\quad\ T$ = brake torque, in.-lb

and the actuating force F_a required at the end of the operating lever is

$$F_a = \frac{aF_2}{L} \tag{15-3}$$

Note that if the direction of rotation is reversed to counter clockwise,

F_1 and F_2 are interchanged; for the same braking torque a larger force F_a is required to operate the brake. For this case

$$F_a = \frac{aF_1}{L} \tag{15-4}$$

The band width w depends upon the maximum unit pressure p_{max} that occurs at the high-tension end and is determined from

$$p_{max} = \frac{F_1}{wr} \tag{15-5}$$

In the differential band brake shown in Fig. 15-2, neither end of the band goes through the fulcrum of the actuating arm. The force F_1 applies a torque onto the arm in the same direction as the actuating force F_a, and hence the friction assists in applying the braking torque. This is called self-energizing. Equations (15-1), (15-2), and (15-5) apply for this brake. For the direction of drum rotation shown,

$$F_a = \frac{F_2a - F_1b}{L} \tag{15-6}$$

For a reversal in the direction of drum rotation, F_1 and F_2 become interchanged, and the required actuating force F_a then becomes

$$F_a = \frac{F_1a - F_2b}{L} \tag{15-7}$$

For the same braking torque, F_a from Eq. (15-7) will be larger than the value from Eq. (15-6); hence it is important to note that for the same braking force F_a, the direction of drum rotation affects the magnitude of the braking torque.

15-4 Self-energizing and Self-locking Brakes. For the differential band brake of Fig. 15-2 with clockwise rotation of the drum, the clockwise moment F_1b applied onto the brake arm by the band is in the same direction as that by the actuating force F_a. When such occurs, the brake is defined as being self-energized, and the moment F_1b is called the

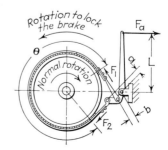

Fig. 15-2 Differential band brake.

self-energizing moment. Equation (15-6) may be revised into

$$F_a = \frac{F_2 a - F_1 b}{L} = \frac{F_2(a - be^{f\theta})}{L} \tag{15-8}$$

Should the quantity $be^{f\theta}$ be larger than a, the algebraic value of F_a would be negative. From a practical standpoint, should this be true, once the friction band came into contact with the drum, the brake would grab or be *self-locking*, and no applied force F_a would be required. For certain applications, a self-locking condition may be desirable; however, in most applications this definitely is not desired, and the designer must check to be sure that the quantity $be^{f\theta}$ does not equal or exceed a in magnitude. Of course, the more nearly a brake approaches the self-locking condition, the smaller is the required applied force F_a, and hence many brakes are designed with self-energization below that required to make the brake self-locking.

15-5 Block Brakes. A simple external block brake is shown in Fig. 15-3. The equations developed in Art. 14-15 for block clutches apply equally well to block brakes. The braking torque T is

$$T = fF_r h \tag{15-9}$$

where F_r = radial force between the drum and each shoe, lb
$\quad\quad f$ = coefficient of friction
$\quad\quad h$ = effective moment arm of the friction force, in.

For a large angle of contact θ, the pressure varies, and from Eq. (14-28),

$$T = \frac{4fF_r r \sin(\theta/2)}{\theta + \sin\theta} = 2fp_{max}wr^2 \sin\theta/2 \tag{15-10}$$

where p_{max} = maximum normal pressure between block and drum, psi
$\quad\quad r$ = radius of the friction surface of the drum, in.
$\quad\quad w$ = axial width of block, in.

and all other symbols are as previously defined.

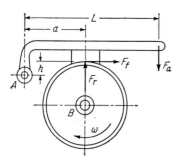

Fig. 15-3 External block brake.

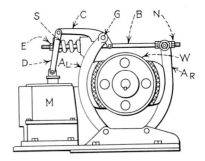

Fig. 15-4 Solenoid-operated block brake.

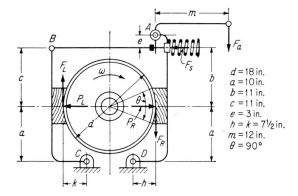

Fig. 15-5   Diagram of a block brake.

An example of a block brake as used for elevator and hoisting service is shown in Fig. 15-4. This brake is positive in action, the brake blocks being pressed against the brake wheel W by means of the spring S, which is compressed between the end of the arm A_L and the spring retainer E on the end of the rod B. The brake is released by a magnet enclosed in the case M, which, when energized by electric current, pulls down on the rod D. As a result, the arm C rotates on its fulcrum G and forces apart the tops of the arms A_R and A_L, thus compressing the spring and releasing the brake. Releasing of both brake blocks is ensured by an adjustable stop which limits the outward movement of the arm A_R. Adjustment for wear is made by means of the adjusting nuts at N.

Example.   A block brake similar to the one described in the preceding paragraph is shown diagrammatically in Fig. 15-5. The dimensions of a typical brake are given in the figure. The coefficient of friction is taken as 0.35 and the spring is assumed to exert a force of 3,000 lb when the brake is set.

The normal force P_R against the right-hand brake shoe is found by taking moments about the lower pin joint D, thus

$$\Sigma M = F_R h + P_R a - F_s(a + b) = 0 \tag{15-11}$$

from which

$$P_R = \frac{(a + b)F_s - F_R h}{a} \tag{15-12}$$

From Eq. (15-10), the relation between the friction force F_R and the total shoe force P_R is

$$F_R = \frac{4fP_R \sin (\theta/2)}{\theta + \sin \theta} \tag{15-13}$$

and, by substituting this value in Eq. (15-12),

$$P_R = \frac{(a+b)F_s}{a + [4hf \sin (\theta/2)/(\theta + \sin \theta)]}$$
$$= \frac{(10+11)3,000}{10 + [4 \times 7.5 \times 0.35 \times 0.707/(1.57+1)]} = 4,890 \text{ lb} \quad (15\text{-}14)$$

In the same way,

$$P_L = \frac{(10+11)3,000}{10 - [4 \times 7.5 \times 0.35 \times 0.707/(1.57+1)]} = 8,860 \text{ lb} \quad (15\text{-}15)$$

Note that in this case the shoe forces P_R and P_L are not equal. If the brake is to act only when the drum is revolving clockwise, the shoe forces P_R and P_L, and hence the friction forces, may be equalized by making the distance b somewhat larger than the distance c, or by making h and k equal to zero.

The total braking capacity of this brake is expressed by

$$T = (F_R + F_L)r = \frac{4f \sin (\theta/2)}{\theta + \sin \theta} (P_R + P_L)r$$
$$= \frac{4 \times 0.35 \times 0.707}{1.57 + 1} (4,890 + 8,860)9 = 47,600 \text{ in.-lb} \quad (15\text{-}16)$$

The force F_a that must be applied to the operating arm by the magnet to release the brake is

$$F_a = 1.1 \frac{eF_{\text{spring}}}{m} = \frac{3 \times 3,000}{12} 1.1 = 825 \text{ lb} \quad (15\text{-}17)$$

The factor 1.1 is used since the spring is further compressed when in the release position. The release force on the spring is usually 10 to 15 per cent greater than the required braking force.

After the forces acting are determined, the individual members of the brake may be considered separately to determine the required size of each. The arms carrying the brake shoes are treated as simple beams, the operating arm as two cantilever beams, and the pins at the various joints as oscillating high-pressure bearings.

15-6   Automotive Shoe Brakes.   Band brakes were used on nearly all of the earliest automobiles, but they were exposed to dirt and water and their heat dissipation capacity was poor. These conditions, together with the tendency toward smaller wheels and larger tires, have forced the use of internal-shoe brakes. Nearly all automotive brakes now are internal-shoe brakes, and at least one shoe per wheel is self-energizing; that is, friction makes the shoe follow the rotating brake drum and wedge itself between the drum and the point at which it is anchored. Tremendous friction thus builds up, giving great braking power without the use of excessive pedal pressures. If such brakes are not properly designed and adjusted, and if the brake facings do not have the proper coefficient of friction, the braking action may be too sensitive.

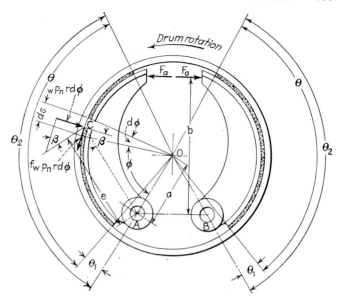

Fig. 15-6 Internal-shoe automotive brake.

Fig. 15-6 is typical of one type of internal-shoe automotive brake in current use. The actuating force F_a is applied equally in magnitude to each shoe, generally by two equal diameter pistons in a common hydraulic cylinder. Two shoes are assumed to be identical in size. For the direction of drum rotation shown, the left shoe is known as the forward or leading shoe and the right shoe as the rear or trailing shoe. Since each shoe is considered as being comparatively rigid in comparison to the friction surfacing, the pressure on the contact surface is a function of the distance of the point on the contact surface away from the pivot point A or B. The normal pressure p_n is given by

$$p_n = K_p \sin \beta = Ka \sin \phi \qquad (15\text{-}18)$$

where K is a brake constant, in pounds per cubic inch, as given by

$$K = \frac{p_n}{a \sin \phi} \qquad (15\text{-}19)$$

If w is the width of each friction surface, in inches, in a direction parallel to the axis of rotation of the drum, the total normal force dF_n and the total frictional force dF_f on the differential area $wr \sin \phi$ are given by

$$dF_n = p_n wr \, d\phi = K_a wr \sin \phi \, d\phi \qquad (15\text{-}20)$$

and

$$dF_f = f \, dF_n = K_a fwr \sin \phi \, d\phi \qquad (15\text{-}21)$$

where f is the coefficient of friction.

Considering the left or forward shoe, summing moments about point A, and employing L as a subscript to designate the left shoe,

$$M_{nL} - M_{fL} - F_a b = 0 \tag{15-22}$$

where M_n and M_f respectively are the moments about point A of the normal and frictional forces acting on the frictional material. Thus

$$M_{nL} = \int_{\theta_1}^{\theta_2} (K_L awr \sin \phi \, d\phi)(a \sin \phi) \tag{15-23}$$

$$= \frac{K_L a^2 wr}{4} (2\theta - \sin 2\theta_2 + \sin 2\theta_1)$$

and

$$M_{fL} = \int_{\theta_1}^{\theta_2} (K_L afwr \sin \phi \, d\phi)(r - a \cos \phi) \tag{15-24}$$

$$= \frac{K_L afwr}{2} [2r(\cos \theta_1 - \cos \theta_2) - a(\cos^2 \theta_1 - \cos^2 \theta_2)]$$

In a similar manner, for the right shoe

$$-M_{nR} - M_{fL} + F_a b = 0 \tag{15-25}$$

$$M_{nR} = \frac{K_r a^2 wr}{4} (2\theta - \sin 2\theta_2 + \sin 2\theta_1) \tag{15-26}$$

$$M_{fR} = \frac{K_r afwr}{2} [2r(\cos \theta_1 - \cos \theta_2) - a(\cos^2 \theta_1 - \cos^2 \theta_2)] \tag{15-27}$$

Solving Eqs. (15-22) through (15-24) for K_L yields

$$K_L = \frac{2F_a b}{awr\{(a/2)(2\theta - \sin 2\theta_2 + \sin 2\theta_1) - f[2r(\cos \theta_1 - \cos \theta_2) - a(\cos^2 \theta_1 - \cos^2 \theta_2)]\}} \tag{15-28}$$

Equations (15-26) and (15-27) give

$$K_R = \frac{2F_a b}{awr\{(a/2)(2\theta - \sin 2\theta_2 + \sin 2\theta_1) + f[2r(\cos \theta_1 - \cos \theta_2) - a(\cos^2 \theta_1 - \cos^2 \theta_2)]\}} \tag{15-29}$$

The total braking torque T on the drum is the moment about the center of the drum of the friction forces acting on both shoes, hence

$$T = \int_{\theta_1}^{\theta_2} K_L afwr^2 \sin \phi \, d\phi + \int_{\theta_1}^{\theta_2} K_R fawr^2 \sin \phi \, d\phi$$

$$= afwr^2(K_L + K_R)(\cos \theta_1 - \cos \theta_2) \tag{15-30}$$

It is interesting to note that for the same applied force F_a on each shoe, the brake constants are not equal, and $K_L > K_R$; the leading shoe contributes more than 50 per cent of the total braking torque on the drum. Since the frictional forces acting on the leading shoe are in the same counterclockwise direction as the moment of the actuating force F_a, the leading shoe is self-energizing. Similarly, the rear shoe is not self-energizing. If the direction of drum rotation were reversed and all other quantities remained the same, the right shoe would be self-energizing, the left shoe non-self-energizing, and the total torque the same as given by Eq. (15-30); however, the magnitudes of braking torque that the right and left shoes apply to the drum would be interchanged.

Examination of Eq. (15-22) shows that friction alone will lock the brake when the friction moment M_{f_L} exceeds the moment of the normal pressure M_{n_L}. The brake should be self-energizing but not self-locking. The amount of self-energizing is measured by the ratio of the friction moment and normal pressure moment. Hence

$$\frac{M_{f_L}}{M_{n_L}} = \frac{2f}{a} \frac{2r(\cos\theta_1 - \cos\theta_2) - a(\cos^2\theta_1 - \cos^2\theta_2)}{2(\theta_2 - \theta_1) - \sin 2\theta_2 + \sin 2\theta_1} \tag{15-31}$$

When the above ratio is equal to or greater than unity, no force F_a is necessary to set the brake, which is therefore self-locking. When the ratio is between unity and zero, the brake will be self-energizing. A well-designed brake should be self-locking only when the coefficient of friction has increased about 0.20 above the highest coefficient that the brake lining will ever have.

It is possible to increase the braking torque of a two-shoe automotive brake of the type shown in Fig. 15-6. If the two shoes are identical in size, the actuating force F_a on the rear, non-self-energizing shoe can be increased by employing a hydraulic piston of larger diameter for it. If actuating forces on each shoe are equal, the braking torque applied by each shoe can be equalized by employing a non-self-energizing shoe having a greater angle of contact θ than that of the self-energizing leading shoe. Some manufacturers employ the braking system indicated in Fig. 15-7. This consists of two separate hydraulic cylinders per braking drum, each applying equal actuating forces F_a onto the two shoes. For the direction

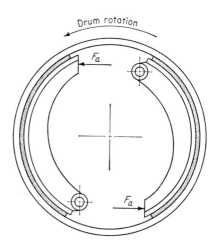

Fig. 15-7 Internal-shoe automotive brake with both shoes self-energizing for the direction of drum rotation indicated, and neither self-energizing for opposite rotation.

of drum rotation indicated in the figure, each of the shoes is self-energizing. For the opposite direction of drum rotation, neither shoe is self-energizing. This explains why automobiles with this type of brake have substantially higher braking torque when moving forward than when moving backward. Usually this is of no great consequence since maximum braking effort is seldom required when the car moves backward.

The brake shoe should, in general, be symmetrical about the line of maximum normal pressure. When the required motion of the force actuator, lining clearances, pressure distribution, and wear are considered, the practical length of shoe is limited to about 120 deg.

The brake shoes here considered have been of rigid construction. When the shoes are somewhat flexible, the pressure distribution will be slightly different. Also, it is evident that the heat of friction will raise the temperature of the outer surface of the shoe above that of the inner surface, causing the shoe to curl up and the pressure to rise near the center of the shoe.

15-7 Disk Brakes. Multiple friction disks arranged in a manner similar to that of a multiple disk clutch but with one set of disks held fixed from rotating constitute what is known as a *disk brake*. Such brakes are used in aircraft, automotive vehicles, and heavy equipment. No analysis of disk brakes is given here since such analysis is identical to that for disk clutches, which was given in Arts. 14-8 through 14-10. As is true for most types of automotive brakes, disk brakes quite often present a heat dissipation problem because of their relatively high energy absorption capacity and comparatively low volumes.

15-8 Spot Brakes. The brake shown in Fig. 15-8 consists of a relatively heavy disk of steel mounted concentrically with the axis of rotation

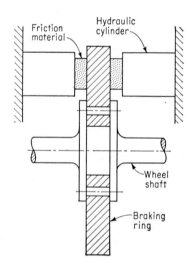

Fig. 15-8 Schematic of spot brake.

of an automotive or aircraft wheel and rotating with the wheel, and two nonrotating hydraulic cylinders, each of which is capable of applying simultaneously a relatively large force onto a cylindrical disk of friction material which in turn contacts the rotating disk when the hydraulic pressure is applied. Such a system applies comparatively high friction forces over a small area or spot on the disk. These brakes often are referred to as *spot brakes*. The forces applied by the two hydraulic cylinders always are collinear. The braking torque capacity T in in.-lb is given by

$$T = 2fF_aR = 2\pi fpr^2R \tag{15-32}$$

where F_a = radial load applied by each cylinder of friction material onto the disk, lb
f = coefficient of friction
p = hydraulic pressure within the actuating cylinders, psig
R = radial distance from the axis of rotation of the wheel to the center of application of F_a, in.
r = radius of each hydraulic cylinder, in.

Because of the high contact pressures possible between the disk and the friction cylinders, this type of brake is capable of absorbing relative high energy rates for its size. In aircraft service it is not uncommon for the disks of this type of brake to become red hot from only one landing of the aircraft. Heat dissipation rates from the disk are relatively high because of the temperature of the disk and the usually rapid airflow across the disk. Disk brakes are finding application in sports cars, but because it must be possible to use the brake several times, or continuously while descending a hill, lower contact pressures are employed in automotive disk brakes than aircraft brakes; hence operating temperatures are lower.

Although the wear of the friction cylinders is comparatively high, usually this is of no great importance since the maximum stroke of each hydraulic cylinder applying the braking force usually is large—1 in. or more. Replacement of the friction cylinders is a comparatively simple process.

15-9   Electric Generator Braking.   Where it is necessary to apply braking over an extended period of time, rather than absorbing energy by means of friction, energy absorption can be accomplished by driving one or more electric generators, the output from which can either be made to do useful work or be dissipated by resistive heating or other means. For a period of time generator braking has been employed on trains and large trucks. This type of braking is easily controlled by any of several different methods of controlling the generator output.

15-10 Eddy-current Brakes. The eddy-current clutches discussed in Art. 14-21 can easily be used as brakes if the output disk is held stationary. The braking torque is a function of the slip between the rotating and the stationary disk, and the current in the coil creating the magnetic field. Various types of construction exist for eddy-current brakes, each of which is easily controlled by the amount of the coil exciting current.

15-11 Hydrodynamic Brakes. A brake utilizing fluid friction instead of mechanical surface friction has been introduced in the oil fields and mines to handle the heavy loads incident to deep hoisting and drilling. These brakes do not replace the regular mechanical brakes. They are used to provide braking when lowering tools, drill pipe, and casing into the hole. The regular mechanical brake is required to stop the descending load and to hold it stationary when drilling operations make this necessary.

15-12 Design Factors for Brakes. The heat generated by the friction must be dissipated to the atmosphere; hence the capacity of a brake may be limited by its ability to dissipate this heat. Stated in a different manner, the work done, which is proportional to the product of the unit pressure and the rubbing velocity, must be kept below certain maximum values that depend upon the type of brake, the efficiency of the heat-dissipating surfaces, and the continuity of service. Brakes operating very infrequently can be much smaller than those which operate almost continuously. The lower the unit pressure at the braking surface, the longer will be the life, and the less the danger of overheating. Some designers recommend the following for different types of service:

$pV = 55,000$ for intermittent operation with long rest periods, and poor heat dissipation, as with wood blocks

$pV = 28,000$ for continuous service with short rest periods, and with poor dissipation

$pV = 83,000$ for continuous operation and with good dissipation as with an oil bath

Table 15-1   **Working pressures for brake blocks**

Rubbing velocity, fpm	Working pressure, psi		
	Wood blocks	Asbestos fabric	Asbestos blocks
200	80	100	160
400	65	80	150
600	50	60	130
800	35	40	100
1,000	25	30	70
2,000	25	30	40

Table 15-2 **Design values for brake facings**

Facing material	Design coefficient of friction	Permissible unit pressure, psi	
		V of 200 fpm	V of 2,000 fpm
Cast iron on cast iron:			
Dry...............................	0.20		
Oily...............................	0.07		
Wood on cast iron.....................	0.25–0.30	80–100	20–25
Leather on cast iron:			
Dry...............................	0.40–0.50	8–15	
Oily...............................	0.15		
Asbestos fabric on metal:			
Dry...............................	0.35–0.40	90–100	25–30
Oily...............................	0.25		
Molded asbestos on metal, dry...........	0.30–0.35	150–175	30–40

where p is the projected area, in pounds per square inch, and V is the rubbing velocity, in feet per minute.

One prominent manufacturer of mine hoists uses allowable pressures as shown in Table 15-1.

Allowable pressures and coefficients of friction for several brake materials operating under ordinary conditions are given in Table 15-2.

Figure 15-9 shows the variation of the coefficient of friction with the pressure and velocity.

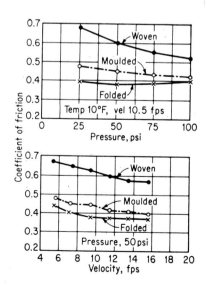

Fig. 15-9 Typical variation of coefficient of friction with pressure and velocity for various types of brake linings.

15-13 Heating of Brakes. Brakes (and many clutches) used continuously often present a difficult heat dissipation problem. The temperature will increase until the rate of dissipation is equal to the rate of heat generation, and the braking area, the radiating surface, and the air circulation must be proportioned so that overheating will be prevented. The heat to be dissipated is

$$H = \frac{fF_r V}{778} \qquad \text{Btu/min} \tag{15-33}$$

For a lowering brake, the heat to be dissipated is

$$H = \frac{Wh}{778} \qquad \text{Btu} \tag{15-34}$$

where W is the weight lowered, in pounds, and h is the total distance, in feet.

The ability of the brake drum to absorb heat is proportional to the mass and to the specific heat of the material. Assuming that all heat generated is absorbed by the brake drum and its supporting flange, the temperature rise t_r, in degrees Fahrenheit, is

$$t_r = \frac{H}{W_r c} \tag{15-35}$$

where W_r is the weight of the brake drum and flange, in pounds, and c is the specific heat of the material, in Btu per pound per degree Fahrenheit (0.13 for cast iron and 0.116 for steel).

Some of the heat generated will be immediately radiated to the air and carried away by air currents, and to a slight degree heat will be conducted away through the contacting parts, so that it is impossible to compute the actual temperature attained by the rim. It is also impossible to compute the time required for the brake to cool, since cooling laws have not been well established, but it is possible to compute the rate of heat loss at a given temperature, and since the heat loss must equal the rate of heat generation, the horsepower capacity of the brake can be determined for any given maximum temperature. The temperature rise should be limited to a maximum of 500°F (350°F for average conditions) for the best grades of asbestos on cast iron, or 150°F for wood or leather. The actual temperature of the drum after a single application of the brake is not so important as the assurance that the rate of cooling equals or exceeds the rate of heat generation.

Since 1 hp is equivalent to 42.4 Btu/min, the heat generated in Btu per minute is

$$H_g = 42.4q(\text{hp}) \tag{15-36}$$

where q is a load factor or the ratio of the actual brake operating time to the total cycle of operation.

Table 15-3   Heat dissipation factors for brakes

Temperature difference t_r	Heat dissipation factor C Btu/(in.²)(min)(°F)	Ct_r Btu/(in.²)(min)
100	0.00060	0.06
200	0.00075	0.15
300	0.00083	0.25
400	0.00090	0.36

The rate of heat dissipation, in Btu per minute, is

$$H_d = Ct_r A_r \qquad (15\text{-}37)$$

where C = heat dissipation factor, Btu/(in.²)(min)(°F) temperature difference

A_r = dissipating surface, in.²

t_r = difference between the temperature of the radiating surface and the surrounding air, °F

Typical values of C, which increases with the temperature difference, may be obtained from Table 15-3.

Equating the heat generated and the heat dissipated, and solving for A_d, the dissipation surface required, in in.², is found to be

$$A_d = \frac{42.4q(\text{hp})}{Ct_r} \qquad (15\text{-}38)$$

The dissipating surface includes all the exposed surface of the brake drum not covered by the friction material, and the solid portion of the center web down to the center of the holes that are usually inserted to lighten the casting. Both sides of the web may be counted provided the air can circulate freely. Dissipation may be increased by providing air passages through the drum and by using fins to increase the surface.

Example. Assume a hoisting engine to be equipped to lower a load of 6,000 lb by means of a band brake having a drum diameter of 48 in. and a width of 8 in. The band width is 6 in., and the arc of contact is 300 deg. The load is to be lowered 200 ft, the hoisting cycle being 1.5 min hoisting, 0.75 min lowering, and 0.3 min loading and unloading.

The generated heat, equivalent to the power developed, is

$$H_g = \frac{6{,}000 \times 200}{0.75 \times 33{,}000} = 48.5 \text{ hp}$$

The load factor is

$$q = \frac{0.75}{1.5 + 0.75 + 0.3} = 0.294$$

Allowing a temperature rise of 300°F, the dissipating surface required is

$$A_r = \frac{42.4 \times 0.294 \times 48.5}{0.25} = 2{,}418 \text{ in.}^2$$

The dissipating surface of the brake drum is

$$A_d = 2(\pi \times 48 \times 8) - (\pi \times 48 \times 6)\tfrac{300}{360} = 1{,}658 \text{ in.}^2$$

which leaves about 760 in.² to be provided for in the flanges and web.

The actual temperature of the drum will vary slightly above and below the 300°F rise assumed, since heat is radiated during the whole cycle but generated during only 29.4 per cent of the cycle. This brake drum will weigh about 600 lb. Hence the temperature rise during the braking operation will be

$$
\begin{aligned}
\Delta t &= \frac{1}{778 W_r c} (Wh - C t_r A_r m \times 778) \\
&= \frac{1}{778 \times 600 \times 0.13} (6{,}000 \times 200 - 0.25 \times 2{,}418 \times 0.75 \times 778) \\
&= 14°\text{F}
\end{aligned}
$$

In this, m is the lowering time in min. This result indicates that the drum temperature will vary about 14°, or 7° above and below the average.

The actual temperature attained by the brake drum and the time required for it to cool cannot be accurately calculated, but the method just outlined may be used for preliminary computations. In the final design of a new brake, heating should be checked by a proportional comparison with a brake already known to give good performance in actual service.

An approximation of the time required for the brake to cool may be made by the formula

$$T_c = \frac{W_r c \log_e t_r}{K a_r} \tag{15-39}$$

where A_r = dissipation surface, in.²
K = a constant varying from 0.4 to 0.8
T_c = cooling time, min

The other symbols have the same meaning as before.

16

Belts

16-1 Flexible Connectors. The uses of belts may be grouped into three general classes: power transmission, conveyor service, and elevator service. Belts for power transmission have been in use for more than a century, competing formerly with rope drives and now with electric drives. Conveyor and elevator belts are a later development.

In Chaps. 11 and 12 gears for connecting shafts for the transmitting of power were discussed. In this chapter and the following chapter belts and chains respectively will be considered. There are several types of belts and chains. The designer will need to consider linear velocities, angular velocities, velocity ratios, pulley diameters, center distances, vibration, noise, environment, space, efficiency, life, and cost in deciding upon the particular drive for a definite application. In some cases more than one type of drive may be used. The final solution will depend upon cost, availability, and maintenance.

Some of the advantages of flat belting are that it can be used with high-speed drives, it can be used in dusty and abrasive environments, it allows long distances between shafts, and it offers long life, high efficiency, low cost, and low maintenance. Flat-belt drives can be made

Table 16-1   **Drive selection data**

Type drive	Max speed ratio per step	Max center distance, ft	Speed of driven shaft, rpm	Pitch line velocity, fpm	Width for equal load, in.
Timing belt..........	12	15	100–10,000	1,000–15,000	3.0
Standard V belt.......	8	15	300–4,000	1,500–8,000	6.7
Leather flat belt.......	...	30	300–4,000	1,500–6,000	
Rubber fabric flat belt	...	30	300–4,000	1,500–6,000	
Standard silent chain...	12	15	75–4,000	1,500–6,000	
Standard roller chain...	10	15	75–4,000	1,500–5,000	1.6
Cast-tooth gear........	6	...	....	0–600	
Cut-tooth gear........	6	...	300–4,000	0–4,000	1.0
Helical and herringbone gear...............	10	...	0–30,000	0–12,000	

Source: E. S. Cheaney, C. L. Paullers, and W. C. Raridan, Mechanical Power Transmission, *Mech. Eng.*, vol. 81, no. 12, December, 1959; W. A. Williams, "Mechanical Power Transmission," Conover-Mast Publications, Inc., New York, 1953; and manufacturers' literature.

competitive with V belts and chains for short-center drives by using pivoted or spring-actuated motor bases. The main disadvantage of flat-belt drives is that the tension must be kept high so as to keep the slip below 2 per cent. This high tension results in high bearing loads and belt stress. Noise is also a disadvantage.

Table 16-1 gives information which should be helpful in selecting the type of drive. It should be understood that the data are not absolute and values may vary. Special drives and types of construction will cause the values to change.

Belts used for power transmission must be strong, flexible, and durable and must have a high coefficient of friction. The most common flat-belt material is oak-tanned leather. Leathers tanned with chestnut bark, vegetable compounds, alum salts, and chrome salts are also used. Fabric, rubber, and balata belts are all commonly used. Steel belts are frequently used in Europe but are very rare in the United States.

16-2   Leather Belts.   The steer hides from which belting is made vary in density and in strength, the fibers near the backbone being shorter but denser than the fibers farther down on the sides, and the belly fibers being the longest and least dense. The backbone leather does not have the highest tensile strength, although it is the best belt material. The best leather is obtained from a strip extending about 15 in. on each side of the backbone and about 54 in. from the tail. Poorer grades of belting are made from strips taken outside, but adjacent to, this region. Short-lap belts are made from the backbone strip, and long-lap belts include the shoulder leather and are therefore inferior to short-lap belts. Double-ply belts are made by cementing two strips of leather together with the hair sides out.

Oak-tanned leather, which is considered to be the standard belt material, is fairly stiff, whereas *chrome leather* is soft and pliable. The manufacturers of chrome leather claim that it grips the pulley better, is more pliable, will not crack when doubled upon itself, and is lighter, stronger, and longer fibered. Because chrome leather is flexible, it is hard to shift with a belt shifter, and hence two-ply belt with the inner ply of chrome leather and the outer ply of oak leather is used where shifting is necessary.

The best leather has an ultimate strength of about 4,000 psi. First-grade belts have a minimum strength of 3,000 psi and an average of 3,750 psi for single ply and 3,500 psi for double ply.

16-3 Fabric and Canvas Belts. This type of belt is made from canvas or cotton duck folded to three or more plies (layers) and stitched together. Woven belts are made of cotton woven to any desired thickness on a loom. Fabric belts are usually impregnated with a filler, largely linseed oil, to make them waterproof and to prevent injury to the fibers. The filler makes the belts rather stiff. These belts are cheap and are used for intermittent service, in hot dry places, and where little attention is given to their upkeep, as in farm machinery. Fabric belts are used to some extent for conveyors.

16-4 Rubber Belts. Rubber belting is made from folded canvas duck with layers of rubber between and surrounding the whole belt structure. The belt is vulcanized under heat and pressure. Such belts are commonly used where exposed to moisture, acids, and alkalies. Neoprene belts also have a high resistance to oil, sun, and heat.

The strength and pulling capacity of the belt are in the duck, the rubber acting only to protect the fibers from internal wear and moisture. For light high-speed service, a lightweight duck, 24 and 26 oz,* is used. For heavy drives in centrifugals, oil fields, and main drives, high tensile strength is required and 36-oz duck is used, inner-stitched to prevent ply separation. Standard rubber belt is usually 32 oz.

Rubber belt is cheaper than leather belt but is affected by light, heat, and oil and deteriorates with age even when in service.

Balata belting is made like rubber belt except that balata gum is substituted for rubber, and it is not vulcanized. Balata does not oxidize and age in the air; it is waterproof, acidproof, and not affected by animal oils or alkalies, but it is seriously affected by mineral oils. When heated,

* Duck is graded by the weight of a strip 36 in. wide by 40 in. long.

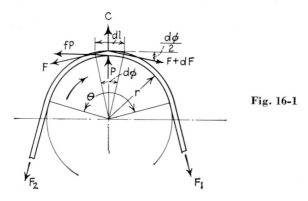

Fig. 16-1

balata becomes soft and sticky, and so it should not be used where the belt temperature exceeds 100 to 120°F. Balata belt is about 25 per cent stronger than rubber belt.

16-5   Belt Design.   There are two methods for the selection of belt drives. The basic or fundamental method is based upon an analysis of the belt tensions caused by the transfer of power from one shaft to the other. The procedure takes into account the strength of the belt material, arc of contact, center distance, effect of centrifugal force, the coefficient of friction, belt thickness and width, and a working stress. An equation which considers these factors is developed in terms of the horsepower and is known as the fundamental belt equation. Adjustments are then made for belt joints, multiple-ply, and inclined belt drives.

The other method of selection is based upon the fundamental equation but uses tables, curves, and nomograms developed by manufacturers and standards associations to take into account the various factors involved and to make solution easier. This method is based upon tests of actual drives including both the belt and the pulleys. In some belt drives, all the quantities needed for the fundamental equation are not available. In this text both methods are used in some cases. Space does not permit a full presentation of the second method. Because materials and construction of belts have been improved the designer should refer to the latest literature of the manufacturer of the particular type of belt.

16-6   Ratio of Belt Tensions.   Referring to Fig. 16-1, consider the forces acting on a short section of belt, dl inches long. These forces are the belt pulls, F and $(F + dF)$, the pulley force P, and the centrifugal force C.

From the conditions of equilibrium

$$P + C - F \sin \frac{d\phi}{2} - (F + dF) \sin \frac{d\phi}{2} = 0 \qquad (16\text{-}1)$$

and

$$(F + dF) \cos \frac{d\phi}{2} - F \cos \frac{d\phi}{2} - fP = 0$$

from which

$$P = \frac{\cos \dfrac{d\phi}{2} \, dF}{f} \tag{16-2}$$

Also

$$C = Ma = \frac{12\rho btr \, d\phi}{g} \frac{v^2}{r} = F_c \, d\phi \tag{16-3}$$

where ρ = weight of the belt, lb/in.³
    b = belt width, in.
    t = belt thickness, in.
    r = pulley radius, in.
    v = belt velocity, fps
    g = 32.2 ft/sec²

By substitution of the values of P and C in Eq. (16-1),

$$\cos \frac{d\phi}{2} \, dF - f(2F + dF) \sin \frac{d\phi}{2} + fF_c \, d\phi = 0$$

In the limit, $\cos d\phi/2$ is unity, and $\sin d\phi/2$ is equal to $d\phi/2$, nearly. By substitution, expansion, and elimination of the product $dF \, d\phi$, since it is negligible, this equation becomes

$$dF - fF \, d\phi + fF_c \, d\phi = 0$$

from which

$$\frac{dF}{F - F_c} = f \, d\phi$$

By integration over the entire arc of contact θ

$$\int_{F_2}^{F_1} \frac{dF}{F - F_c} = f \int_0^\theta d\phi$$

or

$$\log_e \frac{F_1 - F_c}{F_2 - F_c} = f\theta$$

from which

$$\frac{F_1 - F_c}{F_2 - F_c} = e^{f\theta} \tag{16-4}$$

The power transmitted is measured by the difference in tension on the tight and slack sides of the belt, and Eq. (16-4) is more useful in the form

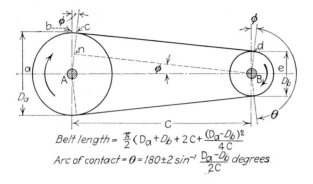

$$\text{Belt length} = \frac{\pi}{2}(D_a + D_b) + 2C + \frac{(D_a - D_b)^2}{4C}$$

$$\text{Arc of contact} = \theta = 180 \pm 2 \sin^{-1} \frac{D_a - D_b}{2C} \text{ degrees}$$

Fig. 16-2 Open belt drive.

$$F_1 - F_2 = (F_1 - F_c) \frac{e^{f\theta} - 1}{e^{f\theta}} \tag{16-5}$$

where F_1 = total tension on the tight side, lb

F_2 = total tension of the slack side, lb

$F_c = 12\rho btv^2/g$ = centrifugal tension, lb

$e = 2.718$

f = coefficient of friction

θ = arc of contact, rad (see Figs. 16-2 and 16-3)

ρ = belt weight, lb/in.[3]: 0.035 for leather; 0.044 for canvas; 0.041 for rubber; 0.040 for balata; 0.042 for single and 0.045 for double woven belt.

The effective belt pull $(F_1 - F_2)$ depends upon the initial tension, i.e., the tension in the belt when the drive is standing idle. The value of the initial tension F_i may be found approximately from the relation

$$F_i = \frac{(F_1^{\frac{1}{2}} + F_2^{\frac{1}{2}})^2}{4} \tag{16-6}$$

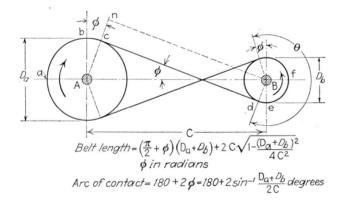

$$\text{Belt length} = \left(\frac{\pi}{2} + \phi\right)(D_a + D_b) + 2C\sqrt{1 - \frac{(D_a + D_b)^2}{4C^2}}$$

$$\phi \text{ in radians}$$

$$\text{Arc of contact} = 180 + 2\phi = 180 + 2\sin^{-1}\frac{D_a + D_b}{2C} \text{ degrees}$$

Fig. 16-3 Crossed belt drive.

Initial tensions should range from 200 to 240 psi for leather belts and from 10 to 12 lb per ply per inch of width for rubber belts.

16-7 Power Transmitted by Belts. The power transmitted by any belt depends on the arc of contact, difference in belt tensions, coefficient of friction, and center distance. The pulley having the lower value of $f\theta$ (usually the smaller pulley) governs the transmitting power. The general expression for power transmitted is

$$\text{hp} = \frac{(F_1 - F_2)v}{550} \tag{16-7}$$

where v is belt velocity, in feet per second.

Combination of Eqs. (16-5) and (16-7) results in

$$\text{hp} = \frac{(F_1 - F_c)v}{550} \frac{e^{f\theta} - 1}{e^{f\theta}} \tag{16-8}$$

The required cross-sectional area of the belt is found by substitution, in this equation, of $12\rho b t v^2/g$ for F_c, and $b t s_w$ for F_1. Hence

$$bt = \frac{550 \text{ hp}}{v(s_w - 12\rho v^2/g)} \frac{e^{f\theta}}{e^{f\theta} - 1} \tag{16-9}$$

where s_w is the maximum working stress in the belt, in pounds per square inch.

The thickness t and the standard widths of belts are given in Table 16-2. In the selection of the proper belt, it is not considered good practice to use single-ply leather belts more than 8 in. wide.

Table 16-2 **Leather belt thickness and minimum pulley diameters**

Weight, plies	Thickness	Tan	Width, in.	Pulley diam., in. Belt velocity, fpm		
				1,000	2,000	3,000 and over
Single..................	$\frac{5}{32}$	Chrome	to 8	2	$2\frac{1}{2}$	3
Light single.............	$\frac{8}{64} \frac{10}{64}$	Oak	to 8	2	$2\frac{1}{2}$	3
Medium single..........	$\frac{10}{64} \frac{12}{64}$	Oak	to 8	3	$3\frac{1}{2}$	4
Heavy single...........	$\frac{12}{64} \frac{14}{64}$	Oak	to 8	4	5	6
Double..................	$\frac{5}{16}$	Chrome	to 12	6	7	8
Light double...........	$\frac{17}{64} \frac{19}{64}$	Oak	to 12	4	5	6
Medium double.........	$\frac{19}{64} \frac{21}{64}$	Oak	to 12	8	10	12
Heavy double...........	$\frac{21}{64} \frac{25}{64}$	Oak	to 12	10	13	14
Medium triple..........	$\frac{30}{64}$	Chrome	to 24	18	24	36
Heavy triple............	$\frac{17}{32}$	Oak	to 24	24	30	36

Note: Standard belt widths increase by $\frac{1}{8}$ in. from $\frac{1}{2}$ to 1 in.; by $\frac{1}{4}$ in. to 3 in.; by $\frac{1}{2}$ in. to 6 in.; by 1 in. to 10 in.; by 2 in. to 56 in.; and by 4 in. to 72 in.

16-8   Working Stress in Belts.   A factor of safety of 10 is usually used with leather belts, making the maximum working stress 300 psi, so that F_1 is $300bt$. Working stresses as low as 250 and as high as 400 psi have been used.

The strength of fabric and rubber belts depends upon the weight of duck used and the number of plies. The ultimate strength is about 300 lb per ply per inch of width for 28-oz duck, 325 lb for 30- and 32-oz duck, and 360 lb for 36-oz duck. Approximate weights of rubber belts are 0.021 lb per ply per inch of width per foot of length for 28-oz duck, 0.024 lb for 32-oz duck, and 0.026 lb for 36-oz duck. Since the weight of the duck is usually unknown, the designer should refer to the manufacturer's horsepower tables before making the final selection. For preliminary computations, the values of effective tensions $(F_1 - F_2)$ given in Table 16-3 can be used, these values agreeing closely with average practice for an arc of contact of 180 deg.

16-9   Belt Joints and Fasteners.   Unless the belts are endless, some type of fastener is required at the joint. The joint is a weak place in the belt, and the permissible working stress should be multiplied by a joint factor from Table 16-4 when computing the power capacity of any belt.

16-10   Multiple-ply Leather Belts.   Theoretical considerations indicate that if the thickness of the belt is doubled by making it two-ply, the horsepower capacity will also be doubled. However, laboratory tests and actual service experience indicate that a two-ply belt will have its capacity reduced to 85 per cent, and a triple-ply belt to 75 per cent, of

Table 16-3   **Effective tensions per inch of belt width**

Number of plies	Belt material				
	Rubber	Balata	Canvas	Woven cotton	Woven camel hair
3	30	48	30	Single 55	Single 60
4	40	65	40	Double 80	Double 85
5	50	80	50	Triple 110	Triple 125
6	60	95	60		
7	70	110	70		
8	80	130	80		
10	100	160	100		

The effective tensions should be increased or decreased $\frac{1}{2}$ per cent for each degree of contact greater or less than 180 deg.

Effective tensions listed are for belts made of 28-oz duck. For 30- and 32-oz duck multiply by 1.10. For 36-oz duck multiply by 1.20.

Table 16-4 **Correction factors for belt joints**

Joint	*Factor*
Cemented by beltmaker	1.00
Cemented	0.98
Wire laced by machine	0.90
Wire laced by hand	0.82
Rawhide laced	0.60
Metal belt hooks	0.35

the capacity computed on the basis of a single-ply belt of the same thickness.

16-11 Inclined Belt Drives. Vertical drives are not so efficient as horizontal drives, since the belt does not cling to the lower pulley. The horsepower transmitted by any inclined belt should be reduced from that of an equivalent horizontal belt by 1 per cent for each degree of inclination exceeding 60 deg.

16-12 Crossed Belts. Crossed belts wider than 8 in. should be avoided. Where wider belts are required, the reversing type of drive with idler pulleys should be used. The capacity of a crossed belt should be reduced to 75 per cent that of an open belt, and if the pulley ratio is 3:1 or more, the reduction should be 50 per cent.

16-13 Belt Slip and Creep. When a belt is transmitting power, there is always a small amount of slip between the belt and the pulleys so that the actual velocity of the belt is slightly less than the surface speed of the driving pulley and slightly greater than that of the driven pulley. Belt slip up to about 3 per cent actually increases the coefficient of friction and under normal conditions $1\frac{1}{2}$ to 2 per cent is present. Creep is a different type of movement of the belt on the pulley surface caused by the fact that any unit length of belt on the tight side decreases in length as it passes around the driving pulley to the slack side. Hence, the belt velocity leaving the driving pulley is slightly less than that of the belt moving onto this pulley. This action adds to the actual belt slip on this pulley. On the driven pulley the creep action is the reverse of that on the driving pulley. Percentage slip usually refers to the combined real slip and creep action.

16-14 Coefficient of Friction for Belts. The coefficient of friction depends on the belt material, the pulley-surface material, the belt slip,

Table 16-5 **Coefficient of friction for belts**

Belt material	Cast iron, steel			Wood	Com-pressed paper	Leather face	Rub-ber face
	Dry	Wet	Greasy				
Leather, oak tanned.....	0.25	0.20	0.15	0.30	0.33	0.38	0.40
Leather, chrome tanned	0.35	0.32	0.22	0.40	0.45	0.48	0.50
Canvas, stitched........	0.20	0.15	0.12	0.23	0.25	0.27	0.30
Cotton, woven..........	0.22	0.15	0.12	0.25	0.28	0.27	0.30
Camel hair, woven......	0.35	0.25	0.20	0.40	0.45	0.45	0.45
Rubber...............	0.30	0.18	....	0.32	0.35	0.40	0.42
Balata................	0.32	0.20	....	0.35	0.38	0.40	0.42

(Pulley material spans the Cast iron/Wood/Compressed paper/Leather face/Rubber face columns)

and the belt speed. Average values of the coefficient as recommended for design purposes are given in Table 16-5.

16-15   Horepower Rating Tables.   The American Leather Belting Association* has adopted a standardized procedure for the determination of the capacity of oak-tanned leather belting. This method of belt selection considers such factors as service conditions, center distance, pulley size, belt speed, and belt thickness. Tables 16-6 to 16-8 give the factors, and the example illustrates the use of these tables.

Example.   Find the width of a light, double-ply, oak-tanned leather belt to be used on a shunt-wound, 25-hp, direct-current motor whose speed is 1,750 rpm and whose fiber pulley diameter is 8 in. The center distance is 15 ft, and the tight side is above. The service is continuous under normal conditions, and the angle of the center line is 70 deg from the horizontal.

Solution.   For a belt speed of 3,660 fpm, Table 16-6 gives 8.4 hp per inch of width. For an 8-in. pulley, a center distance of 15 ft, and tight side above, Table 16-7 modifies this value by the factor 0.72. The service correction factors from Table 16-8 are 1.0 for normal atmospheric conditions, 0.9 for 70-deg angle of center line, 1.2 for fiber pulley, 0.8 for continuous service, and 0.8 for direct-current motor.

$$\text{Width of belt} = \frac{25}{8.4 \times 0.72 \times 1.0 \times 0.9 \times 1.2 \times 0.8 \times 0.8}$$
$$= 5.98 \text{ in.}   \text{(use a standard width of 6 in.)}$$

The student should compare this value with the value obtained from the use of Eq. (16-9).

* Horsepower ratings for oak-tanned flat leather belting adopted **Dec. 7, 1938,** by The American Leather Belting Association.

16-16 Operating Velocity of Belts. An examination of Eq. (16-9) indicates that at high velocities the power-transmitting capacity of the belt is seriously decreased by the centrifugal force; in fact at the higher velocities the power decreases until it is theoretically zero. This does not mean, as is sometimes claimed, that the belt will not transmit power. It simply means that the assumed permissible stresses will be exceeded if power is transmitted. In practice, this decrease in capacity actually occurs but apparently not so rapidly as the equation indicates. In gen-

Table 16-6 **Horsepower per inch of width for oak-tanned leather belts**

Belt speed, fpm	Single-ply		Double-ply			Triple-ply	
	$\frac{11}{64}$ in.*	$\frac{13}{64}$ in.*	$\frac{18}{64}$ in.*	$\frac{20}{64}$ in.*	$\frac{23}{64}$ in.*	$\frac{30}{64}$ in.*	$\frac{34}{64}$ in.*
	Med.	Heavy	Light	Med.	Heavy	Med.	Heavy
600	1.1	1.2	1.5	1.8	2.2	2.5	2.8
800	1.4	1.7	2.0	2.4	2.9	3.3	3.6
1,000	1.8	2.1	2.6	3.1	3.6	4.1	4.5
1,200	2.1	2.5	3.1	3.7	4.3	4.9	5.4
1,400	2.5	2.9	3.5	4.3	4.9	5.7	6.3
1,600	2.8	3.3	4.0	4.9	5.6	6.5	7.1
1,800	3.2	3.7	4.5	5.4	6.2	7.3	8.0
2,000	3.5	4.1	4.9	6.0	6.9	8.1	8.9
2,200	3.9	4.5	5.4	6.6	7.6	8.8	9.7
2,400	4.2	4.9	5.9	7.1	8.2	9.5	10.5
2,600	4.5	5.3	6.3	7.7	8.9	10.3	11.4
2,800	4.9	5.6	6.8	8.2	9.5	11.0	12.1
3,000	5.2	5.9	7.2	8.7	10.0	11.6	12.8
3,200	5.4	6.3	7.6	9.2	10.6	12.3	13.5
3,400	5.7	6.6	7.9	9.7	11.2	12.9	14.2
3,600	5.9	6.9	8.3	10.1	11.7	13.4	14.8
3,800	6.2	7.1	8.7	10.5	12.2	14.0	15.4
4,000	6.4	7.4	9.0	10.9	12.6	14.5	16.0
4,200	6.7	7.7	9.3	11.3	13.0	15.0	16.5
4,400	6.9	7.9	9.6	11.7	13.4	15.4	16.9
4,600	7.1	8.1	9.8	12.0	13.8	15.8	17.4
4,800	7.2	8.3	10.1	12.3	14.1	16.2	17.8
5,000	7.4	8.4	10.3	12.5	14.3	16.5	18.2
5,200	7.5	8.6	10.5	12.8	14.6	16.8	18.5
5,400	7.6	8.7	10.6	12.9	14.8	17.1	18.8
5,600	7.7	8.8	10.8	13.1	15.0	17.3	19.0
5,800	7.7	8.9	10.9	13.2	15.1	17.5	19.2
6,000	7.8	8.9	10.9	13.2	15.2	17.6	19.3

* Average thickness.

For pivoted-base drives, where the tight side of the belt is away from the pivot shaft, do not use these tables.

Table 16-7 **Correction factor for small pulley diameter for oak-tanned leather belts**

Diameter small pulley, in.	Center distance, ft							
	Up to 10 ft		15 ft		20 ft		25 ft and over	
	Tight side		Tight side		Tight side		Tight side	
	Above	Below	Above	Below	Above	Below	Above	Below
2	0.37	0.37	0.38	0.41	0.37	0.43	0.37	0.44
$2\frac{1}{2}$	0.41	0.41	0.43	0.46	0.41	0.48	0.42	0.49
3	0.45	0.45	0.48	0.52	0.48	0.54	0.48	0.55
$3\frac{1}{2}$	0.49	0.49	0.53	0.57	0.53	0.59	0.53	0.60
4	0.53	0.53	0.58	0.63	0.59	0.65	0.59	0.66
$4\frac{1}{2}$	0.56	0.56	0.61	0.66	0.62	0.68	0.62	0.70
5	0.59	0.59	0.65	0.70	0.66	0.72	0.66	0.74
$5\frac{1}{2}$	0.60	0.60	0.66	0.72	0.67	0.74	0.68	0.76
6	0.62	0.62	0.68	0.74	0.69	0.76	0.70	0.78
7	0.64	0.64	0.70	0.76	0.71	0.78	0.72	0.80
8	0.66	0.66	0.72	0.78	0.73	0.80	0.74	0.82
9	0.67	0.67	0.73	0.79	0.74	0.81	0.75	0.83
10	0.68	0.68	0.75	0.81	0.76	0.83	0.77	0.85
11	0.69	0.69	0.76	0.82	0.77	0.84	0.78	0.86
12	0.70	0.70	0.77	0.83	0.78	0.86	0.79	0.88
13	0.71	0.71	0.78	0.84	0.79	0.87	0.80	0.89
14	0.72	0.72	0.79	0.85	0.80	0.88	0.81	0.90
15	0.73	0.73	0.80	0.86	0.81	0.89	0.82	0.91
16	0.74	0.74	0.80	0.87	0.81	0.89	0.82	0.91
17	0.74	0.74	0.81	0.88	0.82	0.90	0.83	0.92
18	0.75	0.75	0.82	0.89	0.83	0.91	0.84	0.93
20	0.75	0.75	0.83	0.90	0.84	0.92	0.85	0.94
22	0.76	0.76	0.84	0.91	0.85	0.93	0.86	0.95
24	0.77	0.77	0.85	0.92	0.86	0.94	0.87	0.96
30	0.79	0.79	0.87	0.94	0.88	0.96	0.89	0.98
36	0.80	0.80	0.88	0.95	0.89	0.98	0.90	1.00

eral factory practice with line shafting and machine belts, moderate velocities of 1,000 to 3,000 fpm are most satisfactory, higher speeds requiring the use of excessively large pulleys. For large power transmission with the pulley bearings mounted on solid foundations, velocities of 5,000 and 6,000 fpm are used, although high velocities always increase the belt troubles.

Belts, like other mechanical equipment, have critical speeds, which are indicated by the belt riding from side to side on the pulley face and by violent flapping of the slack side. The critical speed depends upon the belt tannage, thickness, and center distance and can be remedied by altering the center distance, the load, or the belt velocity.

16-17 Pulley Sizes. The minimum pulley diameters recommended for use with various belts are given in Table 16-2. The pulley face should be about 1 in. wider than the belt for belts up to 12 in., 2 in. wider for 12- to 24-in. belts, and 3 in. wider for belts over 24 in. wide. The face should be *crowned* to assist in keeping the belt centered. Standard pulleys are crowned by making the center diameter $\frac{1}{8}$ in. per foot of face width larger than the edge diameter. Flanges, single or double, may be used on pulleys to retain the belt.

16-18 Short-center Drives. High speed ratios and short center distances decrease the arc of contact on the smaller pulley until the power-transmitting capacity of the drive is seriously reduced. The arc of contact should never be less than 155 deg, and in practice it is found that arcs less than 165 deg require high belt tensions. The proper arc of contact

Table 16-8 **Service correction factors for oak-tanned leather belts**

Select the one appropriate factor from each of the five divisions

Atmospheric condition:
Clean, scheduled maintenance on large drives............................. 1.2
Normal.. 1.0
Oily, wet, or dusty... 0.7
Angle of center line:
Horizontal to 60 deg from horizontal...................................... 1.0
60 to 75 deg from horizontal.. 0.9
75 to 90 deg from horizontal.. 0.8
Pulley material:
Fiber on motor and small pulleys.. 1.2
Cast iron or steel.. 1.0
Service:
Temporary or infrequent... 1.2
Normal.. 1.0
Important or continuous... 0.8
Peak loads:
Steady belt loads as obtained with steam engines, turbines, diesel and multi-cylinder gas engines, fans, centrifugal pumps, and steady line shaft loads..... 1.0
Jerky belt loads as obtained with large induction motors; compensator-started, shunt-wound, direct-current motors; single-cylinder gas engines; reciprocating machines; and machines developing series of peak loads, such as compressors, rock crushers, and punch presses.. 0.8
Shock and reversing belt loads on all motors 10 hp and under, all cross-the-line start motors, wound-rotor (slip-ring) motors, synchronous motors, and reversing loads such as printing presses, elevator service, and laundry washers..... 0.6

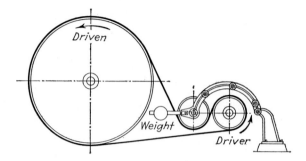

Fig. 16-4

may be obtained with short-center distances by the use of spring-loaded or gravity idlers, as shown in Figs. 16-4 and 16-5. The idler should be located so that the wrap on the small pulley is 225 to 245 deg, and the clearance between the idler and the pulley should be not more than $1\frac{1}{2}$ to 2 in. The idler should always be flat-faced, never crowned, and should always be located next to the small pulley (whether driving or driven) and on the slack side. A fixed idler is suited only for drives in which the small pulley is driven, and in which the drive is steady. Spring-loaded idlers are not so satisfactory as gravity idlers, since the increase in the slack on the belt when loaded relieves the spring tension when it is most needed.

With the use of properly designed and balanced idlers, it is possible, but not always advisable, to use extremely short center distances. Many operating men are prejudiced against short drives and idlers in any form because of the troubles they have experienced, such as short belt life, insufficient slack in belt to permit the belt stretch to absorb shock loads, and the necessity of frequent adjustments. However, the decreased cost of the original belt and the saving in space may often justify the use of these drives.

16-19   Rockwood Belt Drive.   The Rockwood drive, or pivoted mount, lessens some of the objections common to idlers. In this drive, a part of the weight of the driving motor is balanced against the belt pulls as shown in Fig. 16-6. The leverage and the belt tensions may be adjusted

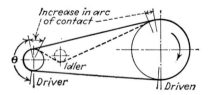

Fig. 16-5   Effect of an idler on a short-center drive.

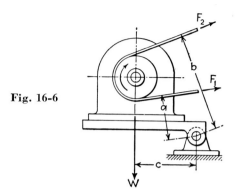

Fig. 16-6

to any desired value by shifting the motor along the hinged support. When the motor has been properly balanced to maintain the required belt pull at the heaviest peak loads, no further adjustment is necessary. Belt stretch, whether caused by effective belt tension or by centrifugal force, is taken up by the movement of the motor about the hinge point. An important feature of this drive is that the pressure between the belt and pulley is fixed by the motor weight and its leverage, and the sum of the belt tensions decreases as the power load transmitted is decreased. Hence at all loads, except the extreme peak loads, the belt tension is much less than that required with an ordinary open belt drive. This increases the life of the belt.

Whenever possible, the tight or pulling side of the belt should pass between the motor pulley and the hinge point. This will reduce the effective moment of the belt pulls about the hinge point and permit the motor to be mounted closer to the hinge than would be possible if the slack side were placed nearest the hinge.

Once balanced, the Rockwood drive limits the maximum load that may be imposed on the belt and hence on the motor. At the maximum load the belt will slip and protect the motor from further overloading. Of course this slippage may shorten the life of the belt, but it prevents the motor from burning out. The action of the drive may be better indicated by an example.

Example. Assume that, at maximum load, F_1 is 300 lb, F_2 is 100 lb, and the effective tension is 200 lb. The motor weight is 150 lb and the distances a and b are 3 and 10 in., respectively. See Fig. 16-6.

To determine the required motor moment arm, take moments about the hinge point when maximum power is being transmitted.

$$F_1 a + F_2 b = Wc$$

and

$$c = \frac{F_1 a + F_2 b}{W} = \frac{300 \times 3 + 100 \times 10}{150} = 12.67 \text{ in.}$$

When operating at one-half the maximum load,

$$F_1 - F_2 = 100$$

and

$$3F_1 + 10F_2 = 150 \times 12.67 = 1,900$$

from which

$$F_1 = 223 \text{ lb}$$

and

$$F_2 = 123 \text{ lb}$$

When operating under no load, the belt tensions will be practically equal. Hence

$$F_1 = F_2$$

and

$$F_1(3 + 10) = 1,900$$

from which

$$F_1 = F_2 = 146 \text{ lb}$$

The total belt pull or load on the motor bearing ($F_1 + F_2$) will be 400 lb at maximum load, 346 lb at half load, and 292 lb when idling. With an open belt drive with fixed centers, the sum of the belt tensions will remain practically constant at all loads. Hence the bearing load will always be 400 lb, a value only reached with the Rockwood drive when the belt is operating at its maximum load.

16-20   V-belt Drives.   A V belt runs over grooved pulleys or sheaves. However, when speeds do not exceed 6,000 fpm, one pulley of the drive may be flat faced as used with flat belts. The sheave may be single or multiple grooved. The wedging action against the sides of the grooves increases the friction and results in a greater amount of power transmission in a smaller space. Each manufacturer has one or more types of construction, but in general the belt consists of a central layer of cords to carry the load, surrounded by rubber to transmit the pressure of the cords to the side walls of the belt, and both rubber and cords are enclosed in an elastic wearing cover. A typical section construction is shown in Fig. 16-7(a). Most manufacturers produce at least three grades of standard V belts: "common," premium, and super, each grade having a different horsepower rating per belt.

The advantages of V-belt drives are (1) the belts cannot come out of the grooves; (2) the wedging action permits a smaller arc of contact; (3) shorter center distances can be used; (4) the gripping action results in lower belt tension; (5) the drives are quiet at high speeds; (6) the drive is capable of absorbing high shock; and (7) standardization results in better initial installation and replacement.

The absolute minimum center distance is $D + d/2$ plus minimum

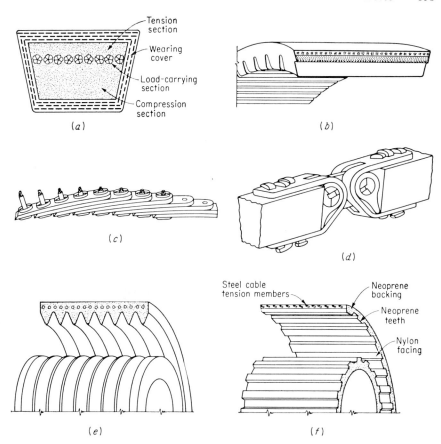

Fig. 16-7 (*a*) Typical V belt section. (*b*) Cogged or notched V belt. (*c*) Link-type or quick detachable V belt. (*d*) Open-end V belt. (*e*) V-ribbed belt. (*f*) Positive drive or timing belt.

allowance for installation. The center distance should be at least equal to the diameter of the larger sheave. The preferred minimum center distance is a distance equal to the sum of the pitch diameters.

Figure 16-7(*b*) shows a cogged or notched V belt which provides transverse rigidity, lengthwise flexibility, and increased heat dissipation.

A link-type or quick-detachable V belt is shown in Fig. 16-7(*c*). This belt simplifies the assembly of multiple drives where it is difficult to match the length of each belt, especially of large fixed-center drives. This type should not be used for speeds over 5,000 fpm and is suited for continuous heavy-duty service.

Figure 16-7(*d*) shows an open-end V belt which is obtainable in rolls. The capacity of this belt is about 75 per cent of that of an endless

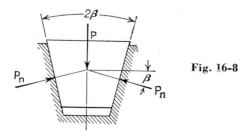

Fig. 16-8

belt. This belt may be used for emergency replacement and for drives where endless V belts are not physically suitable. It is obtainable in B, C, and D sections (see below) and should not be run at high speeds.

V-ribbed belts as shown in Fig. 16-7(e) combine the strength and simplicity of the flat belt with the more nearly positive action of a V belt.

A recent addition to the V-belt family is the narrow or wedge V belt. It is similar to the standard V belt but uses a narrower cross section to give greater wedging action. With improved materials, high speeds, and this increased wedging action approximately three times the horsepower capacity may be carried in the same space as required for standard V belts. These belts may be run at speeds up to 7,000 fpm.

There are five belt sections for the older standard regular- and deep-groove V belts: A, $\frac{1}{2}$ in. outside width by $\frac{5}{16}$ in. depth; B, $\frac{21}{32}$ by $\frac{13}{32}$ in.; C, $\frac{7}{8}$ by $\frac{17}{32}$ in.; D, $1\frac{1}{4}$ by $\frac{3}{4}$ in.; and E, $1\frac{1}{2}$ by $\frac{29}{32}$ in. The newer narrow belts have three sections: 3V, $\frac{3}{8}$ by $\frac{5}{16}$ in.; 5V, $\frac{5}{8}$ by $\frac{17}{32}$ in.; and 8V, 1 by $\frac{29}{32}$ in. Both groups are available in standard lengths.

The groove dimensions for the older type are given in Standard Specifications for Multiple V-Belt Drives (ASA B55.1-1961) and for the newer narrow belts in Engineering Standard Specifications for Drives Using Narrow V-Belts (Mechanical Power Transmission Association and the Rubber Manufacturers Association, January, 1962). The following are groove angles for the older sections: A and B sections—34 and 38 deg; C and D—34, 36, and 38 deg; and E—36 and 38 deg. Grooves for the newer sections: 3V—36, 38, 40, and 42 deg; 5V—38 and 40 deg; and 8V—38, 40, and 42 deg.

From Fig. 16-8, the normal force on the groove face is

$$P_n = \frac{P}{2 \sin \alpha} \tag{16-10}$$

Then the tractive force is

$$F = 2fP_n = \frac{2fP}{2 \sin \alpha} = f_e P \tag{16-11}$$

where f = coefficient of friction

f_e = equivalent or effective coefficient of friction

$= \dfrac{f}{\sin \alpha}$

The design coefficient of friction on flat surfaces is taken as 0.13, making f_e equal to 0.45 for 34-deg, 0.42 for 36-deg, 0.40 for 38-deg, 0.38 for 40-deg, and 0.36 for 42-deg grooves.

The regular power-transmission equations for belts may be applied to V belts when f_e is substituted for the regular coefficients of friction f. For the older standard V belts, the weight of the belt is approximately $0.235b^2$ lb/ft, and the maximum permissible working load $145b^2$ lb, where b is the belt width at the outer surface. Using these values, Eq. (16-5) becomes

$$F_1 - F_2 = nb^2(145 - 0.0073v^2)\,\frac{e^{f_e\theta} - 1}{e^{f_e\theta}} \qquad (16\text{-}12)$$

where n = number of individual belts in the drive

d = pitch diameter of the small pulley, in.

v = belt velocity, fps

When the driven pulley is comparatively large, a grooved surface may not be necessary. This will be true if $e^{f\theta}$ for the large flat-faced pulley exceeds $e^{f_e\theta}$ for the small grooved pulley.

Since V belts are made endless and in standard lengths, the pulley center distances must be arranged to suit the available belt lengths. In computing belt lengths, the pitch diameters of the pulleys should be used. These may be obtained from the outside diameters of grooved or flat pulleys by making the correction for belt thickness. Belt length is given by the equation

$$L = 2C + 1.57(D + d) + \frac{(D - d)^2}{4C} \qquad (16\text{-}13)$$

where C = center distance, in.

D = pitch diameter of the large pulley, in.

d = pitch diameter of the small pulley, in.

Equation (16-12) may be used for a rough estimate of the power transmitted. The V-belt industry has developed horsepower rating formulas derived from a stress-fatigue theory. In many years of V-belt testing, a correlation was found between peak tensions and average number of tension cycles to failure. Tests which were run on "standard drives" with equal diameter pulleys, ideal center distances, belt lengths, etc., provide data for obtaining a curve similar to an endurance curve for steel. The effect of variations in horsepower transmitted, sheave diameter, center distance, belt length, speed ratio, and arc of contact were determined. A recent publication by E. O. Michael discusses fatigue effects in V belts.* Horsepower tables, service factors for various installations, correction factors, and other design information are available in manufacturers' design manuals.†

16-21 Positive Drive Belts. A recent development in mechanical power transmission is the positive drive or timing belt shown in Fig. 16-7(f). The molded teeth of the belt make positive engagement with the mating sprocket teeth in a smooth rolling manner similar to the action of a gear. The action is also similar to a silent chain. The transmission of power does not depend upon friction. The timing belt combines the advantages of the chain and gear with those of a belt. Up to 600 hp may be transmitted. Speeds from 100 to 10,000 fpm are obtainable.

The advantages are (1) positive grip eliminates slippage and speed variation; (2) low initial tension; (3) uniform speed due to the absence of chordal effect; (4) high horsepower to weight ratio; (5) low maintenance costs; (6) wide speed range; (7) elimination of friction, low initial tension, and thin belt construction prevents heat build-up; (8) low noise level; (9) small sprockets, short center distances, narrow belts, and high capacity reduce space requirements; (10) smooth running due to minimum backlash; (11) economical drive; and (12) back of belt may be run over a flat-faced pulley.

The belts are available in the following pitches and widths: $\frac{1}{5}$-in. pitch with widths from $\frac{1}{4}$ to $\frac{3}{4}$ in.; $\frac{3}{8}$-in. pitch with $\frac{1}{2}$- to $1\frac{1}{2}$-in. widths; $\frac{1}{2}$-in. pitch with $\frac{3}{4}$- to 6-in. widths; $\frac{7}{8}$-in. pitch with 2- to 8-in. widths; and $1\frac{1}{4}$-in. pitch with 2- to 10-in. widths. The number and size of increments of width depend on the pitch.

The pulleys or sprockets are made of cast iron, steel, light metals, and nonmetallic materials. At least one pulley (preferably the smaller) is flanged. Both pulleys should be flanged when the center distance is greater than eight times the smaller pulley diameter.

The literature of manufacturers‡ provides horsepower rating tables and other information for the design of the drives. The design procedure is similar to that for V belts.

16-22 Speed Changers. There are many devices for speed reducing and increasing. Manual, remote, automatic, and mechanical controls are

---

* E. O. Michael, How Fatigue Affects V-belt Life Expectancy, Reprint from February–March, 1962, issue of *Power Drive Engineering*, Huebner Corp., Cleveland, Ohio.

† Dayton Industrial Products Co., Melrose Park, Ill.; Dodge Manufacturing Corp, Mishawaka, Ind.; The Gates Rubber Co., Denver, Colo.; Worthington Corp., Oil City, Pa.

‡ Worthington Corp., Oil City, Pa.; Morse Chain Co., Ithaca, N.Y.

used. The speed variation is obtained by the use of V belts on variable pitch sheaves whose pitch diameters are changed by adjusting movable flanges. The pulley may be fixed directly to the motor pulley shaft. Enclosed units may be driven by a motor or connected to a drive shaft with a power take-off from the unit. Either fixed- or variable-speed ratios are obtainable.

Recently a speed changer employing a triple reduction with timing belts was placed on the market.

17

Power Chains

17-1 Power Chains. Various types of chains are used for the transmission of power. For low speeds where the loads are not great, detachable chain (Fig. 17-1) is used. The links are usually made of malleable cast iron or manganese steel, cast in one piece with no separate bushings or pins at the joints. Because of the open construction at the joint, links are readily removed from, or added to, the chain. For conveyor and elevator service, this type of chain is furnished with special flanges and other attachments to provide connections to the other parts of the conveyors. Detachable chain is suitable for power transmission up to speeds of 350 fpm and 20 hp.

Fig. 17-1 Ewart detachable chain.

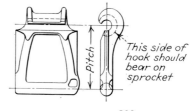

This side of hook should bear on sprocket

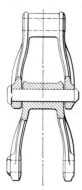

Fig. 17-2   Closed-end pintle chain.

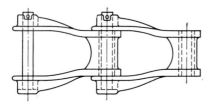

Fig. 17-3   Bushed pintle chain.

When the chains are exposed to grit, closed-end pintle chains (Fig. 17-2) are preferable to the detachable chains. The two types are interchangeable. The pins at the joints are either riveted over or made removable so that chain links may be removed. Closed-end pintle chains are slightly stronger than detachable chains and are suitable for speeds up to 450 fpm and 40 hp.

Bushed closed-end pintle chains (Fig. 17-3) are made of malleable iron with casehardened steel pins and bushings, the pins being either riveted over or made removable. The bushed chain is suitable for power transmission under severe service conditions at speeds up to 450 fpm and 40 hp.

All of the chains mentioned are cast and operate on cast tooth sprockets. Their pitch is only approximate. Where more accurate chains are required, roller and silent chains are used. Roller chains are made

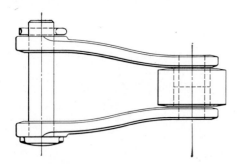

Fig. 17-4   Malleable-iron roller chain.

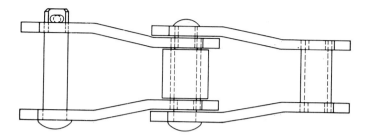

Fig. 17-5 Steel roller chain, offset links.

with steel side bars with hardened steel bushings and pins. In the cheaper grades, the rollers are malleable iron or cast steel. This type (Fig. 17-4) is suitable for power transmission at speeds up to 600 fpm. The chains shown in Figs. 17-5 and 17-6 have alloy-steel side plates and heat-treated steel rollers and are suitable for heavy duty up to 75 hp and shock loading at speeds up to 700 fpm.

The highest grade roller chains, usually referred to as finished-steel roller chains (Figs. 17-7 through 17-9) are used for accurate timing, for the transmission of power, and as the basic members for many types of conveyors. Dimensions of these chains are shown in Table 17-1. The individual parts of these chains—pins, bushings, rollers, and link plates—

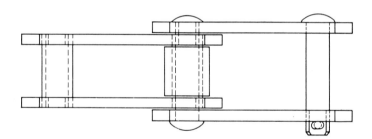

Fig. 17-6 Steel roller chain, straight links.

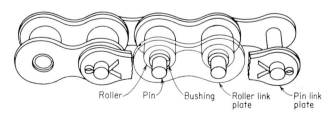

Roller   Pin   Bushing   Roller link plate   Pin link plate

Fig. 17-7 Construction of roller chain. (*Courtesy Diamond Chain and Manufacturing Company.*)

Fig. 17-8   Transmission roller chain. (*Courtesy Diamond Chain and Manufacturing Company.*)

are made of special alloy steels, selected, processed, and heat-treated so that each part may function with maximum efficiency in its own field as a tension member or as a bearing member. It is characteristic of roller chains that no single part is required to resist both tension and wear.

17-2   Speed and Sprocket Limits.*   Linear speeds of these chains are not limited except by the number of teeth and the rotative speed of the smallest sprocket in the drive. The shorter the pitch, the higher the permissible rotative speed, with speeds as high as 8,800 rpm approved for $\frac{1}{4}$-in.-pitch chains.

Sprockets with fewer than 16 teeth may be used for relatively slow speeds, but 18 to 24 teeth are recommended as the minimum for the higher speeds. Ordinarily, sprockets with fewer than 25 teeth, running at speeds above 500 rpm, should be heat-treated to have a tough, wear-resistant surface of 35 to 45 Rockwell C hardness.

If the speed ratio requires a larger sprocket with more than 120 teeth, or with more than eight times the number of teeth on the small sprocket, it is advisable to make the desired speed change in two or more steps. Comparatively small numbers of teeth account for a faster rate of wear in the articulating joints of the chain, and comparatively large numbers of teeth allow the chain to top the teeth before the chain is elongated enough to be unsuitable for further service on sprockets with

---

*Material on finished-steel roller chains, chain sprockets, and chain selection furnished by H. G. Taylor, Diamond Chain and Manufacturing Company, Indianapolis, Ind.

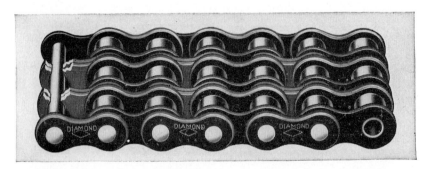

Fig. 17-9   Multiple-strand roller chain. (*Courtesy Diamond Chain and Manufacturing Company.*)

Table 17-1 **Finished steel roller chains**

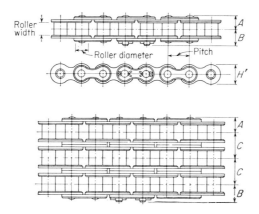

Chain No.		Roller			Roller link plate		Transverse dimensions			Tensile strength, lb/ strand	Weight, lb/ft	Recommended maximum rpm		
ASA	Dia-mond	Pitch p	Width w	Diam-eter d	Pin diam-eter	Thick-ness t	Height H'						Teeth	
								A	B	C			12	18	24
25	89	$\frac{1}{4}$	$\frac{1}{8}$	0.130	0.0905	0.030	0.226	0.149	0.188	0.254	875	0.084	5,800	7,800	8,880
35	82	$\frac{3}{8}$	$\frac{3}{16}$	0.200	0.141	0.050	0.344	0.224	0.290	0.400	2,100	0.21	2,380	3,780	4,200
41	65*	$\frac{1}{2}$	$\frac{1}{4}$	0.306	0.141	0.050	0.383	0.256	0.315	...	2,000	0.26	1,750	2,725	2,850
40	66	$\frac{1}{2}$	$\frac{5}{16}$	0.312	0.156	0.060	0.452	0.313	0.358	0.563	3,700	0.41	1,800	2,830	3,000
50	449*	$\frac{5}{8}$	$\frac{3}{8}$	0.400	0.200	0.080	0.594	0.384	0.462	...	6,100	0.68	1,300	2,030	2,200
50	148	$\frac{5}{8}$	$\frac{3}{8}$	0.400	0.200	0.080	0.545	0.384	0.462	0.707	6,600	0.65	1,300	2,030	2,200
60	433	$\frac{3}{4}$	$\frac{1}{2}$	0.469	0.234	0.094	0.679	0.493	0.567	0.892	8,500	0.99	1,025	1,615	1,700
80	434	1	$\frac{5}{8}$	0.625	0.312	0.125	0.903	0.643	0.762	1.160	14,500	1.73	650	1,015	1,100
100	470	$1\frac{1}{4}$	$\frac{3}{4}$	0.750	0.375	0.156	1.128	0.780	0.910	1.411	24,000	2.51	450	730	850
120	472	$1\frac{1}{2}$	1	0.875	0.437	0.187	1.354	0.977	1.123	1.796	34,000	3.69	350	565	650
140	474	$1\frac{3}{4}$	1	1.000	0.500	0.220	1.647	1.054	1.219	1.929	46,000	5.00	260	415	500
160	478	2	$1\frac{1}{4}$	1.125	0.562	0.250	1.900	1.250	1.433	2.301	58,000	6.53	225	360	420
200	480	$2\frac{1}{2}$	$1\frac{1}{2}$	1.562	0.781	0.312	2.275	1.533	1.850	2.800	95,000	10.65	170	260	300

Courtesy of Diamond Chain and Manufacturing Company.

* Not made in multiple strands. All dimensions are in inches.

fewer teeth. The rate of wear and resultant elongation of chains vary directly with the angle of articulation in the pin-bushing joint. This angle is inversely proportional to the number of teeth and is equal to 360 deg divided by the number of teeth.

The American Standards Association tooth profile (Fig. 17-10) allows the chain to adjust itself to a larger effective pitch polygon as the chain elongates. To determine the allowable elongation, it is assumed that the diameter of the maximum effective pitch circle followed by the elongated

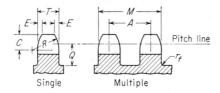

Single   Multiple

Fig. 17-10 Standard tooth section profile. (*Courtesy Diamond Chain and Manufacturing Company.*)

n = number of strands in multiple-strand chain

p = pitch of chain, in.

t = nominal thickness of link plate

T = thickness of tooth

$T = 0.93W - 0.006$, for single-strand chain

$T = 0.90W - 0.006$, for double- and triple-strand chain

$T = 0.88W - 0.006$, for four- and five-strand chain

$T = 0.86W - 0.006$, for six-strand chain or wider

r_f = max fillet radius with max hub diameter = $0.04p$

M = overall width of tooth-profile section

W = inside width of single-strand roller chain or length of roller

$C = 0.5p$

$E = 0.125p$

$R = 1.063p$

$Q = 0.5p$ for link plates with figure-of-eight contour

$A = W + 4.15t + 0.003$

$M = A(n - 1) + T$

Tolerance on T or

$$M = \pm(0.02W + 0.002)$$

chain is equal to the nominal pitch diameter plus the diameter of the chain roller. The ASA roller diameter is five-eighths of the pitch. Based upon these figures, the theoretical allowable elongation, in percentage, will be approximately 200 divided by the number of teeth. For 200 teeth, therefore, the percentage of allowable elongation is 1 per cent; for 133 teeth, $1\frac{1}{2}$ per cent; for 100 teeth, 2 per cent; and for 57 teeth, $3\frac{1}{2}$ per cent. Chains which have elongated $3\frac{1}{2}$ per cent should be replaced.

In combining the effects of small and large numbers of teeth, it may be concluded that the service life of a chain on 12- and 60-tooth sprockets, for example, will be about the same as for 24 and 120 teeth if the rotative speeds and chain pulls were the same in each case, the greater wear and elongation due to the 12-tooth sprocket being counterbalanced by the greater allowable elongation on the 60-tooth sprocket. A multiple chain, four strands wide on the smaller sprockets, would have practically the same power capacity as a double-strand chain of the same pitch on the larger sprockets. The four-strand drive would require less space radially, but the double-strand drive would have all other advantages—less axial width, smoother action, quieter operation, less load on shaft bearings because there is only half the chain pull, higher permissible rotative speed, equivalent or slightly lower cost, and a lower rate of wear on the larger number of teeth, which offsets the lower allowable percentage of elongation.

Idler sprockets may be meshed with either face of standard roller chains, to take up slack, to guide the chain clear of obstructions, to obtain directions of rotation opposite to that of the driver sprocket, or to provide a longer arc of chain wrap on another sprocket. Idlers should not turn faster than the rotative speeds recommended as maximum for other sprockets with the same number of teeth. It is desirable that the idlers have at least three teeth in mesh with the chain, preferably with the idle span of the chain. Idler sprockets should have accurately machined teeth, the same as other sprockets. Plain disks, plain or grooved cylinders, and flanged pulleys should not be used, especially those which would allow contact with the edges of the link plates.

17-3   Chain Selection.   Horsepower ratings for two-sprocket drives are based upon the number of teeth and the rotative speed of the smaller sprocket, either driver or follower. The pin-bushing bearing area, as it affects the allowable working load, is the important factor for medium- and high-speed drives. Because of the relationship between lightness and tensile strength of roller chains, the effect of centrifugal force does not need to be considered. Even at the unusual speed of 6,000 fpm, the added tension in the chain due to centrifugal force is only 3 per cent of the ultimate tensile strength. The permissible working load is based upon a static bearing pressure of 5,100 psi.

For standard roller chain, the pin diameter is five-sixteenths of the pitch, or one-half of the roller diameter, and the bushing length is seven-eighths of the pitch, or roller length (five-eighths of pitch) plus two times the link plate thickness (one-eighth of pitch). Therefore

$$A = \frac{5p}{16}\frac{7p}{8} = 0.273p^2 \tag{17-1}$$

The horsepower at normal speeds and medium numbers of teeth is

$$\text{hp} = \frac{F_w \times V}{33,000} \tag{17-2}$$

where F_w = permissible load, lb

$V = \dfrac{Tnp}{12}$ = chain velocity, fpm

T = number of teeth
n = speed of sprocket, rpm
p = chain pitch, in.

Table 17-2   **Horsepower ratings for single-strand roller-chain drives**

ASA No. 25, $\frac{1}{4}$-in. pitch, Diamond No. 89

Teeth	rpm of sprocket											
	200	400	600	800	1,200	1,800	2,400	3,000	4,000	5,000	6,000	8,000
12	0.10	0.19	0.27	0.33	0.45	0.58	0.67	0.73	0.78	0.73		
15	0.12	0.24	0.35	0.43	0.60	0.79	0.94	1.04	1.14	1.20	1.14	
18	0.15	0.29	0.42	0.53	0.74	0.98	1.17	1.32	1.49	1.56	1.56	
21	0.18	0.36	0.50	0.62	0.86	1.16	1.38	1.57	1.77	1.86	1.88	1.66
24	0.21	0.40	0.57	0.71	0.98	1.31	1.58	1.78	2.02	2.12	2.14	1.89

ASA No. 35, $\frac{3}{8}$-in. pitch, Diamond No. 82

Teeth	rpm of sprocket											
	200	400	800	1,200	1,600	2,000	2,400	2,800	3,200	3,600	4,000	4,500
12	0.34	0.60	1.01	1.31	1.53	1.66	1.72	1.73				
15	0.43	0.78	1.35	1.78	2.12	2.37	2.54	2.65	2.70	2.69		
18	0.52	0.96	1.65	2.21	2.65	2.98	3.24	3.43	3.52	3.57	3.55	
21	0.61	1.12	1.95	2.61	3.14	3.53	3.86	4.08	4.22	4.28	4.28	
24	0.70	1.28	2.22	2.98	3.57	4.04	4.38	4.65	4.81	4.86	4.87	4.75

ASA No. 40, $\frac{1}{2}$-in. pitch, Diamond No. 66

Teeth	rpm of sprocket											
	200	400	600	800	1,000	1,200	1,600	1,800	2,000	2,400	2,800	3,200
12	0.77	1.34	1.81	2.16	2.46	2.71	2.99	3.07	3.10			
15	0.99	1.76	2.40	2.93	3.38	3.77	4.32	4.52	4.67	4.81		
18	1.20	2.15	2.94	3.63	4.21	4.71	5.48	5.76	5.97	6.27	6.35	
21	1.41	2.52	3.47	4.27	4.97	5.57	6.50	6.86	7.13	7.50	7.63	
24	1.60	2.88	3.96	4.87	5.67	6.35	7.40	7.80	8.12	8.51	8.68	8.57

ASA No. 50, $\frac{5}{8}$-in. pitch, Diamond No. 449 (single), No. 148 (multiple)

Teeth	rpm of sprocket											
	100	200	300	400	600	800	1,000	1,200	1,400	1,600	1,800	2,200
12	0.80	1.44	1.99	2.48	3.26	3.86	4.3	4.6	4.8			
15	1.02	1.87	2.61	3.27	4.39	5.31	6.0	6.8	7.0	7.3	7.5	
18	1.23	2.27	3.19	4.01	5.41	6.58	7.5	8.3	8.9	9.4	9.7	
21	1.45	2.66	3.75	4.70	6.38	7.77	8.9	9.8	10.6	11.1	11.6	11.9
24	1.65	3.05	4.27	5.37	7.28	8.85	10.2	11.2	12.1	12.6	12.1	13.6

Table 17-2 **Horsepower ratings for single-strand roller-chain drives** (*Continued*)

ASA No. 60, $\frac{3}{4}$-in. pitch, Diamond No. 433

Teeth	rpm of sprocket											
	50	100	200	300	400	600	800	1,000	1,200	1,400	1,600	1,800
12	0.73	1.34	2.41	3.30	4.05	5.2	6.1	6.6	6.9			
15	0.92	1.72	3.14	4.34	5.39	7.1	8.5	9.5	10.2	10.6		
18	1.12	2.10	3.82	5.31	6.63	8.9	10.6	12.0	13.0	13.7	14.1	
21	1.31	2.46	4.49	6.24	7.80	10.4	12.5	14.1	15.4	16.3	16.9	
24	1.50	2.80	5.11	7.12	8.90	11.9	14.3	16.1	17.6	18.6	19.2	19.5

ASA No. 80, 1-in. pitch, Diamond No. 434

Teeth	rpm of sprocket											
	50	100	150	200	300	400	500	600	700	800	1,000	1,160
12	1.68	3.07	4.28	5.3	7.2	8.7	9.8	10.7	11.4			
15	2.14	3.95	5.57	7.0	9.6	11.8	13.6	15.1	16.3	17.3		
18	2.59	4.81	6.79	8.6	11.8	14.5	16.9	18.9	20.5	21.9	24.0	
21	3.03	5.62	7.96	10.1	13.9	17.1	19.9	22.3	24.3	26.0	28.5	
24	3.46	6.43	9.10	11.5	15.8	19.5	22.7	25.4	27.7	29.6	32.5	33.9

ASA No. 100, $1\frac{1}{4}$-in. pitch, Diamond No. 470

Teeth	rpm of sprocket											
	25	50	100	200	300	400	500	650	700	750	800	870
12	1.72	3.19	5.8	9.9	13.0	15.6	17.2					
15	2.19	4.10	7.5	13.1	17.5	21.3	24.0	27.2	28.1			
18	2.55	4.97	9.1	16.0	21.6	26.6	30.2	34.5	35.7	36.8		
21	3.08	5.80	10.7	18.9	25.5	31.4	35.7	40.9	42.3	43.5	44.6	
24	3.52	6.62	12.2	21.5	29.2	35.4	40.5	46.5	48.1	49.5	50.6	52.0

ASA No. 120, $1\frac{1}{2}$-in. pitch, Diamond No. 472

Teeth	rpm of sprocket											
	25	50	75	100	150	200	250	300	350	400	500	600
12	2.90	5.4	7.6	9.6	13.2	16.2	18.7	21.0	22.8	24.3		
15	3.71	6.9	9.8	12.5	17.3	21.6	25.3	28.6	31.4	33.9	38.0	
18	4.74	8.4	12.0	15.3	21.3	26.6	31.3	35.4	39.2	42.4	47.9	
21	5.24	9.9	14.0	17.9	24.9	31.2	36.8	41.7	46.2	50.0	56.7	61.7
24	5.99	11.3	16.0	20.4	28.5	35.7	41.9	47.6	52.6	57.1	64.6	70.3

Table 17-2   **Horsepower ratings for single-strand roller-chain drives**
(*Continued*)

ASA No. 140, 1¾-in. pitch, Diamond No. 474

Teeth	rpm of sprocket											
	20	30	50	100	150	200	250	300	350	400	450	475
12	3.72	5.4	8.4	14.8	20.1	24.5	28.1	31.0				
15	4.73	6.9	10.8	19.3	26.6	32.8	38.2	42.8	46.7			
18	5.73	8.3	13.1	23.7	32.7	40.5	47.3	53.2	58.4	62.9		
21	6.70	9.7	15.3	27.7	38.4	47.6	55.7	62.8	69.0	74.5	79.0	
24	7.65	11.1	17.5	31.7	43.7	54.3	63.6	71.6	78.7	84.8	89.9	92.4

ASA No. 160, 2-in. pitch, Diamond No. 478

Teeth	rpm of sprocket											
	10	20	40	80	120	160	200	240	280	320	360	400
12	2.9	5.5	10.1	18.0	24.6	30.1	34.8	38.6				
15	3.7	7.0	13.0	23.5	32.4	40.2	47.0	52.9	58.0	62.4		
18	4.4	8.5	15.8	28.7	39.7	49.5	58.1	65.7	72.4	78.3		
21	5.2	9.9	18.5	33.6	46.7	58.1	68.3	77.5	85.5	92.5	99	
24	5.9	11.3	21.1	38.4	53.5	66.5	78.0	88.3	97.4	105.4	112	118

ASA No. 200, 2½-in. pitch, Diamond No. 480

Teeth	rpm of sprocket											
	10	20	40	60	80	100	120	160	200	240	260	280
12	5.6	10.5	19.1	26.8	33.6	39.6	45.1	54.4				
15	7.1	13.4	24.7	34.8	43.9	52.2	59.8	73.4	85			
18	8.6	16.2	30.0	42.4	53.7	64.1	73.7	90.7	105	118		
21	10.0	18.9	35.1	49.7	63.1	75.3	86.6	106.9	124	139	146	
24	11.4	21.6	40.2	56.8	71.9	86.0	98.8	121.8	142	159	166	173

These tables are abbreviated. Intermediate values may be interpolated and some values for greater numbers of teeth may be extrapolated. Blank spaces indicate that these numbers of teeth are not approved for these speeds. Ratings for multiple-strand chains are proportional to the number of strands. The recommended numbers of strands for multiple-strand chains are 2, 3, 4, 6, 8, 10, 12, 16, 20, and 24, with a maximum overall width of 24 in. The horsepower ratings are conservatively based upon a satisfactory service life of 15,000 hr and a chain length of 100 pitches. These ratings also take for granted that reasonable care will be given to installation and maintenance, including adequate lubrication at all times. Theoretically, a drive operating 24 hr a day should have double the horsepower capacity of a drive required to operate only 10 to 12 hr per day, if satisfactory service is specified for the same number of years.

The effects of chordal action and chain velocity reduce the permissible load. By considering these factors, the equation for the horsepower capacity* is

$$\text{hp} = p^2 \left\{ \frac{V}{23.7} - \frac{V^{1.41}[26 - 25 \cos{(180/T)}]}{1050} \right\} \qquad (17\text{-}3)$$

The horsepower ratings given in Table 17-2 are based upon this formula.

For very slow speeds and favorable operating conditions, including intermittent service, the chain selection may be based upon the ultimate tensile strength of the chain rather than durability. For chain speeds of 25 fpm or less, the chain pull may be as much as one-fifth of the ultimate strength; for 50 fpm, one-sixth; for 100 fpm, one-seventh; for 150 fpm, one-eighth; for 200 fpm, one-ninth; and for 250 fpm, one-tenth of the ultimate tensile strength.

17-4   Center Distance and Chain Length.   If a center distance is to be nonadjustable after installation, it should be selected for an initially snug fit for an even number of pitches of chain. For the average application a center distance equivalent to 30 to 50 pitches of chain represents good practice, and it must be greater than half the sum of the outside diameters of the sprockets. Extremely short center distances should be avoided, if possible, especially for ratios greater than 3:1. It is desirable to have at least 120-deg wrap in the arc of contact on a power sprocket. For ratios of 3:1 or less, the wrap will be 120 deg or more for any center distance or numbers of teeth. To have a wrap of 120 deg or more, **for** ratios greater than $3\frac{1}{2}$:1, the center distance must not be less than the difference between the outside diameters of the two sprockets.

The chain length is

$$L = 2C \cos \alpha + \frac{T_1 p(180 + 2\alpha)}{360} + \frac{T_2 p(180 - 2\alpha)}{360} \qquad (17\text{-}4)$$

$$L_p = 2C_p \cos \alpha + \frac{T_1}{2} + \frac{T_2}{2} + \frac{\alpha(T_1 - T_2)}{180} \qquad (17\text{-}5)$$

$$L_p = 2C_p + \frac{T_1}{2} + \frac{T_2}{2} + \frac{K(T_1 - T_2)^2}{C_p} \qquad (17\text{-}6)$$

where L = chain length, in.
L_p = chain length, pitches
C = center distance, in.
C_p = center distance, pitches
p = pitch of chain, in.

---

* "Design Manual for Roller and Silent Chains," pp. 46–47, Association of Roller and Silent Chain Manufacturers, 1955.

R = pitch radius of large sprocket, in.
r = pitch radius of small sprocket, in.
α = angle between tangent and center line, deg = $\sin^{-1}\left(\dfrac{R-r}{C}\right)$
T_1 = number of teeth on large sprocket
T_2 = number of teeth on small sprocket
$180 + 2\alpha$ = angle of wrap on large sprocket, deg
$180 - 2\alpha$ = angle of wrap on small sprocket, deg
K = a variable, its value depending upon the value of $\dfrac{T_1 - T_2}{C_p}$

Table 17-3   **Condensed values of K in Eq. (17-6)**

$\dfrac{T_1 - T_2}{C_p}$	0.1	1.0	2.0	3.0	4.0	5.0	6.0
K	0.02533	0.02538	0.02555	0.02584	0.02631	0.02704	0.02828

Formulas for chain length on multisprocket drives are too cumbersome to be useful. Multisprocket drives should be laid out accurately to scale, preferably full size, or larger.

Since standard roller chains are made up of alternate roller links and pin links, it is preferable to use chain lengths in even numbers of pitches. An odd number of pitches requires an offset link one pitch long. The link plates are offset to affect half a roller link at one end and half a pin link at the other end. Offset links should be avoided if possible.

Since roller chain cannot slip on a sprocket, it is advisable to avoid nonproductive pull on the chain, which is due to the unnecessary tension in the slack span. The permissible amount of slack depends upon several factors—length of span, weight of chain, character of load (whether impulsive or jerky), and slope of center line. Extremely long horizontal center distances for comparatively heavy chains should be avoided. The relationships between depth of sag and tension due to weight of chain in the catenary are approximately

$$h = 0.433(S^2 - L^2)^{\frac{1}{2}} \quad \text{and} \quad T = w\left[\frac{S^2}{8h} + \frac{h}{2}\right] \tag{17-7}$$

where h = depth of sag, in.
L = distance between points of support, in.
S = catenary length of chain, in. (approximately equal to the length tangent to the sprockets)
T = tension or chain pull, lb
w = weight of chain, lb/in.

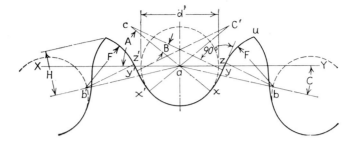

Fig. 17-11 Design of standard sprocket tooth for roller chains. (*Courtesy Diamond Chain and Manufacturing Company.*)

p = pitch; d = nominal roller diameter; T = number of teeth; D = pitch diameter = $\dfrac{p}{\sin (180/T)}$; H = height of tooth above center line between two adjacent seating-curve centers.

$d' = 1.005d + 0.003''$; $A = 35 \deg + \dfrac{60 \deg}{T}$; $B = 18 \deg - \dfrac{56 \deg}{T}$; $ac = 0.8d$; $C = \dfrac{180 \deg}{T}$.

Draw line XY. Locate point a, and with that as center and radius ax equal to $\frac{1}{2}d'$, draw circular arc for the "seating curve" xx'.

Draw line xac making angle A with line XY and locate point c so that $ac = 0.8d$. Draw line cy making angle B with line cx. With center at c and radius cx, draw arc xy for the "working curve."

Draw line yz perpendicular to line cy. Draw line ab making angle C with line XY, and locate point b so that $ab = 1.24d$. Draw line bz parallel to line yc. With b as center and radius bz, draw the "topping curve," arc zu tangent to line zy.

A similar construction for the other half will complete the tooth outline.

Outside diameter of sprocket when tooth is pointed = $p \cot \dfrac{180 \deg}{T} + 2H$.

The recommended value for H is $0.3p$; and when this value is chosen, the outside diameter of the sprocket will be

$$p \left(0.6 + \cot \frac{180 \deg}{T} \right)$$

The pressure angle, when the chain is new, is $xab = 35 \deg - \dfrac{120 \deg}{T}$.

The minimum pressure angle is abz or $17 \deg - \dfrac{64 \deg}{T}$.

The average pressure angle is $26 \deg - \dfrac{92 \deg}{T}$.

The standard tooth form is designed to give maximum efficiency throughout the life of the drive. Because of the large pressure angle and the distribution of the load over a number of teeth the tendency of the teeth to wear hook-shaped is greatly reduced. The reason for this is that the chain rides higher on the teeth as it elongates, thus accommodating itself to a larger pitch circle. This is the standard approved by the ASME, SAE, and AGMA.

17-5   Chain Sprockets.   The tooth form adopted by ASA for roller chain is shown in Fig. 17-11. One of the most important requirements of a sprocket cutter is that it cut the space (roller seat) between teeth slightly oversize, to allow rollers to seat without being pinched.

The ASA tooth-section profile for standard-width chains is shown in Fig. 17-10 with all dimensions in inches.

The sprocket selected must be large enough to accommodate the shaft. Under normal conditions the shaft diameter and keyway dimensions determine the minimum hub diameter of a sprocket, usually equal to the shaft diameter plus four times the keyway depth. If there is to be a setscrew over the key, the minimum hub diameter is the shaft diameter plus six times the keyway depth, assuming a square key. See Table 5-1 for standard keys.

To provide clearance between ends of adjacent link plates, the end radius must be less than half of the pitch, or the height of the link plate must be less than the pitch. Roughly, therefore, the maximum hub diameter to allow clearance under the edges of the link plates with figure-eight contour would be the pitch diameter minus pitch (see Q under Fig. 17-10). The extra clearance needed under standard offset link plates and straight-edge link plates requires the use of a smaller maximum hub diameter, which can be obtained from the equation

$$\text{Maximum hub diameter} = D \times \cos\frac{180}{T} - (H' + 0.05) \qquad (17\text{-}8)$$

where D = pitch diameter, in.
   T = number of teeth
   H' = height of link plate, in. (Table 17-1)

In some forms of cast sprockets, clearance is provided between the rollers and teeth by a slight increase of the pitch diameter of the driving sprocket and a decrease in that of the driven sprocket. This construction causes one tooth to carry all the load and is not desirable except with

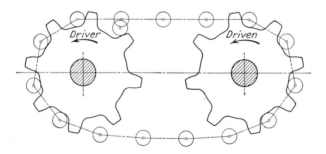

Fig. 17-12   Roller-chain drive showing tooth action when clearance is provided.

Table 17-4 **Service factors for roller chains**

Operating characteristics	Intermittent: few hours per day, few hours per year	Normal: 8–10 hr per day, 300 days per year	Continuous: 24 hr per day
Easy starting, smooth, steady load....	0.6–1.0	0.9–1.5	1.2–2.0
Light to medium shock or vibrating load..........................	0.9–1.4	1.2–1.9	1.5–2.2
Medium to heavy shock or vibrating load..........................	1.2–1.8	1.5–2.3	1.8–2.5

cast teeth and cast chain in which the pitch cannot be accurately maintained. See Fig. 17-12.

The theoretical number of teeth on a sprocket should be large so as to reduce the wear in the pin joints of the chain, to minimize the variation in chain speed, and to reduce the tooth load to produce the same power at the same rotative speed. On the other hand, the greater the number of teeth, the larger the sprocket, the greater the cost, and the greater the centrifugal force and lubrication difficulties. A compromise must be made. Whenever possible, the number of teeth on the sprockets should be an odd number so as to reduce wear and speed variation. This is not too important when a large number of teeth are used on a sprocket.

17-6 Service Factors. The horsepower ratings in Table 17-2 should be modified for various service conditions by using service factors as given in Table 17-4. The designer should consult manufacturers' catalogs for load classification information.

The following conditions justify service factors in the higher range: sprocket ratios greater than 6:1; more than two sprockets in the drive; less than 120-deg wrap on sprockets other than idlers; very short center distances, especially with few teeth; vertical or near-vertical center lines, small sprocket below; frequent starting and stopping, especially under heavy load.

17-7 Lubrication. Chain drives suffer more from lack of proper lubrication than from many years of normal service. Light-bodied oil of good quality, fluid at the prevailing temperature, is the best general-purpose lubricant for both high- and low-speed drives; SAE 20 or SAE 30 for the shorter pitches and SAE 40 for the heavier chains. Winter oil should be used when temperatures are very low. Heavy oils and greases applied at ordinary temperatures cannot pass through the narrow clearance spaces leading to the pin-bushing surfaces.

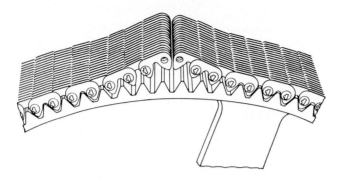

Fig. 17-13   Rocker joint silent chain before ends are assembled.

Three types of lubrication are recommended: (1) drip (4 to 10 drops per minute), shallow bath or manual, used for chain speeds up to 600 fpm; (2) rapid drip (20 or more drops per minute), or continuous with shallow bath, disk, or slinger, used for chain speeds of 600 to 1,500 fpm; (3) continuous with disk, slinger, or circulating pump, used for speeds over 1,500 fpm. For other than splash and bath systems, the oil should be delivered to the upper edges of link plates in the bottom span of chain. Oil applied to the upper span of chain or to the chain rollers is of little value in retarding wear. Enough oil should be applied to prevent overheating and drawing of the hardened bearing surfaces; the amount of oil depends upon operating conditions.

17-8   Silent Chains.   Silent chains (Fig. 17-13) are made of alternating steel links whose tooth edges are made straight so that they bear along the entire length of the sprocket teeth. There are two general types of silent chains, depending on the method of joining the links. The Reynolds type uses pins similar to those used in roller chains. The Morse type uses a rocker joint. They are suitable for speeds up to 1,600 fpm in general service, and up to 6,000 fpm if enclosed and well lubricated. The selection of silent chains is essentially the same as for roller chains.

Silent chains are used ·on sprockets having straight-sided teeth. Sprockets* for silent chains have been standardized to provide interchangeability. The chains are held on the sprockets by center or double plates riding in grooves in the sprocket, or outside links extending down over the outer ends of the teeth.

Example.   Select a roller-chain drive to transmit 40 hp from a 1,200-rpm motor to a line shaft at 250 rpm. The motor shaft diameter is $2\frac{3}{8}$ in., and the center distance is adjustable from 24 in. Service will be 10 hr/day, 6 days/week, and good lubrication will be provided.

* Inverted Tooth (Silent) Chains and Sprocket Teeth, *ASA B29.2*, 1957.

Solution. As shown in Table 17-1, the pitch of the chain is governed by the number of teeth and the speed of the small sprocket. For economy in price, the longest approved pitch is usually selected unless extreme quietness is desirable.

The longest pitch chain for 1,200 rpm is $\frac{3}{4}$ in.

The minimum hub diameter for $2\frac{3}{8}$ in. bore is $3\frac{5}{8}$ in.

The minimum trial pitch diameter is $3\frac{5}{8}$ in. $+$ $\frac{3}{4}$ in. $= 4\frac{3}{8}$ in. (see Fig. 17-10).

Since the perimeter of the pitch polygon is approximately equivalent to the circumference of the pitch circle, the minimum number of teeth will be

$$\frac{4.375\pi}{0.75} = 18.35, \text{ say } 19$$

The pitch diameter $D = \dfrac{p}{\sin 180/T} = \dfrac{0.75}{\sin 180/19} = 4.557$ in.

The outside diameter $= p\left(0.6 + \cot\dfrac{180}{T}\right) = 0.75\left(0.6 + \cot\dfrac{180}{19}\right) = 4.945$ in.

The maximum hub diameter for 19 teeth, $\frac{3}{4}$ in. pitch, from Eq. (17-8) is found to be $3\frac{49}{64}$ in. This is adequate for $2\frac{3}{8}$ in. bore, being more than the recommended $3\frac{5}{8}$ in. minimum hub diameter.

The reduction ratio is 4.8:1, and the nearest ratio which can be provided with a 19-tooth driver is 91:19, or 4.79:1.

For 91 teeth the pitch diameter is 21.729 in. and the outside diameter is 22.166 in.

One-half of the sum of the outside diameters is 13.555 in., providing 10.445 in. clearance between the two sprockets.

The difference between the pitch diameters is 17.172 in. and will provide 138 deg of wrap, which is more than the desired minimum of 120 deg.

The speed ratio is 4.8:1, and a combination of 19 and 91 teeth represents a ratio of 4.79:1. If the exact ratio is more important than the minimum number of teeth, 20 and 96 teeth should be used.

From Table 17-2, the rating for a $\frac{3}{4}$-in.-pitch chain on 19 teeth at 1,200 rpm is 13.8 hp per strand, making it necessary to use triple-strand chain, with a rating of 41.4 hp, assuming a service factor of 1.

The chain speed is $\dfrac{19 \times 1,200 \times 0.75}{12} = 1,425$ fpm.

The chain pull for 40 hp is $\dfrac{40 \times 33,000}{1,425} = 925$ lb.

With an ultimate tensile strength of 25,500 lb (Table 17-1) for this chain, the safety factor will be slightly more than 27. Tensile strengths are always adequate, and usually more than adequate, for chain having requisite power capacity.

The center distance, 24 in., divided by the pitch, $\frac{3}{4}$ in., is equivalent to 32 pitches. Using Eq. (17-6), the chain length in pitches is

$$L_p = 2 \times 32 + \frac{91}{2} + \frac{19}{2} + \frac{0.02562(91 - 19)^2}{32} = 123.15 \text{ pitches}$$

A center distance slightly less than 24 in. would require exactly 123 pitches of chain, but this length would include an offset link. It is advisable, therefore, to specify a length of 124 pitches. The 0.85 pitch, or approximately $\frac{5}{8}$ in. of slack, can be taken up by adjusting the center distance after the chain is installed.

To give an initially snug-fitting chain on a fixed center distance, C_p in Eq. (17-6) must be increased a little more than half of the 0.85 pitch slack. (The rate of increase for values of K becomes less as the center distance is increased.)

Substituting 32.5 for C_p in Eq. (17-6), the exact chain length would be 124.05 pitches, giving a value for C of $24\frac{3}{8}$ in., for which 124 pitches would be entirely satisfactory.

18

Wire Ropes

18-1 Wire Ropes. Wire ropes were once favored for long-distance transmission of power. Electric transmission has made this use practically obsolete. The chief use of wire ropes at the present time is in elevators, mine hoists, cranes, oil-well drilling, aerial conveyors, tramways, haulage devices, and suspension bridges.

As the requirements for strength and service increased, there followed in order wrought-iron, cast-steel, extra-strong cast-steel, and plow-steel ropes. For extra-high strength, alloy-steel ropes (known by various trade names) are now available. For certain purposes, ropes are made of aluminum alloys, copper, bronze, and stainless steel.

The various grades of steel wires have minimum ultimate strengths approximately as follows: iron, 85,000; cast steel, 170,000; extra-strong cast steel, 190,000; plow steel, 210,000; and alloy steel, 230,000 psi. The smaller wire sizes have strengths from 10 to 20 per cent higher. The wires are laid in curved form in the rope; hence it is impossible to develop the full strength of the metal in the finished rope, the loss in strength amounting to 5 to 20 per cent. The ultimate strengths* of plow-steel

* Tables of rope strengths and other properties are given in "Machinery's Handbook," 16th ed., pp. 454, 460, The Industrial Press, New York, 1959.

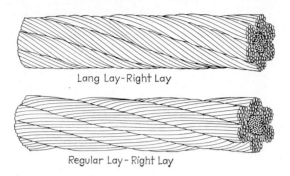

Lang Lay-Right Lay

Regular Lay- Right Lay

Fig. 18-1

ropes may be approximated by the formulas

$$F_u = 76{,}000d^2 \text{ lb} \qquad \text{for 6 by 7 and 6 by 19 ropes}$$
$$ = 75{,}000d^2 \text{ lb} \qquad \text{for 6 by 37 ropes} \qquad\qquad (18\text{-}1)$$

where d = diameter of the rope, in.

The designation 6 by 7 indicates that the rope is made of six strands each containing seven wires.

The nominal size of wire rope is the diameter of a circle that just encloses the rope. Ropes are available in the following sizes: 6 by 7 ropes vary by $\frac{1}{16}$-in. increments from $\frac{1}{4}$ to $\frac{5}{8}$ in., and $\frac{1}{8}$-in. from $\frac{5}{8}$ to $1\frac{1}{2}$ in.; 6 by 19 vary by $\frac{1}{16}$-in. increments from $\frac{1}{4}$ to $\frac{5}{8}$ in., $\frac{1}{8}$-in. from $\frac{5}{8}$ to $2\frac{1}{4}$ in., and $\frac{1}{4}$-in. from $2\frac{1}{4}$ to $2\frac{3}{4}$ in.; and 6 by 37 vary by $\frac{1}{16}$-in. increments from $\frac{1}{4}$ to $\frac{5}{8}$ in., $\frac{1}{8}$-in. from $\frac{5}{8}$ to $2\frac{1}{4}$ in., and $\frac{1}{4}$-in. from $2\frac{1}{4}$ to $3\frac{1}{2}$ in.

The weight of wire ropes is approximately $1.58d^2$ lb/ft.

18-2   Wire-rope Construction.   The individual wires are first twisted into strands, and then the strands are twisted around a hemp or steel center to form the rope. The ropes are right- or left-lay, depending on whether the strands form right- or left-hand helices. Most rope is right-lay. Regular-lay rope has the wires twisted opposite to the strands and is standard construction in this country. Lang-lay rope has the wires and strands twisted in the same direction, giving a rope with a better wearing surface, but it is harder to splice and twists more easily when loaded (see Fig. 18-1).

The most common rope constructions are illustrated in Fig. 18-2, each construction being designed for particular properties. By varying the construction, the metal may be distributed to give maximum wear, maximum flexibility, or any desired intermediate quality. The standard constructions are 6 by 7 or coarse lay; 6 by 19 or flexible; and 6 by 37, together with 8 by 19, or extra-flexible.

In making a selection, one must consider flexibility, wear resistance, strength, reserve strength, core strength, and corrosion resistance. For equal diameters, the use of a large number of small wires gives a rope of high flexibility. Increasing the wire diameter and decreasing the number of wires reduce the flexibility. When extreme flexibility without extreme strength is required, tiller rope (6 by 6 by 7) is used.

Large wires give better wear resistance. Laying two sizes of wire alternately in the outer layer increases the wearing qualities and retains the flexibility. For severe service, steel-clad rope is used. This consists of a regular rope with each strand wrapped with a thin flat strip of steel to protect the wires. The strength of the rope is not increased, but its life is lengthened, since after the covering is worn through, the regular rope is still intact, and the metal cover, being forced down between the strands, presents more wearing surface.

The strength does not depend on the rope construction but is due entirely to the material from which the wires are made. Of course, the flexibility that is obtained with small wire sizes increases the fatigue strength. By *reserve strength* is meant the ratio of the strength of all the inside wires to the strength of all the wires in the rope. The outside wires are subject to service wear and the inside wires are protected, forming the reserve capacity that can be relied upon throughout the life of the rope. This consideration is important in choosing ropes for service in which human life is involved. The coarse-lay ropes, being made of large wires with few inner or protected wires, have the least reserve strength, and ropes like the modified Seale and the Warrington have the highest reserve strength.

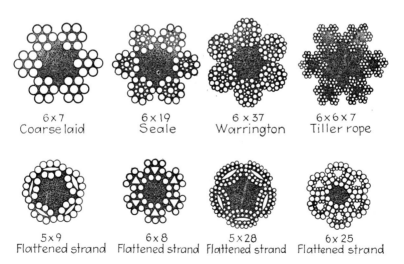

| 6 x 7 Coarse laid | 6 x 19 Seale | 6 x 37 Warrington | 6 x 6 x 7 Tiller rope |

| 5 x 9 Flattened strand | 6 x 8 Flattened strand | 5 x 28 Flattened strand | 6 x 25 Flattened strand |

Fig. 18-2 Types of wire-rope construction.

18-3 Corrosion Resistance. Corrosion is chiefly due to moisture and stray electric currents and may be effectively overcome by using suitable rope lubricants. Galvanized ropes are suitable for some purposes; but since the zinc flakes off easily, galvanizing is not suitable for ropes running over sheaves or those which are frequently bent. Copper, bronze, and stainless steels are sometimes used where corrosion resistance is a major consideration.

18-4 Bending Stresses.* As the rope passes around the drums and sheaves, the wires on the outside increase in length and those on the inside decrease, producing additional tension in the outer wires. The rope does not bend as a solid bar, but there is a movement and rearrangement of the wires, a movement that varies with the rope construction, the wire size, the type of center, and the amount of pinching or restraint in the grooves. It is evident that these items vary with the rope and installation, and that no mathematical formula based on a series of assumptions at static conditions can give the true bending stress under all operating conditions.

There are several bending-stress formulas in use, giving a wide variety of results. Only the simplest of these is given here, as it seems to agree fairly well with experimental data available. An experienced designer of rope installations will modify the results according to his experience with similar installations.

By combining the elastic curve equation $M = EI/\rho$ and the beam equation $M = s_b I/c$,

$$s_b = \frac{Ec}{\rho}$$

But $c = d_w/2$ and $\rho = D/2$.

Then the bending stress is

$$s_b = \frac{E_r d_w}{D} \tag{18-2}$$

where s_b = bending stress, psi
E_r = modulus of elasticity of the rope, psi of wire area
d_w = wire diameter, in.
D = drum or sheave diameter, in.

The value E_r is not the modulus of elasticity of the wire material, but of the entire rope. Tests and theoretical investigations by J. F. Howe

---

* For complete discussions on bending stresses the reader is referred to James F. Howe, Determination of Stresses in Wire Rope as Applied to Modern Engineering Problems, *Trans. ASME*, vol. 40, p. 1043, 1918; and to "Wire Engineering," John A. Roebling's Sons, October, 1932.

indicate that for steel ropes of the ordinary constructions the value of E_r may be taken as 12,000,000 psi.

The value of the wire diameter d_w depends on the rope construction. For preliminary computations, the wire diameter and the total cross-sectional area of the metal in a rope may be obtained from Table 18-1.

Table 18-1 **Approximate wire diameters and areas**

Rope	d_w	A
6 × 7	0.106d	0.38d^2
6 × 19	0.063d	0.38d^2
6 × 37	0.045d	0.38d^2
8 × 19	0.050d	0.35d^2

In most cases, it is more convenient to convert the bending stress into the equivalent bending load, i.e., the direct tension load that would produce the same wire stress. This equivalent bending load, in pounds, is

$$F_b = A\,\frac{E_r d_w}{D} \tag{18-3}$$

18-5 Starting Stresses. When starting and stopping, the rope and the supported load must be accelerated by a force transmitted through the rope. In hoisting and elevator service where the acceleration may be as high as 10 ft/sec² the additional load becomes an important item. If, when the rope drum begins to rotate, there is any slack rope to be taken up before the load is moved, there will be considerable impact load on the rope. This impact may be determined by the usual impact equations if the acceleration of the rope is known, so that the velocity at the instant of impact can be determined. However, a rope will stretch much more than a solid bar, because of the twisted lay of the wires, and this condition relieves the impact effect to some extent. Computations for the impact load are of no practical value because of the unknown factors, and the designer must use his judgment based on experience in cases of this kind.

18-6 Fatigue Strength. From the experimental data obtained by Drucker and Tachau[*] with 6 by 19 and 6 by 37 regular-lay ropes, there seems to be a correlation between fatigue failure and the total working stress, including the bending stress. However, they found that the bearing

[*] D. C. Drucker and H. Tachau, A New Design Criterion for Wire Rope, *Trans. ASME*, vol. 67, p. A-33, 1945.

pressure of the rope on the sheave gives a better criterion for the fatigue strength of the rope. For a contact angle of approximately 180 deg, the nominal bearing pressure $p = 2F_t/Dd$, where F_t is the tensile load on the rope (total working load minus the bending load), D is the diameter of the sheave in inches, and d is the diameter of rope in inches. They give curves for p/s_{tu} as the number of bends to failure, where s_{tu} is the ultimate tensile strength of the rope in pounds per square inch. These curves give the following approximate values for p/s_{tu}: 0.0014 to 0.0018 for 500,000 to 1,000,000 bends, 0.0018 to 0.0028 for 200,000 to 500,000 bends, and 0.0028 to 0.006 for 50,000 to 200,000 bends. The reader should consult the curves of Drucker and Tachau for more exacting values. The

Table 18-2   **Common wire-rope applications**

Type of service	Type of rope	Sheave diameter	
		Recommended	Minimum
Haulage rope: Mine haulage Factory-yard haulage Inclined planes Tramways Power transmission Guy wires	6×7	$72d$	$42d$
Standard hoisting rope: (most commonly used rope) Mine hoists Quarries Ore docks Cargo hoists Car pullers Cranes Derricks Dredges Tramways Well drilling Elevators	6×19	$45d$ 60–$100d$ 20–$30d$	$30d$
Extra-flexible hoisting rope	8×19	$31d$	$21d$
Special flexible hoisting rope: Steel-mill ladles Cranes High-speed elevators Service where sheave diameters are limited	6×37	$27d$	$18d$

fatigue strength may be estimated by the expression

$$p = \frac{2F_t}{Dd} = cs_{tu} \tag{18-4}$$

where c is a constant depending on the desired rope life and having the above values.

18-7 Sheave and Drum Diameters. Because of the bending stresses where the rope wraps around the drum, it is important that the drum diameters be kept fairly large. This is especially true with high-speed ropes in continuous service, where the fatigue action will materially affect the life of the rope. Practice varies in regard to the proper diameter, but the values recommended as standard practice are given in Table 18-2. Where larger diameters are possible, their use will give better and more economical service. Space requirements may lead to the use of sheaves smaller than those given in the table.

18-8 Sheave Grooves. The contour of the sheave groove has a great influence on the life and service of the rope. If the groove has a bottom radius much larger than the rope, there will be insufficient support for the rope, which will flatten out from its normal circular section. This tends to increase the fatigue effects. With too small a bottom radius, the rope will be wedged into the groove and the normal rotation of the rope will be prevented. This concentrates the wear along two lines parallel to the axis instead of distributing it around the entire circumference.

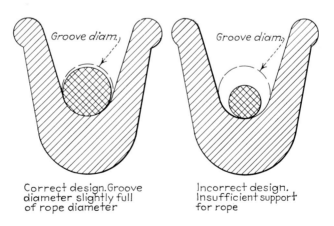

Correct design. Groove
diameter slightly full
of rope diameter

Incorrect design.
Insufficient support
for rope

Fig. 18-3 Sheave grooves. Correct groove diameter = rope diameter $+ \frac{1}{64}$ in. for $\frac{1}{4}$ and $\frac{5}{16}$ ropes; $+\frac{1}{32}$ in. for $\frac{3}{8}$ to $\frac{3}{4}$ ropes; $+\frac{3}{64}$ in. for $\frac{3}{16}$ to $1\frac{1}{8}$ ropes; $+\frac{1}{16}$ in. for $1\frac{3}{16}$ to $1\frac{1}{2}$ ropes; $+\frac{3}{32}$ in. for $1\frac{9}{16}$ to $2\frac{1}{4}$ ropes; and $+\frac{1}{8}$ in. for $2\frac{5}{16}$ and larger ropes.

The rope center will also be distorted, and premature breaking of the wires will occur in the valleys between the strands. The correct design of the groove is shown in Fig. 18-3. This groove gives support to the rope on nearly half its circumference.

18-9 Winding the Rope.   There are four common methods of imparting motion to the rope: winding drums, winding machines, friction spools, and grip wheels.

When winding drums are used, the end of the rope is attached to the drum. The rope should always wrap once or more around the drum so that the rope is held by friction instead of by direct pull on the fastener, where the rope will not sustain its full breaking strength. Cylindrical drums, either plain or with guide grooves, are the most commonly used in all classes of service, and have the advantage that several layers of rope may be wound on them, thus reducing the size required. Conical drums and combined conical and cylindrical drums are used with deep mine hoists to balance the torque change due to the decreasing weight of the rope as the hoist rises.

Since the rope is wound under tension, the drum is subjected to external crushing loads as well as bending loads, and the body should be designed as a thick cylinder subjected to external pressure. Hoists fitted with conical and plate clutches may also impose axial loads on the drums. Drum flanges are usually subjected to tangential loads by the brake, axial or tangential loads by the clutch, and axial loads by the side pressure of the ropes when they are wound in several layers. The complete design of drums cannot be discussed here.

18-10 Idlers and Guide Sheaves.   Rollers and sheaves used to support and guide the rope should be of generous proportions to reduce bending and fatigue effects. Since the forces involved are usually less than on the driving sheaves, the diameters may be reduced. They should be grooved at the center and wide enough to prevent the rope from moving off sideways.

18-11 Rope Fasteners.   Spliced wire rope should be avoided, but, when it is used, the permissible working load should be reduced to about 75 per cent of the working load on unspliced ropes.

Several types of rope fasteners are shown in Fig. 18-4. Of these, the wire-rope socket, Fig. 18-4e, is the only one that will develop 100 per cent of the rope strength. The wires at the end are separated, the hemp center is removed, and the wires are cleaned, dipped in muriatic acid, inserted in the socket, evenly distributed, and anchored in place by filling the socket with high-grade molten zinc. If properly made, the joint will be as strong as the rope. Babbitt, lead, and other antifriction metals are

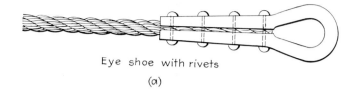

Eye shoe with rivets

(a)

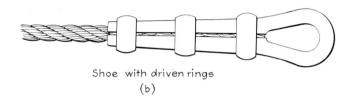

Shoe with driven rings

(b)

Thimble and clamps

(c)

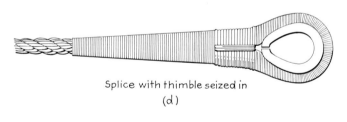

Splice with thimble seized in

(d)

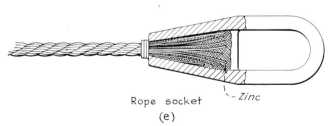

Rope socket   Zinc

(e)

Fig. 18-4

Table 18-3   Efficiencies of wire-rope fasteners

| | Per cent of total |
Method of fastening	rope strength
Wire-rope socket with zinc......................	100
Thimble with four or five wire tucks.............	90
Special offset thimble with clips.................	90
Regular thimble with clips......................	85
Three-bolt wire clamps.........................	75

sometimes substituted for the zinc, but these do not bond perfectly with the wires and such a joint will not develop full rope strength.

When thimbles or eye splices are used, the full rope strength cannot be developed, such joints having the efficiencies indicated in Table 18-3.

18-12   Factor of Safety.   Some authorities base the factor of safety on the external load, not including the bending load. The magnitude and method of using the factor of safety are usually included in the code of the city, state, or other agency where the installation is to be used. The factors of safety given in Table 18-4 are the ultimate tensile strength of the rope divided by the total working load, including the bending load.

18-13   Examples of Wire-rope Selection.   To bring out the principles of rope applications, several examples with solutions are presented.

Example 1.   Vertical Shaft Hoist.   Select a wire rope for a vertical mine hoist to lift 1,200 tons of ore in each 8-hr shift from a depth of 2,400 ft. Assume a two-compartment shaft with the hoisting skips in balance.

Solution.   Several combinations of rope velocity, loading, etc., could be used, and, in practice, several combinations would be worked out and the proper combination selected after due consideration of the effect of each on the hoist design and power requirements. Calculations for only one combination are given here.

A maximum rope velocity of 2,500 fpm with acceleration and deceleration

Table 18-4   Factors of safety for wire ropes (based on ultimate strength)

Service	Factor of safety
Elevators......................	8–12
Mine hoists....................	2.5–5
Cranes, motor driven...........	4–6
hand powered................	3–5
Derricks......................	3–5

periods of 12 sec each, and a rest period of 10 sec for discharging and loading are assumed.

During the acceleration period, the skip travels

$$S_a = \frac{vt}{2} = \frac{2,500 \times 12}{60 \times 2} = 250 \text{ ft}$$

with an acceleration of

$$a = \frac{2,500}{60 \times 12} = 3.47 \text{ ft/sec}^2$$

The total distance traveled at full speed is 2,400 ft less $2S_a$, or 1,900 ft, and the time required is

$$t = \frac{1,900 \times 60}{2,500} = 45.6 \text{ sec}$$

The time required for one trip is

	Seconds
Acceleration	12
Full speed	45.6
Deceleration	12
Discharging and loading	10
Total	79.6

This allows 45 trips per hour and requires a skip of 3.3 tons, say 3.5 tons capacity. A hoisting skip weighs approximately 0.6 of its load capacity, or, in this case, about 2 tons, making a total load of 5.5 tons to be hoisted.

The rope selected must have strength sufficient to support the load, the rope weight, the accelerating load, and the bending load with a factor of safety of approximately 5. As there are several unknown quantities involved, trial diameters must be assumed, and later modified if not satisfactory. Try a $1\frac{1}{4}$-in. 6 by 19 plow-steel rope, weighing 2.47 lb/ft and having an ultimate strength of 59.4 tons. Average mine-hoist practice is to use drums 60 to 100 times the rope diameter. Assume the rope drum to be 9 ft 6 in. in diameter.

Then the loads are

	Tons
Useful load	3.5
Weight of skip	2.0
Weight of rope, $\dfrac{2,400 \times 2.47}{2,000}$	2.96
Acceleration of load, $\dfrac{(3.5 + 2)3.47}{32.2}$	0.59
Acceleration of rope, $\dfrac{2.96 \times 3.47}{32.2}$	0.32
Equivalent bending load	2.47
Total	11.84

The factor of safety for this rope is

$$FS = \frac{59.4}{11.84} = 5.02$$

which is slightly greater than the desired value; thus the rope is satisfactory.

$$p = \frac{2F_t}{Dd} = \frac{2(11.84 - 2.47)2,000}{114 \times 1.25} = 263 \text{ psi}$$

Assuming $s_{tu} = 210,000$ psi, $263/210,000 = 0.00125 < 0.0014$; therefore, the life should be satisfactory.

Example 2.   Inclined Shaft Hoist.   Select a wire rope for an inclined shaft whose length is 2,400 ft, at a 60 per cent slope, with a loaded skip weighing 22,000 lb, a rope velocity of 2,000 fpm, an acceleration period of 10 sec, and a factor of safety of 5.

Solution.   Assume a $1\frac{1}{8}$-in. 6 by 19 plow-steel rope, weighing 2.00 lb/ft and having an ultimate strength of 48 tons. Assume the rope drum to be 8 ft in diameter.

Since the hoist operates on an incline, the friction of the skip and of the wire supported on track rollers must be overcome by the rope pull. The car friction will amount to about 50 lb/ton of weight normal to the incline, using bronze bushings in the wheels. The friction of the rope on proper track rollers is about 100 lb/ton of weight normal to the incline.

From the data

$$a = \frac{2,000}{60 \times 10} = 3.33 \text{ ft per sec}^2$$
Rope weight $= 2,400 \times 2 = 4,800$ lb

and the angle of incline is 30 deg 58 min.

The loads on the rope are

		Tons
Weight of skip and load,	$\dfrac{22,000 \sin \theta}{2,000}$	5.66
Weight of rope,	$\dfrac{4,800 \sin \theta}{2,000}$	1.23
Skip friction,	$\dfrac{50 \times 22,000 \cos \theta}{2,000 \times 2,000}$	0.24
Rope friction,	$\dfrac{100 \times 4,800 \cos \theta}{2,000 \times 2,000}$	0.10
Acceleration,	$\dfrac{(22,000 + 4,800)3.33}{2,000 \times 32.2}$	1.39
Equivalent bending load....................		2.13
Total......................................		10.75

The factor of safety with this rope is

$$FS = \frac{48}{10.75} = 4.46$$

which is less than the required factor of 5. A $1\frac{1}{4}$-in. rope should now be assumed, and the calculations repeated. This calculation is left to the student.

$$p = \frac{2F_t}{Dd} = \frac{2(10.75 - 2.13)2,000}{96 \times 1.125} = 319.3 \text{ psi}$$

Assuming $s_{tu} = 210,000$ psi, $319.3/210,000 = 0.00152$ and therefore the life should be satisfactory.

Example 3.   High-speed Passenger Elevator.   Select a wire rope for the elevator in a building where the total lift is 600 ft, the rope velocity 900 fpm, and the full speed is to be reached in 40 ft. The lifting sheaves are to be of the traction type. The cage will weigh 2,500 lb and the passengers 2,000 lb.

Solution.   State laws generally require at least 4 ropes and a factor of safety of 8 on passenger elevators. The total weight to be lifted is 4,500 lb, and allowing the same amount for the weight of the rope, acceleration, and bending loads, the total required strength is $2 \times 4,500 \times 8$ or 72,000 lb, which is 18,000 lb/rope.

Elevator ropes are usually 6 by 19 construction, although constructions like the Special Seale are being used on the longest high-speed elevators. For this elevator, assume $\frac{1}{2}$-in. 6 by 19 plow-steel ropes, weighing 0.39 lb/ft and having an ultimate strength of 10 tons. The head-shaft sheaves for a $\frac{1}{2}$-in. rope should be about 36 in. in diameter.

The rate of acceleration is

$$a = \frac{v^2}{2S} = \frac{900^2}{2 \times 40 \times 3,600} = 2.81 \text{ ft/sec}^2$$

Then the rope loads are

	Tons
Elevator and passengers, $\dfrac{4,500}{4 \times 2,000}$	0.563
Weight of rope, $\dfrac{0.39 \times 600}{2,000}$	0.117
Acceleration of load, $\dfrac{4,500 \times 2.81}{4 \times 2,000 \times 32.2}$	0.049
Acceleration of rope, $\dfrac{0.117 \times 2.81}{32.2}$	0.010
Equivalent bending load....................	0.497
Total......................................	1.236

The factor of safety with this rope is

$$FS = \frac{10}{1.236} = 8.10$$

which is slightly greater than the factor desired.

$$p = \frac{2F_t}{Dd} = \frac{2(1.236 - .497)2,000}{36 \times 0.5} = 164.2 \text{ psi}$$

If $s_{tu} = 210,000$ psi, $164.2/210,000 = 0.000792$ and should last indefinitely.

Since this is a traction-type elevator, the number of wraps of rope around the driving sheaves must be determined. The driving-sheave layout will be similar to Fig. 18-5. The counterweight is usually made sufficient to balance the elevator plus one-third the live weight, making the counter-weight, in this case, 3,167 lb. When the elevator is at the bottom, and the counterweight at the top,

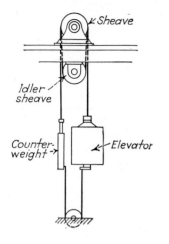

Fig. 18-5

the rope tensions are

$$F_1 = \frac{2,500 + 2,000}{4} + 234 + 98 + 20 = 1,477 \text{ lb/rope}$$

$$F_2 = \frac{3,167}{4}\left(1 - \frac{2.81}{32.2}\right) = 723 \text{ lb}$$

and

$$F_1 - F_2 = 754 \text{ lb}$$

which must be supplied by sheave friction. The arc of contact is 180 deg and the coefficient of friction for greasy wire rope on round-bottom cast-iron sheaves is about 0.18 to 0.23 with hard-rubber groove linings. Hence

$$\frac{F_1}{F_2} = e^{f\theta} = 2.718^{0.18 \times 3.14} = 1.76$$

and

$$F_2 = \frac{F_1}{e^{f\theta}} = \frac{1,477}{1.76} = 843 \text{ lb}$$

from which

$$F_1 - F_2 = 634 \text{ lb} \qquad \text{for one wrap around the sheave}$$

Then the required number of wraps is

$$n = \frac{754}{634} = 1.19 \qquad \text{say 2 wraps}$$

The number of wraps required should also be checked with the elevator in its top position.

19

Fits, Tolerances, Limits, and Surface Roughnesses

19-1 Need for Machine Fits and Tolerances. In order for each machine element to fit together and operate properly with its mating members, each part must be manufactured according to a relatively narrow range of acceptable dimensions. Journals, pistons, and other sliding parts must be made so that they are capable of moving relative to other machine elements but without so much freedom that they will not function properly. On the other hand, keys, gears on shafts, and similar members mounted by press or shrink fits must be so dimensioned that the desired interference is maintained without being so large as to make assembly impossible or the resulting stresses too high. In the case of sliding or free fits, the male or inner member is slightly smaller than the female or outer member in which it fits; i.e., the parts are made with a small clearance or noninterfering difference in mating dimensions. In the case of force and shrink fits, the male member is made slightly larger than the mating hole into which it finally is assembled; i.e., the parts have a slight interference or overlap of dimensions before assembly; after assembly, these mating dimensions are the same. In order for the assembled machine parts to have their proper clearance or interference, desired dimensions must be established for the manufacture of these parts; in addition, reasonable tolerances (allowable dimensional variations) must be established in order to determine whether or not the produced parts pass inspection even though they may not have precisely the desired dimensions.

19-2 Definitions.   Consider a 2-in.-diam journal and oil-lubricated journal bearing. In order for the bearing to operate, an operating clearance is necessary; that is, the bore of the bearing must be slightly larger than the diameter of the journal. The 2-in. dimension is a *nominal dimension* that denotes the approximate size of both the bearing bore and the journal diameter; it is probable that neither of these two diameters actually would be exactly 2 in. The *nominal size* of a part is the approximate dimension of the part or of mating parts without any indication of an operating clearance or overlap or the degree of accuracy required for the machine work. The *basic dimension* is that exact theoretical size desired for a part before a variation of dimension is allowed for manufacturing. *Allowance* is the intentional difference in the basic dimensions of two mating parts; it is the amount of clearance or overlap desired without variation due to machine work. In the case of sliding members with a free fit, allowance is referred to as a *clearance;* for force and shrink fits, as an *interference.* Sometimes clearance is called a positive allowance and interference a negative allowance. *Tolerance* is the maximum acceptable size variation from the basic dimension; it may be plus, minus, or zero. *Limits* are the maximum and minimum acceptable dimensions of a produced part; they are obtained by the algebraic addition of the basic dimension and the positive or negative tolerance. The larger limit is referred to as the *upper limit*, the smaller as the *lower limit*.

19-3 Basic Hole and Basic Shaft Systems.   In the *basic hole system*, the basic dimension of the hole or female member of two mating parts is the nominal dimension. The basic dimension of the mating male member would be the nominal dimension plus the allowance. In the *basic shaft system*, the basic dimension of the male member would be the nominal dimension; the basic dimension of the mating female member would be the nominal dimension minus the allowance. The basic hole system is used almost exclusively. Consider the 2-in.-nominal-diameter journal and bearing mentioned in Article 19-2 and assume an allowance (total diametral clearance) of 0.002 in. According to the basic hole system, the basic dimension of the bearing and journal would be 2.000 and 1.998 in. respectively; according to the basic shaft system, the basic dimensions of the bearing and the journal would be 2.002 and 2.000 in. respectively.

19-4 Standard Classes of Fits. The amount of allowance and tolerance of produced parts depend upon the service for which they are intended, their method of manufacture, and other considerations based mainly upon experience. Many manufacturers have their own set of so-called standard fits; however, these standards for one manufacturer may be completely different and nontranslatable into the standards of any other manufacturer. In order that all manufacturers may have a common set of standard classes of fits, the American Standards Association has established the set of eight different classes of fits indicated in Table 19-1 together with related recommended allowance and tolerances for each class of fit.

The *loose fit* is intended for use where accuracy is not essential and where considerable freedom is permissible, such as in agricultural, mining, and general-purpose machinery.

The *free fit* is suitable for running fits where the speeds are in excess of 600 rpm, and the pressures in excess of 600 psi. This fit is suitable for shafts of generators, motors, engines, and some automotive parts.

Table 19-1 **ASA standard classes of fits, with formulas for recommended allowances and tolerances for the basic hole system**

Class of fit	Method of assembly	Allowance	Hole tolerance	Shaft tolerance
1. Loose.........	Strictly inter-changeable	$-0.0025 \sqrt[3]{D^2}$	$0.0025 \sqrt[3]{D}$	$-0.0025 \sqrt[3]{D}$
2. Free..........	Strictly inter-changeable	$-0.0014 \sqrt[3]{D^2}$	$0.0013 \sqrt[3]{D}$	$-0.0013 \sqrt[3]{D}$
3. Medium.......	Strictly inter-changeable	$-0.0009 \sqrt[3]{D^2}$	$0.0008 \sqrt[3]{D}$	$-0.0008 \sqrt[3]{D}$
4. Snug.........	Strictly inter-changeable	-0.0000	$0.0006 \sqrt[3]{D}$	$-0.0004 \sqrt[3]{D}$
5. Wringing......	Selective assembly	0.0000	$0.0006 \sqrt[3]{D}$	$0.0004 \sqrt[3]{D}$
6. Tight.........	Selective assembly	$0.00025D$	$0.0006 \sqrt[3]{D}$	$0.0006 \sqrt[3]{D}$
7. Medium force..	Selective assembly	$0.0005D$	$0.0006 \sqrt[3]{D}$	$0.0006 \sqrt[3]{D}$
8. Heavy force or shrink	Selective assembly	$0.001D$	$0.0006 \sqrt[3]{D}$	$0.0006 \sqrt[3]{D}$

D = nominal diameter = basic diameter of hole. Basic diameter of shaft = D plus algebraically the allowance. Hole limits = D, and D plus the hole tolerance; shaft limits = basic shaft diameter, and basic shaft diameter plus algebraically the shaft tolerance. Allowances and tolerances usually are rounded off to the nearest 0.0001 in.

The *medium fit* is suitable for running fits where the speeds are under 600 rpm, and the pressures under 600 psi. This fit is used for the more accurate machine-tool and automotive parts.

The *snug fit* is a zero-allowance fit and is the closest fit that can be assembled by hand. It is suitable where no perceptible shake is permissible and also where the parts are not to slide freely when under load.

The *wringing fit* is practically a metal-to-metal fit and is selective, not interchangeable. Light tapping with a hammer is necessary to assemble the parts.

The *tight fit* has a slight positive allowance or metal interference, and light pressure is required to assemble the parts. This fit is suitable for semipermanent assembly, for long fits in heavy sections, and for drive fits in thin sections. It also is suitable for shrink fits in light sections.

The *medium force fit* requires considerable force to assemble. This fit is suitable for press fits on locomotive wheels, car wheels, generator and motor armatures, and crank disks. It is also suitable for shrink fits on medium sections. This is the tightest fit recommended for cast-iron external members since it often stresses the cast iron to its yield stress.

The *heavy force* and *shrink fit* are used for steel external members that have a high yield stress. They will overstress cast-iron external members. When the force fit requires impractical assembly pressure, the shrink fit should be used.

The shrink fits are used in the assembly of steel rims on cast-iron wheels, high-grade steel rims on cast gear spiders, aluminum-alloy heads on steel cylinders of aircraft engines, and built-up large-bore guns. For heavy power transmission, keys are used in addition to the force fits, especially with shafts over 3 in. in diameter. The key is used as a locating guide during assembly and also to maintain a tight connection, since the slight twist of the shaft may cause creep between the shaft and hub and allow the joint gradually to work loose.

19-5 Unilateral and Bilateral Tolerance Systems. A *unilateral tolerance system* is one in which one of the tolerances is zero; the resulting limits are the basic dimension and the basic dimension plus the nonzero tolerance algebraically combined. A *bilateral tolerance system* is one in which no tolerances are zero; that is, the basic dimension is not a limit. The limits are the basic dimension plus the nonzero positive tolerance and the basic dimension plus the negative tolerance. In the bilateral tolerance system, the two tolerances are often of equal magnitude, but such is not necessary. The unilateral tolerance system appears to be the prevalent system in use.

Example of Fits, Tolerances, and Limits. By use of an ASA class 3 fit, determine the limits for a 2-in. journal and bearing. Indicate the journal and bearing diameters as they are calculated by the systems of limits, unilateral tolerance, and bilateral tolerance.

Solution. Since the basic hole system is to be used, the basic bore of the bearing is 2.0000 in., the nominal diameter of the mating members. From Table 19-1, the allowance is $-0.0009 \sqrt[3]{2.0^2} = -0.0014$ in.; thus the basic diameter of the journal is $2.0000 - 0.0014 = 1.9986$ in. Excluding algebraic sign, the nonzero tolerances of the bearing and journal each are $0.0008 \sqrt[3]{2.0} = 0.0010$ in., or 0.0010 in. and -0.0010 in. respectively taking into account the algebraic sign. This means that the second basic dimensions of the bearing bore and journal diameter are 2.0010 in. and 1.9976 in. respectively. In the systems of limits, unilateral tolerances, and bilateral tolerances, the limiting bores of the bearing would be indicated by 2.0000 in. and 2.0010 in., $2.0000^{+0.0010}_{-0.0000}$ in., and $2.0005^{\pm 0.005}$ in. respectively. In a similar manner, the limiting diameters of the journal would be indicated by 1.9986 in. and 1.9976 in., $1.9986^{+0.0000}_{-0.0010}$ in., or 1.9981 ± 0.0005 in. These notations would describe the limits equally well. Proper indication of these limiting dimensions is shown in Fig. 19-1. Note that when limits are used, the limit first encountered in the production of the part appears above the dimension line and the other limit appears below the dimension line. For male members manufactured by turning down their diameter from larger stock than finally required, the upper limit would be first encountered in production. For holes produced by boring out a hole smaller than finally required, the lower limit would first be encountered in the production of the bearing. Although for many decades unilateral or bilateral tolerances have been successfully used on production drawings, there appears to be an increasing preference for limits rather than tolerances. Perhaps this is because the use of limits helps prevent mistakes when limiting dimensions are being obtained by adding or subtracting tolerances from

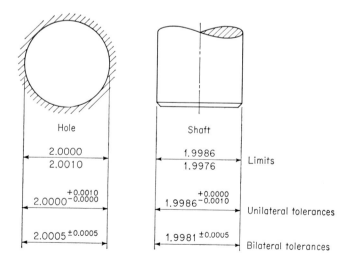

Fig. 19-1 Indication of the use of limits, unilateral tolerances, and bilateral tolerances for female and male dimensions.

basic dimensions. Limits directly present the dimensions necessary for the gaging and inspection of manufactured parts.

19-6   Need for Recognizing Surface Roughness.   Although machine parts may be manufactured to conform to certain standard fits, they may be unsuitable for use because of surface irregularities which affect the matability, appearance, color, luster, corrosion and wear resistance, and other properties of the surface. Perhaps the most important of these items is the *surface roughness*, which refers to the relatively finely spaced surface irregularities produced by production techniques. Until recent times, it has been difficult to specify surface roughness adequately, mainly because surface-roughness standards common to a majority of manufacturers were lacking.

19-7   Theory of Surface Roughness.   The irregularities of any surface usually are so minute that often it is difficult if not impossible to make direct measurements of surface roughness. Figure 19-2 depicts a profile through a typical finished metallic surface; in this figure, the vertical dimension has been magnified approximately 10 times more than the horizontal in order to exaggerate the surface roughness. The *mean surface* is that straight line for plane surfaces, or that circle for cylindrical or spherical surfaces, which represents the desired surface of the machine part if it could be produced without surface irregularities of any sort. If two lines are drawn parallel to the mean surface through the highest peak and the lowest valley in the profile under inspection, the distance between these two lines is the *maximum peak-to-valley surface roughness,* sometimes called the *maximum overall surface roughness.* This roughness dimension, as well as all other roughness indications, usually is given

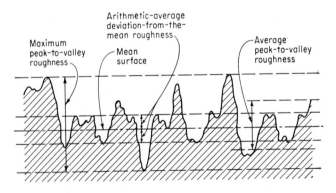

Fig. 19-2   Greatly magnified profile through a metallic surface (vertical scale magnified approximately 10 times the horizontal scale).

in *microinches* (μin.) or millionths of an inch. If two lines are drawn, one through the average of the peaks and one through the average of the valleys of the surface, the distance between these two lines is the *average peak-to-valley surface roughness*, frequently called the *predominant-peak surface roughness*. There is a personal factor of interpretation concerning exactly where these two lines should be placed. For a given length of mean surface, if the absolute values of the areas of the peaks above the mean surface and the areas of the valleys or voids below the mean surface are added together and then divided by the length of the mean surface, the *average deviation from the mean surface* is obtained. The *arithmetic average deviation from the mean surface* is the arithmetic average of the deviations of the actual surface from the mean surface and is calculated by making numerous equally spaced readings along the surface. With the equally spaced deviations, plus the square root of the quotient of the sum of the squares of the individual deviations divided by the number of readings, one can obtain the *root-mean-square average deviation from the mean surface*, usually called the *rms surface roughness*. Although the average deviation from mean, arithmetic average from mean, and rms surface roughness are not identical, the disagreement between them almost always is less than the usual error involved in making actual surface-roughness measurements.

The rms surface roughness is more commonly used for surface-roughness designation than any of the other defined roughnesses. Equipment such as the Brush Surface Analyzer or the Profilometer is available to give indications of rms surface roughness automatically. The use of such equipment together with standard surface-roughness calibration specimens enables a manufacturer to monitor the surface roughness of produced machine parts.

19-8   Typical Surface Roughness of Metallic Finishes.   Typical average values of rms surface roughness of machined and natural metallic finishes are given in Table 19-2. The expected range of roughness by the usual production techniques also is given.

For some applications, it is more desirable to use average peak-to-valley surface roughness than rms surface roughness. Unfortunately, good data concerning average peak-to-valley roughness are scarce. Table 19-3 lists typical ratios of average peak-to-valley to rms surface roughness for several machined finishes. Typical average peak-to-valley surface roughness therefore may be obtained by multiplying the rms roughness obtained from Table 19-2 by the appropriate dimensionless roughness ratios from Table 19-3.

19-9   Suggested Standard Values of Roughness.   Under sponsorship of ASA, ASME, and SAE, the suggested values of standard rough-

Table 19-2 **Typical rms roughness of metallic finishes**

Metallic finish	Average value, rms μin.	Typical range, rms μin.
Bored........................	30	16–250
Buffed.......................	10	2–16
Cut with torch................	750	500–1,000
Die cast......................	50	32–125
Drilled.......................	80	63–250
Extruded.....................	80	32–250
Filed.........................	100	63–500
Forged.......................	150	63–250
Ground:		
Cylindrical.................	10	6–63
Disk......................	80	63–500
Hand......................	400	250–1,000
Honed.......................	8	2–16
Lapped.......................	8	2–16
Lathe-turned.................	30	10–500
Milled.......................	50	32–500
Permanent-mold cast..........	40	32–250
Polished:		
Electropolished.............	3	1–5
Mechanically...............	8	2–16
Reamed......................	30	16–250
Rolled.......................	25	16–250
Sand cast....................	600	250–1,000
Sawed.......................	100	32–250
Shaped......................	50	32–500
Superfinished................	4	1–8

Table 19-3 **Typical ratios of average peak-to-valley roughness to rms roughness for several metallic finishes**

Metallic finish	Average value of ratio	Typical range of ratio
Ground........................	4.5	3.5–5.0
Hydrolapped...................	6.5	5.5–7.5
Loose-abrasive lapped...........	10.0	7.5–13.0
Sandpapered...................	7.0	5.5–9.0
Superfinished..................	7.0	5.5–9.5
Turned........................	5.0	3.5–8.0

Table 19-4 Suggested standard values of roughness (μin.)

0.25	4*	13	40	125*	400
0.50	5	16*	50	160	500*
1	6	20	63*	200	600
2*	8*	25	80	250*	800
3	10	32*	100	320	1000

* It is suggested that preference be given to these values because flat standard roughness specimens for these roughnesses are already readily available which permit actual comparison by use of the fingernail.

ness given in Table 19-4 were devised. Although there is no real reason why a specified roughness should have to conform to any of those values listed in the table, it is believed that the tabulated values present a variety sufficient for most purposes.

19-10 Waviness, Lay, and Flaws. *Waviness* refers to those irregularities of greater spacing than the roughness irregularities. Waviness is usually specified in maximum peak-to-valley height in inches and is usually measured with a dial indicator.

Lay refers to the predominant direction of the marks visible on the surface. The standard types of lay are parallel to a designated direction, perpendicular to a designated direction, in both directions, multiangular, circular, and radial relative to the surface indicated.

Flaws are those irregularities occurring at one or more places on the surface which are not characteristic of the surface and cannot be designated by surface roughness, waviness, or lay.

19-11 Surface-roughness Symbols. Formerly the finish symbol f was placed on surfaces or drawings to indicate that these surfaces were to be machined; the symbol was often accompanied by a note such as "high-speed lathe finish" or "smooth grind." In order to indicate the rms roughness in microinches, one currently provides the check or v surface-roughness mark with accompanying indication of maximum allowable roughness. Such a mark, indicating a maximum surface roughness of 32 rms μin., is shown in Fig. 19-3.

19-12 Description of Various Degrees of Surface Roughness. Below are listed 12 degrees of rms surface roughness in common use today in American industry. Typical methods of producing the surfaces and the applications for the surfaces are given.*

* J. A. Broadston, Surface-finish Requirements in Design, *Metals Engineering–Design*, pp. 291–292, ASME.

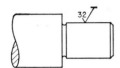

Fig. 19-3 Surface finish indication of 32-rms-μin. roughness.

2000 rms μin. An unusually rough, low-grade surface produced by removing excess stock on heavy work to provide clearance or improved appearance.

1000 rms μin. A very rough surface made by removing stock to nominal dimensions.

500 rms μin. A very rough surface resulting from heavy cuts and coarse feeds. Suitable only for clearance surfaces that are not subject to vibration, fatigue, or stress concentration, and that do not make contact with other parts. In heavy work, this roughness may be used where close tolerances are not required, for bolted or riveted assembly, and for large bearings.

250 rms μin. This surface definitely shows tool marks from rapid feeds in producing a medium machine finish, or may be produced by very coarse surface grinding, rough filing, coarse disk grinding, etc. It is used as a rough finish for soft alloys and for steel, and for hard alloys where moderate notch sensitivity exists but where a rough finish is otherwise tolerable. It also is suitable for parts where an average machine finish is acceptable, and on deep bores that do not require special finishing. It may be used on heavy work when good surface contact is essential.

125 rms μin. This is a good machine finish resulting from high-grade machine work using high speeds and fine feeds taking light cuts with sharp cutters or tools. It may be produced by all methods of direct machining under proper conditions, including coarse surface and cylindrical grinding, average disk grinding, and ordinary hand filing. It is the coarsest finish suitable for rough bearing surfaces where loads are light and infrequent. It also may be used on moderately close fits. It is suitable for parts requiring a good finished appearance.

63 rms μin. This is a smooth machine finish suitable for ordinary bearings and ordinary machine parts where fairly close dimensional tolerances must be held. It may be used for highly stressed parts that are not subject to severe stress reversals. It is just about as smooth a finish as can be produced economically by turning and milling without subsequent operations. This finish can be produced on a flat surface by a power grinder.

32 rms μin. This surface corresponds to a fine machine finish, rough commercial carbide or diamond boring, medium surface grinding, coarse cylindrical grinding, reaming, and similar operations. In the case of turned parts, this finish may be produced by subsequent hand work with emery cloth. *In using this finish and finer finishes, careful consideration must be given to increased production costs.* This finish may be used on parts subject to stress concentration and vibration. While this finish is difficult to produce directly on a lathe, mill, or by other direct machining operations, it is relatively easy to produce with a centerless grinder, cylindrical grinder, or surface grinder; it is the most suitable general finish for hardened steel parts. It may be used for bearings where accuracy is essential, where motion is continuous, and where loads are light, particularly if the lay of the surface is in the direction of motion.

16 rms μin. This fine surface may be produced by cylindrical grinding, very smooth reaming, fine surface grinding, smooth emery buffing, coarse honing, coarse lapping, etc. It is seldom used except where surface finish is of primary importance for the proper functioning of machine parts. Typical applications are rapidly rotating shaft bearings, heavily loaded bearings, and extreme tension members. It also is required in hydraulic applications for static sealing rings, at the bottom of sealing-ring grooves, etc. It is suitable for close tolerance work, journals, journal bearings, for commercial-grade bearing balls and rollers as well as the mating races, the outer diameters of such bearings, the outer diameters of shafts inside such bearings, and the bores in which such bearings are housed.

8 rms μin. This may be produced by a fine cylindrical grind, microhone, hone, lap, buff, etc. It is for use only when coarser finishes are known to be inadequate. It is used for work to very close tolerances where scratches cannot be tolerated. It is suitable for the interior surfaces of hydraulic struts and similar applications where properly lubricated seals must slide. It is suitable for raceways, balls, and rolls of rolling-element bearings when the loads are nominal. This finish and finer finishes may be dull or bright in appearance, depending mainly on how they are produced, so appearance should not be considered in judging quality.

4 rms μin. This surface may be produced by honing, lapping, superfinishing, very fine buffing, or bright polishing. It is a mirrorlike surface without tool or file marks of any kind. Because of their high cost, this and finer finishes should not be used unless essential to design. It is suitable for use where packings and rings must slide across the direction of the finish lay, such as on buffed chrome-plated piston rods and for hydraulic components. It is suitable for ball and roller bearings carrying heavy loads.

2 rms μin. This may be produced by precision lapping, superfinishing, and similar processes. It is suitable for gages, refrigerator shaft seals, etc. It is expensive to produce.

1 rms μin. This very fine finish may be produced by precision lapping, rouge lapping, cloth wheel and tripoli polishing, hand lapping, superfinishing, and similar processes. It is very expensive unless done with modern precision-finishing machines. It is used for micrometer anvils, high-grade gages, and similar applications.

Table 19-5   **Typical surface roughness of machine elements***

Roughness, rms μin.	*Typical applications in machine elements*
250	Clearance surfaces; rough machine parts
125	Mating surfaces without any relative motion; drilled holes
63	Gear locating faces; gear shafts and bores; cylinder head face; cast-iron gearbox faces; piston crowns
32	Brake drums; reamed holes; broached holes; bronze bearings; precision parts
16	Splined shafts; O-ring grooves where no motion occurs; gear teeth; motor shaft bearings; ground roller and ball bearings
12	Piston outside diameters; cylinder bores
8	Crankshaft bearings; camshaft bearings; connecting-rod bearings; valve stems; cam faces; hydraulic cylinder bores; lapped roller and ball bearings
4	Vernier caliper faces; wrist pins; O-ring grooves where relative motion occurs; hydraulic piston rods; railroad axles; precision tools; honed roller and ball bearings
2	Shop gage faces; comparator anvils
1	Micrometer anvils; mirrors; gages

* J. A. Broadston, Surface-finish Requirements in Design, *Metals Engineering–Design*, p. 293, ASME.

19-13   Typical Surface Roughness of Machine Elements.   Ten degrees of surface roughness commonly used in industry and typical uses of each in machine elements are listed in Table 19-5.

20

Cylinders, Pipes, Tubes, and Plates

20-1 Introduction. The variety of applications of cylinders in machine design and the wide range of materials used have led to different methods of design, depending upon the material, the type of cylinder, the service conditions, and many factors which must be considered in addition to the strength requirements. For small cylinders, strength, wear, and corrosion resistance are probably the most important considerations. For large cylinders, distortion caused by their own weight may be serious; hence rigidity is a vital requirement. Initial distortion and variations in wall thickness are more serious in cylinders subjected to external pressure than in cylinders subjected to internal pressure only. In steam apparatus, stills, cookers, and similar apparatus, the effects of temperature on the permissible working stress and on the distortion of the cylinder become important. The design of cylinders for many uses has been more or less standardized. The following paragraphs present the more important theories and standard formulas.

20-2 Thin Cylinders. A cylinder in which the ratio of the wall thickness to the inside diameter is less than 0.07 may be called a thin

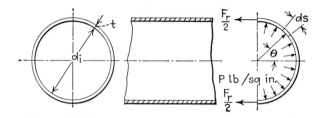

Fig. 20-1

cylinder. In Fig. 20-1, such a cylinder is shown cut by an imaginary plane through its axis. When an internal pressure is applied, the total force acting on the half cylinder and tending to rupture it along the cutting plane is

$$F = \int pL \, ds \cos \theta = \int_{-\pi/2}^{\pi/2} pLr \cos \theta \, d\theta = 2pLr = pLd_i$$

and the total resisting force in the cylinder walls cut by the plane is

$$F_r = 2tLs_t$$

For equilibrium, the rupturing force and the resisting forces must be equal. Hence, assuming uniform stress distribution

$$pLd_i = 2tLs_t$$

and

$$s_t = \frac{pd_i}{2t} \tag{20-1}$$

where p = internal pressure, psi
   d_i = inside diameter, in.
   t = wall thickness, in.
   L = length, in.
   s_t = tensile stress, psi

When there is a seam or joint in the cylinder, the joint efficiency e must be considered, and

$$s_t = \frac{pd_i}{2et} \tag{20-2}$$

Imagine a plane passed through the cylinder perpendicular to the axis. The total force tending to rupture the cylinder along this plane and the corresponding resisting force are

$$F = \frac{\pi d_i^2}{4} p$$

and

$$F_r = \pi d_i t s_t$$

from which

$$s_t = \frac{p d_i}{4t} \tag{20-3}$$

This is the stress in the cylinder wall parallel to its axis, and is one-half the tangential stress, as found in Eq. (20-1).

Internal pressure on a thin vessel that is not truly cylindrical tends to make it become cylindrical. However, if the material is fairly rigid it will resist this action and bending stresses of unknown magnitude will be induced in the wall.

20-3   Thin Spheres with Internal Pressure.   The stress in the wall of a sphere is the same as the stress in a cylinder in the direction parallel to the axis; hence Eq. (20-3) applies to spheres.

20-4   Long Thin Tubes with Internal Pressure.   Small tubes and pipes generally have much thicker walls than are required by the thin-cylinder equation, since with threaded joints the effective metal below the threads is only about one-half the wall thickness. Also, some allowance must be made for flexural stresses due to out-of-round, for nonuniform thickness, for corrosion, and for wear under service conditions. In some conditions of service, flexural stresses are also set up when the supports are far apart and when the installation is such that temperature changes may cause bending.

Iron and steel pipes are made from plates, formed while hot, with the longitudinal joint either butt-welded or lap-welded. Seamless tubing is formed by drawing and has no longitudinal joint. Iron and steel pipes have been thoroughly standardized with respect to wall thickness and with respect to the steam or hydrostatic pressure that may be safely carried. The nominal diameter of standard weight pipe up to and including 12 in. is the approximate inside diameter; for 14 in. and above, the nominal diameter is the outside diameter. The extra metal required in extra-strong and double extra-strong pipes is added to the inside, the outside diameters of all pipes of the same nominal diameter being the same. Dimensions of pipes and pipe fittings, and the permissible pressures for each size may be found in tabulated form in any mechanical engineer's handbook or pipe manufacturer's catalog.

The American Society of Mechanical Engineers' Boiler Code, the American Petroleum Institute Standards, and the American Standards Association provide formulas, information on materials, and allowable working stresses for obtaining the thickness of pipes and tubes. These formulas are in general based upon Barlow's thick cylinder equation, Eq. (20-32), with additive constants to provide for the number of threads and pipe sizes, grooves, plain ends, expansion of tubes into tube sheets and welds. The literature of these agencies should be consulted.

20-5 Openings in Cylindrical Drums.   Many cylindrical vessels subjected to internal pressure must have openings provided in the shell. A rule providing for such openings in cylinders of ductile material, based on data secured through experience with a large number of installations having unreinforced openings and giving satisfactory service over long periods of time, has been proposed by D. S. Jacobus.* The largest permissible diameter of opening in inches is given by the formula

$$d' = 2.75 \sqrt[3]{d_o t(1.0 - K)} \tag{20-4}$$

where K is the ratio of the stress in the solid plate to one-fifth the minimum tensile strength of the steel used in the shell, or $(pd_o/2t)(5/s_{tu})$; and d_o is the outside diameter of the drum, in inches. The maximum diameter of the unreinforced hole should be limited to 8 in. and should not exceed $0.6d_o$.

20-6 Thin Tubes and External Pressure.   The equations developed in the preceding paragraphs are not applicable to cylinders subjected to external pressure, since these will collapse at apparent stresses much below those required for direct compression failure of the material. For this reason, experimental data must be used as the basis of empirical design formulas. Such data indicate that the tube length between transverse joints which tend to maintain the circular form has no influence upon the collapsing pressure of tubes so long as the length is not less than six times the tube diameter. Very short tubes are subject to the supporting effect of the end connections.

Experiments conducted by A. P. Carman† and R. T. Stewart‡ agree quite closely. Carman's formulas for the collapsing pressure of tubes follow:

For seamless steel tubes,

$$p_{cr} = 50,200,000 \left(\frac{t}{d_o}\right)^3 \tag{20-5}$$

---

* D. S. Jacobus, Openings in Cylindrical Drums, *Mech. Eng.*, p. 368, May, 1932.

† A. P. Carman, Resistance of Tubes to Collapse, *Univ. Illinois Eng. Exp. Sta. Bull.* 5, 1906.

‡ R. T. Stewart, Collapsing Pressures of Bessemer Steel Lap-welded Tubes, *Trans. ASME*, vol. 27, p. 731, 1906.

when t/d_o is less than 0.025; and

$$p_{cr} = 95,520 \frac{t}{d_o} - 2,090 \qquad (20\text{-}6)$$

when t/d_o is greater than 0.03.
For lap-welded steel tubes,

$$p_{cr} = 83,290 \frac{t}{d_o} - 1,025 \qquad (20\text{-}7)$$

when t/d_o is greater than 0.03.
For brass tubes,

$$p_{cr} = 25,150,000 \left(\frac{t}{d_o}\right)^3 \qquad (20\text{-}8)$$

when t/d_o is less than 0.025; and

$$p_{cr} = 93,365 \frac{t}{d_o} - 2,474 \qquad (20\text{-}9)$$

when t/d_o is greater than 0.03.

All the formulas are for the collapsing pressure, i.e., the ultimate strength of the tubes, and must be modified to suit the service conditions. The factor of safety to be used should not be less than 3 for the most favorable conditions and should be increased to 6 when there is a possibility of loss of life. When corrosion, weakening due to heating, and other service conditions reduce the collapsing resistance of the tube, the factor of safety should be increased in proportion to the weakening effect. For example, external pressure tubes in boilers are designed with an apparent factor of safety of from 8 to 10.

20-7 Short Tubes and External Pressure. The cylinder heads and the end connections tend to stiffen the tubes and prevent their collapse. When the length is less than six diameters, the strengthening effect must be considered. The best known experimental work on the crushing strength of short tubes is that of Sir William Fairbairn who, in 1858, developed the following formula from which many more recent formulas have been deduced:

$$p_{cr} = 9,675,600 \frac{t^{2.19}}{Ld_o} \qquad (20\text{-}10)$$

For very short tubes, it is possible that the material will fail by direct compression or crushing before the collapsing pressure is reached; hence the crushing stress should be computed from the equation

$$s_c = \frac{p_o d_o}{2t} \qquad (20\text{-}11)$$

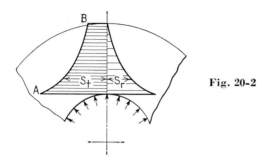

Fig. 20-2

20-8   Unfired Pressure Vessels with External Pressure.   There are a number of formulas in use for the design of cylindrical vessels for steam cookers, digesters, stills, vacuum tanks, and similar applications, and the results have not always been reliable. Probably the most reliable formula* for the critical pressure or collapsing pressure of large cylinders is

$$p_{cr} = \frac{2.6E(t/d_o)^{2.5}}{(L/d_o) - 0.45(t/d_o)^{0.5}} \qquad (20\text{-}12)$$

where L is the maximum distance between supports or stiffening rings, in inches.

To obtain safe working pressures, the critical pressure should be at least five times the working pressure.

20-9   Thick Cylinders.   In the discussion of thin cylinders, the stress was assumed to be uniformly distributed through the entire wall thickness. This assumption is approximately true for very thin walls only. In the case of a cylinder with internal pressure, the stress varies from a maximum at the inner surface to a minimum at the outer surface, as indicated in Fig. 20-2. Several theories of stress distribution have been developed, and several of those most commonly used are discussed in the following paragraphs. The following notation will be used:

d_i = inside diameter, in.
d_o = outside diameter, in.
d_c = diameter of the contact surface in a compound cylinder, in.
r = radius at any point in the wall, in.
s_a = axial stress, psi
s_r = radial stress, psi
s_t = tangential stress, psi
p_i = internal pressure, psi
p_o = external pressure, psi

* H. F. Saunders and D. F. Windenburg, Strength of Thin Cylindrical Shells under External Pressure, *Trans. ASME*, vol. 53, 1931.

p_c = pressure at the contact surface of a compound cylinder, psi
μ = Poisson's ratio of lateral contraction
t = wall thickness, in.

20-10   Lamé's Equations.   In the general case, there will be pressure applied to the cylinder both on the inside and on the outside. In Fig. 20-3, consider the cylinder to be cut by a plane perpendicular to the axis and at some distance from the end wall so as to eliminate the constraining action of the end wall. In this plane, consider an annular ring of radius r and thickness dr. The axial stress may be considered to be uniform over the wall thickness. Hence

$$s_a = \frac{(\pi d_i^2/4)p_i - (\pi d_o^2/4)p_o}{\pi d_o^2/4 - \pi d_i^2/4} = \frac{p_i d_i^2 - p_o d_o^2}{d_o^2 - d_i^2} \tag{20-13}$$

Now consider a small portion of wall material of unit axial length, of radial thickness dr, and of tangential thickness $r\,d\theta$, as shown in Fig. 20-3. This small body is held in equilibrium by axial, radial, and tangential stresses, s_a, s_r, and s_t as shown. The total deformation in the axial direction due to these stresses is

$$\Delta_a = \frac{s_a + \mu s_r - \mu s_t}{E}$$

But Δ_a, s_a, μ, and E are constant, hence

$$s_r - s_t = \text{a constant} = C \tag{20-14}$$

The radial stresses on the small body under consideration are s_r at the inner surface, and $(s_r + ds_r)$ at the outer surface.

The tangential stress on this body is found from the thin-cylinder formula. Hence

$$s_t = \frac{pd}{2t} = \frac{-(s_r + ds_r) \times 2(r + dr) + 2rs_r}{2dr}$$

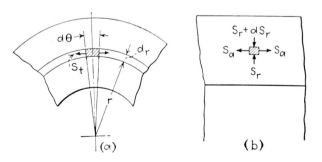

Fig. 20-3

Expanding, collecting the terms, and neglecting the infinitesimal product $ds_r \times d_r$,

$$\frac{dr}{r} = \frac{-ds_r}{s_r + s_t} = \frac{-ds_t}{2s_t + C}$$

since

$$s_r = C + s_t \qquad \text{and} \qquad ds_r = ds_t$$

by integration,

$$2 \log r = \log \frac{B}{2s_t + C}$$

and

$$r^2 = \frac{B}{2s_t + C}$$

where B and C are constants. Then

$$s_t = -\frac{C}{2} + \frac{B}{2r^2} \qquad (20\text{-}15)$$

and

$$s_r = \frac{C}{2} + \frac{B}{2r^2}$$

The constants B and C may be eliminated, since s_r equals p_i and s_t equals $C - p_i$, when r equals $d_i/2$, and s_r equals p_o and s_t equals $C - p_o$, when r equals $d_o/2$. By substitution of these values in Eq. (20-15), the tangential stress in the cylinder wall is

$$s_t = \frac{p_i d_i^2 - p_o d_o^2}{d_o^2 - d_i^2} + \frac{d_i^2 d_o^2}{4r^2} \frac{p_i - p_o}{d_o^2 - d_i^2} = a + \frac{b}{r^2} \qquad (20\text{-}16)$$

and the radial stress is

$$s_r = \frac{p_i d_i^2 - p_o d_o^2}{d_o^2 - d_i^2} - \frac{d_i^2 d_o^2}{4r^2} \frac{p_i - p_o}{d_o^2 - d_i^2} = a - \frac{b}{r^2} \qquad (20\text{-}17)$$

where a and b are constants for any given values of pressure and diameter.

Equations (20-16) and (20-17) are Lamé's general equations for the tangential and radial stresses at any radius r in the wall of a thick cylinder. These stresses are the maximum normal stresses in the wall at the radius r.

20-11   Lamé's Equations for Internal Pressure.   The most common case dealt with in machine design is a cylinder subjected to internal pressure only. In this case, p_o is zero, and

$$s_t = \frac{p_i d_i^2}{4r^2} \frac{4r^2 + d_o^2}{d_o^2 - d_i^2} \tag{20-18}$$

and

$$s_r = \frac{p_i d_i^2}{4r^2} \frac{4r^2 - d_o^2}{d_o^2 - d_i^2} \tag{20-19}$$

Inspection of these equations shows that both stresses increase as the radius decreases; hence the maximum stresses are at the inner surface where r equals $d_i/2$. Then

$$s_{t\ max} = \frac{p_i(d_o^2 + d_i^2)}{d_o^2 - d_i^2} \tag{20-20}$$

and

$$s_{r\ max} = -p_i \tag{20-21}$$

the negative sign indicating that s_r is a compressive stress.

The usual design problem is to determine the wall thickness when the allowable working stress and the internal pressure are known. Substitution of $d_i + 2t$ for d_o in Eq. (20-20) leads to the more convenient form

$$t = \frac{d_i}{2} \left(\sqrt{\frac{s_t + p_i}{s_t - p_i}} - 1 \right) \tag{20-22}$$

This equation is based on the maximum normal stress in the wall and is therefore applicable to brittle materials.

Maximum Shear in the Cylinder Wall. The maximum shear stress in the cylinder wall is found by the substitution of the values of s_t and s_r from Eqs. (20-20) and (20-21) in Eq. (3-36). Hence

$$s_{s\ max} = \frac{s_t - s_r}{2} = \frac{p_i d_o^2}{d_o^2 - d_i^2} \tag{20-23}$$

Since the maximum-shear theory holds that the shear stress is one-half the tensile stress, the equivalent maximum tensile stress in the cylinder wall is

$$s_{t\ max} = \frac{2p_i d_o^2}{d_o^2 - d_i^2} \tag{20-24}$$

and

$$t = \frac{d_i}{2} \left(\sqrt{\frac{s_t}{s_t - 2p_i}} - 1 \right) \tag{20-25}$$

This equation is applicable to ductile materials but is not considered to be so accurate as Clavarino's and Birnie's equations, since the effect of lateral contraction is neglected.

20-12   Clavarino's Equations for Closed Cylinders.   In the preceding article, the effects of lateral deformations of the wall material were neglected, but according to the maximum-strain theory, these lateral deformations affect the load-carrying capacity of the material. By the introduction of the effect of the lateral deformation, the equivalent axial, radial, and tangential stresses are

$$s_a' = s_a + \mu s_r - \mu s_t$$
$$s_r' = s_r + \mu s_a + \mu s_t$$

and

$$s_t' = s_t - \mu s_a + \mu s_r$$

When these values are used as the stresses acting on the body shown in Fig. 20-3, an analysis similar to that for Lamé's equations leads to the following general equations for the equivalent stresses:

$$s_t' = (1 - 2\mu)a + \frac{(1 + \mu)b}{r^2} \tag{20-26}$$

and

$$s_r' = (1 - 2\mu)a - \frac{(1 + \mu)b}{r^2} \tag{20-27}$$

when μ is Poisson's ratio, and a and b have the same meaning as in Eq. (20-16). When a and b are evaluated, and when $d_i/2$ is substituted for r, the equation for the wall thickness is

$$t = \frac{d_i}{2} \left[\sqrt{\frac{s_t' + (1 - 2\mu)p_i}{s_t' - (1 + \mu)p_i}} - 1 \right] \tag{20-28}$$

where s_t' is the permissible working stress in tension, in pounds per square inch.

These equations are known as Clavarino's equations for closed cylinders and are applicable to cylinders having the ends closed or fitted with heads so that axial stress is produced in the wall.

20-13   Birnie's Equations for Open Cylinders.   If the cylinder is considered to be open at the end so that no direct axial stress is possible and if the same notation is used as in Art. 20-10, an analysis similar to that for Clavarino's equations leads to the following equations for equivalent stress:

$$s_t' = (1 - \mu)a + (1 + \mu)\frac{b}{r^2} \tag{20-29}$$

and

$$s_r' = (1 - \mu)a - (1 + \mu)\frac{b}{r^2} \tag{20-30}$$

The equation for the wall thickness is

$$t = \frac{d_i}{2}\left[\sqrt{\frac{s_t' + (1 - \mu)p_i}{s_t' - (1 + \mu)p_i}} - 1\right]$$ (20-31)

These equations are applicable to certain types of pump cylinders, rams, cannon, and similar cylinders where no axial stress is present. An examination of Eqs. (20-28) and (20-31) shows that Birnie's equation always gives greater values of t than does Clavarino's equation; therefore, if there is any doubt as to whether the cylinder is open or closed, Birnie's equation should be used.

20-14 Barlow's Equation. In Lamé's equation for cylinders subjected to internal pressure only, substitute $d_i + 2t$ for d_o, and Eq. (20-21) becomes

$$s_t = \frac{p_i(2d_i^2 + 4td_i + 4t^2)}{4(d_i t + t^2)}$$

When t is small compared to d_i, the term t^2 may be neglected, and

$$s_t = \frac{p_i d_o}{2t}$$ (20-32)

This is Barlow's equation, which is similar to the thin-cylinder equation except that d_o replaces d_i. It is slightly more accurate and is commonly used in computing wall thicknesses for high-pressure oil and gas pipes.

20-15 Changes in Cylinder Diameter Due to Pressure. Although the changes are relatively small, there are cases, such as press and shrink fits, where the changes in cylinder diameter must be known. The unit deformation is equal to the unit stress divided by the modulus of elasticity. Hence the total increase in cylinder diameter is

$$\Delta d = \frac{s_t}{E} d$$ (20-33)

The value of s_t may be found from Eq. (20-29). When the cylinder is subjected to internal pressure only, p_o becomes zero, and the maximum s_t is at r equal to $d_i/2$. Hence the increase in cylinder diameter due to internal pressure is

$$\Delta d_i = \frac{d_i}{E}\frac{(1 - \mu)p_i d_i^2}{d_o^2 - d_i^2} + \frac{(1 + \mu)p_i d_i^2 d_o^2}{d_i^2(d_o^2 - d_i^2)}$$

$$= \frac{p_i d_i}{E}\left(\frac{d_o^2 + d_i^2}{d_o^2 - d_i^2} + \mu\right)$$ (20-34)

Table 20-1 Values of s_t to be used in Eq. (20-32) for nonferrous seamless tubes and pipes

Material	Spec. no.	Condition	For temperatures in °F not to exceed									
			Sub-zero to 150	250	300	350	400	450	500	550	600	700
Seamless copper pipe and tube	SB-13, 42 and 75	Annealed	6,000	5,000	4,750	4,500	3,000					
	SB-42 and SB-75	Light-drawn	7,200	6,000	5,000	4,500	3,000					
		Hard-drawn	10,000	9,000	7,000	5,000	3,000					
Brass pipe, Muntz	SB-43	Annealed	10,000	9,000	7,000	2,000	1,500					
Brass pipe, high brass	SB-43	Annealed	9,000	9,000	9,000	6,000	3,000	1,500	800			
Brass pipe, admiralty	SB-43	Annealed	9,000	9,000	9,000	6,000	5,500	4,500	800			
Brass pipe, red brass	SB-43	Annealed	8,000	8,000	7,000	6,000	3,000	1,500	800			
Copper–Ni 70-30	SB-111	Annealed	13,300	12,700	12,300	12,000	11,700	11,300	11,000	10,500	10,000	9,000
Copper–Ni 80-20	SB-111	Annealed	12,000	11,300	11,000	10,700	10,300	10,000	9,500	9,000	8,500	7,500
Copper–Ni 90-10	SB-111	Annealed	10,000	9,500	9,300	9,000	8,700	8,300	7,500	6,700	6,000	

Stresses in this table when used with Eq. (20-32) are applicable only to diameters $\frac{1}{2}$ to 6 in. outside diameter inclusive, and for wall thicknesses not less than No. 18 BWG (0.049 in.). Additional wall thickness should be provided where corrosion or wear due to cleaning operations is expected. Where tube ends are threaded, additional wall thickness of $\dfrac{0.8}{\text{number of threads}}$ is to be provided. Requirements for rolling or otherwise setting tubes in tube plates may require additional wall thickness.
    From ASME Boiler Code.

Similarly, the decrease in external diameter of a cylinder subjected to external pressure only is

$$\Delta d_o = \frac{p_o d_o}{E}\left(\frac{d_o^2 + d_i^2}{d_o^2 - d_i^2} - \mu\right) \tag{20-35}$$

20-16 Compound Cylinders. When high internal pressures are to be sustained, it is good practice to make the cylinder from two or more annular cylinders with the outer ones shrunk onto the inner ones. This puts the inner cylinder in compression before the internal pressure is applied, and when the internal pressure is applied the resulting stress in the cylinder wall is much lower than if a single cylinder had been used.

Consider a cylinder open at the ends so that no axial stress is imposed. The tangential stress is the larger and is the stress to be considered in the design. The value of this tangential stress at any radius r from Birnie's equation is

$$s_t = (1 - \mu)\frac{p_i d_i^2 - p_o d_o^2}{d_o^2 - d_i^2} + (1 + \mu)\frac{d_i^2 d_o^2 (p_i - p_o)}{4r^2(d_o^2 - d_i^2)} \tag{20-36}$$

In Fig. 20-2 showing a simple cylinder subjected to internal pressure only, the curve AB shows the variation of the tangential stress over the cross section of the wall. The equation of this stress, found by making p_o equal to zero in Eq. (20-36), is

$$s_t = (1 - \mu)\frac{p_i d_i^2}{d_o^2 - d_i^2} + (1 + \mu)\frac{p_i d_i^2 d_o^2}{4r^2(d_o^2 - d_i^2)} \tag{20-37}$$

Figure 20-4 shows a compound cylinder with the outer cylinder shrunk on. In this case the inner cylinder is subjected to an external pressure caused by the shrinking of the outer cylinder, and the outer cylinder is subjected to an internal pressure. Call the unit pressure between the cylinders p_c. In Eq. (20-36) substitute for p_i, p_o, d_o, and r the respective values zero, p_c, d_c, and $d_i/2$. Then the tangential stress at the inner surface of the inner cylinder is

$$s_{tii} = \frac{-2p_c d_c^2}{d_c^2 - d_i^2} \tag{20-38}$$

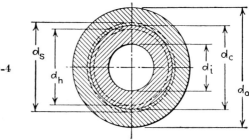

Fig. 20-4

Similarly, the tangential stress at the outer surface of the inner cylinder is

$$S_{toi} = -p_c \left(\frac{d_c^2 + d_i^2}{d_c^2 - d_i^2} - \mu \right) \tag{20-39}$$

the tangential stress at the inner surface of the outer cylinder is

$$S_{tio} = p_c \left(\frac{d_o^2 + d_c^2}{d_o^2 - d_c^2} + \mu \right) \tag{20-40}$$

and the tangential stress at the outer surface of the outer cylinder is

$$S_{too} = \frac{2p_c d_c^2}{d_o^2 - d_c^2} \tag{20-41}$$

Equations (20-38) through (20-41) give the stresses due to shrinking only. If internal pressure is applied to the cylinder there are stresses set up in addition to the shrinkage stresses. These additional stresses may be found for both cylinders when the proper values of r and p are substituted in Eq. (20-36).

The final tangential stress distribution in the cylinder wall is shown in Fig. 20-5. Curve A shows the distribution due to shrinkage pressure only. Curve B shows the distribution due to internal pressure only. Curve C shows the final stress distribution, the stresses in curve C being the sum of the stresses in curves A and B.

20-17   Radial Pressure between the Cylinders.   The radial pressure between the cylinders at the surface of contact depends on the modulus of elasticity of the materials and on the difference between the outer diameter of the inner cylinder and the bore of the outer cylinder, before they are shrunk together.

Changes in the diameters of the two cylinders, in Fig. 20-4, due to pressure at their contact surface, are

$$\Delta d_s = d_s - d_c$$

and

$$\Delta d_h = d_c - d_h$$

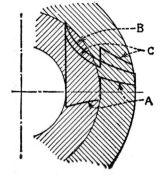

Fig. 20-5

from which

$$\Delta d_s + \Delta d_h = d_s - d_h = B$$

where B is the total shrinkage allowance, in inches.

The values of Δd_s and Δd_h in terms of the pressure p_c have been found in Eqs. (20-34) and (20-35). Hence

$$B = \frac{p_c d_s}{E_s}\left(\frac{d_s^2 + d_i^2}{d_s^2 - d_i^2} - \mu_s\right) + \frac{p_c d_h}{E_h}\left(\frac{d_o^2 + d_h^2}{d_o^2 - d_h^2} + \mu_h\right)$$

Without appreciable error, d_c may be substituted for d_s and d_h, since they differ by only a few thousandths of an inch. Then

$$\frac{B}{d_c} = p_c\left[\frac{d_c^2 + d_i^2}{E_s(d_c^2 - d_i^2)} + \frac{d_o^2 + d_c^2}{E_h(d_o^2 - d_c^2)} - \frac{\mu_s}{E_s} + \frac{\mu_h}{E_h}\right] \tag{20-42}$$

From this equation, the pressure between the cylinders, p_c, can be determined when the total shrinkage allowance is known, or the total allowance to produce the required pressure can be found.

In most compound cylinders, the outer and inner cylinders are made of the same material, and E_s and E_h are equal, and μ_s and μ_h are equal, and the last two terms in the brackets cancel out. After expansion of the terms and simplification

$$p_c = \frac{BE(d_c^2 - d_i^2)(d_o^2 - d_c^2)}{2d_c^3(d_o^2 - d_i^2)} \tag{20-43}$$

20-18 Selection of the Thick-cylinder Equation.

The particular equation to be applied depends upon the material utilized. Equation (20-16) may be used for the brittle materials such as hard steel, cast iron, and cast aluminum, inasmuch as the maximum-normal-stress theory applies. In general, the maximum-shear theory applies to ductile materials such as the low-carbon steels, brass, bronze, and aluminum alloys. Research indicates, however, that the maximum-strain theory agrees with experimental results on thick cylinders more closely than do the equations based on the maximum-shear theory. Hence engineers favor the maximum-strain theory for all materials used in thick cylinders.

20-19 Working Stress in Thick Cylinders.

In thick cylinders and compound cylinders, the maximum stress is present at the inner surface only, and the stress decreases rapidly toward the outer surface. Momentary overstresses are therefore not so serious as in some other machine members, since the material at the inner surface may flow slightly and readjust the stress distribution without causing failure. It is therefore permissible to use relatively high stresses, and if no shock is present, 85 per cent of the yield stress may be considered satisfactory. Thick-walled cylinders of cast iron or cast steel are liable to contain defects and to be more unreliable than thin-walled cylinders; hence with cast cylinders

Table 20-2 **Maximum stresses and deflections in flat plates**

Form of plate	Type of loading	Type of support	Eq.	Total load F	Maximum stress s_{max}	Location of s_{max}	Maximum deflection y_{max}
	Distributed over the entire surface	Edge supported	(1)	$\pi r_0^2 p$	$s_r = s_t = \dfrac{-3F(3m+1)}{8\pi mt^2}$	Center	$\dfrac{3F(m-1)(5m+1)r_0^2}{16\pi Em^2t^3}$
		Edge fixed	(2)	$\pi r_0^2 p$	$s_t = \dfrac{3F}{4\pi t^2}$	Edge	$\dfrac{3F(m^2-1)r_0^2}{16\pi Em^2t^3}$
	Distributed over a concentric circular area of radius r	Edge supported	(3)	$\pi r^2 p$	$s_r = s_t = \dfrac{-3F}{2\pi t^2}\left[(m+1)\log_e\dfrac{r_0}{r} - (m-1)\dfrac{r^2}{4r_0^2} + m\right]$	Center	$\dfrac{3F(m^2-1)}{16\pi Em^2t^3}\left[\dfrac{(12m+4)r_0^2}{m+1} - 4r^2\log_e\dfrac{r_0}{r} - \dfrac{(7m+3)r^2}{m+1}\right]$
		Edge fixed	(4)	$\pi r^2 p$	$s_r = \dfrac{3F}{2\pi t^2}\left(1 - \dfrac{r^2}{2r_0^2}\right)$ $s_r = s_t = \dfrac{-3F}{2\pi mt^2}\left[(m+1)\log_e\dfrac{r_0}{r} + (m+1)\dfrac{r^2}{4r_0^2}\right]$	Edge Center	$\dfrac{3F(m^2-1)}{16\pi Em^2t^3}\left[4r_0^2 - 4r^2\log_e\dfrac{r_0}{r} - 3r^2\right]$ when r is very small (concentrated load) $\dfrac{3F(m^2-1)r_0^2}{4\pi Em^2t^3}$
	Distributed on circumference of a concentric circle of radius r	Edge supported	(5)	$2\pi r p$	$s_r = s_t = \dfrac{-3F}{2\pi mt^2}\left[\dfrac{m-1}{2} + (m+1)\log_e\dfrac{r_0}{r} - (m-1)\dfrac{r^2}{2r_0^2}\right]$	All points inside the circle of radius r	$\dfrac{3F(m^2-1)}{2\pi Em^2t^3}\left[\dfrac{(3m+1)r_0^2}{2(m+1)} - \dfrac{(m-1)r^2}{2(m+1)} - r^2\left(\log_e\dfrac{r_0}{r}+1\right)\right]$
		Edge fixed	(6)	$2\pi r p$	$s_r = s_t = \dfrac{-3F}{4\pi mt^2}\left[(m+1)\left(2\log_e\dfrac{r_0}{r}\right) + \dfrac{r^2}{r_0^2} - 1\right]$ $s_r = \dfrac{3F}{2\pi t^2}\left(1 - \dfrac{r^2}{r_0^2}\right)$	Center when $r < 0.31r_0$ Edge when $r > 0.31r_0$	$\dfrac{3F(m^2-1)}{2\pi Em^2t^3}\left[\dfrac{1}{2}(r_0^2 - r^2) - r^2\log_e\dfrac{r_0}{r}\right]$

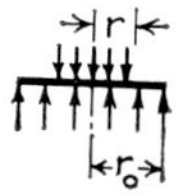	Distributed over a concentric circular area of radius r	Uniform pressure over entire lower surface	(7)	$\pi r^2 p$	$s_r = s_t = \frac{-3F}{2\pi m t^2}\left[(m+1)\log_e \frac{r_o}{r} + \frac{m-1}{4}\left(1 - \frac{r^2}{r_o^2}\right)\right]$	Center	$\frac{3F(m^2-1)}{16\pi E m^2 t^3}\left[4r^2 \log_e \frac{r_o}{r} + 2r^2\left(\frac{3m+1}{m+1}\right) - r_o^2\left(\frac{7m+3}{m+1}\right) + \frac{(r_o^2 - r^2)r^4}{r^2 r_o^2} + \frac{r^4}{r_o^2}\right]$ when r is very small (concentrated load) $\frac{3F(m-1)(7m+3)r_o^2}{16\pi E m^2 t^3}$
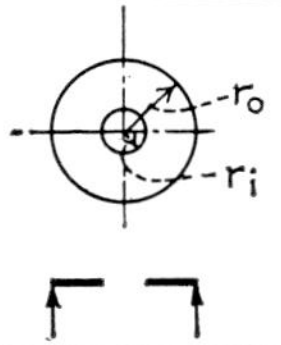	Distributed over the entire surface	Outer edge supported	(8)	$F = \pi(r_o^2 - r_i^2)p$	$s_t = \frac{-3p}{4mt^2(r_o^2 - r_i^2)}\left[r_o^4(3m+1) + r_i^4(m-1) - 4mr_o^2r_i^2 - 4(m+1)r_o^2r_i^2 \log_e \frac{r_o}{r_i}\right]$	Inner edge	$\frac{3p(m^2-1)}{2Em^2t^3}\left[\frac{r_o^4(5m+1)}{8(m+1)} + \frac{r_i^4(7m+3)}{8(m+1)} - \frac{r_o^2r_i^2(3m+1)}{2(m+1)} + \frac{r_o^2r_i^2(3m+1)}{2(m-1)}\log_e \frac{r_o}{r_i} - \frac{2r_o^2r_i^4(m+1)}{(r_o^2 - r_i^2)(m-1)}\left(\log_e \frac{r_o}{r_i}\right)^2\right]$
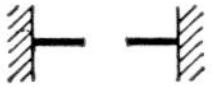	Distributed over the entire surface	Outer edge fixed and supported	(9)	$F = \pi(r_o^2 - r_i^2)p$	$s_t = \frac{-3p(m^2-1)}{4mt^2}\left[\frac{r_o^4 - r_i^4 - 4r_o^2r_i^2 \log_e \frac{r_o}{r_i}}{r_o^2(m-1) + r_i^2(m+1)}\right]$	Inner edge	
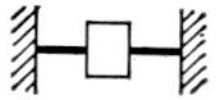	Distributed over the entire surface	Outer edge fixed and supported, inner edge fixed	(10)	$F = \pi(r_o^2 - r_i^2)p$	$s_r = \frac{-3p}{4t^2}\left[(r_o^2 + r_i^2) - \frac{4r_o^2r_i^2}{r_o^2 - r_i^2}\left(\log_e \frac{r_o}{r_i}\right)\right]$	Inner edge	$\frac{3p(m^2-1)}{16Em^2t^3}\left[r_o^4 + 3r_i^4 - 4r_o^2r_i^2 - 4r_o^2r_i^2 \log_e \frac{r_o}{r_i} + \frac{16r_o^2r_i^4}{r_o^2 - r_i^2}\left(\log_e \frac{r_o}{r_i}\right)^2\right]$
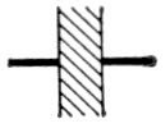	Distributed over the entire surface	Inner edge fixed and supported	(11)	$F = \pi(r_o^2 - r_i^2)p$	$s_r = \frac{3p}{4t^2}\left[\frac{4r_o^4(m+1)\log_e(r_o/r_i) - r_o^4(m+3) + r_i^4(m-1) + 4r_o^2r_i^2}{r_o^2(m+1) + r_i^2(m-1)}\right]$	Inner edge	

Table 20-2 **Maximum stresses and deflections in flat plates** (*Continued*)

Form of plate	Type of loading	Type of support	Eq.	Total load F	Maximum stress s_{max}	Location of s_{max}	Maximum deflection y_{max}
	Uniform over entire surface	All edges supported	(12)	$F = abp$	$s_b = \dfrac{-0.75 b^2 p}{t^2 \left(1 + 1.61\, \dfrac{b^3}{a^3}\right)}$	Center	$\dfrac{0.1422 b^4 p}{E t^3 \left(1 + 2.21\, \dfrac{b^3}{a^3}\right)}$
	Uniform over entire surface	All edges fixed	(13)	$F = abp$	$s_b = \dfrac{0.5 b^2 p}{t^2 \left(1 + 0.623\, \dfrac{b^6}{a^6}\right)}$	Center of long edge	$\dfrac{0.0284 b^4 p}{E t^3 \left(1 + 1.056\, \dfrac{b^5}{a^5}\right)}$
	Uniform over entire surface	Short edges fixed, long edges supported	(14)	$F = abp$	$s_a = \dfrac{0.75 b^2 p}{t^2 \left(1 + 0.8\, \dfrac{b^4}{a^4}\right)}$	Center of short edge	
	Uniform over entire surface	Short edges supported, long edges fixed	(15)	$F = abp$	$s_b = \dfrac{-b^2 p}{2 t^2 \left(1 + 0.2\, \dfrac{a^4}{b^4}\right)}$	Center of long edge	

Abstracted from S. J. Roark, "Formulas for Stress and Strain," 3d ed., McGraw-Hill Book Company, New York, 1954.

Positive sign for s indicates tension at upper surface and equal compression at lower surface; negative sign indicates reverse condition.

it is often better to use high stresses in order to obtain better castings with thinner walls.

20-20 Flat Heads. In many machines flat plates or slightly dished plates are used as diaphragms, pistons, cylinder heads, boiler heads, and the sides of rectangular tanks. These plates may be simply supported at the edges or at the center, or they may be more or less rigidly connected to the supporting member by means of bolts, welds, or rivets or by being integral with the support. The rigorous stress analysis of such plates is difficult, and many of the equations developed have not been entirely verified by experimental evidence. The formulas generally used are based on the work of Bach and Grashof, and since the derivations are complex and may be found in texts on stress analysis, the derivations will not be presented here. Typical formulas employed are presented in Table 20-2. Results obtained by the use of these formulas are in general conservative.

The following symbols are used with the formulas in Table 20-2:

F = total load supported, lb

p = load per unit area, psi

s = unit stress, psi

E = modulus of elasticity, psi

a = length of the long side of a rectangular plate, in.

b = length of the short side of a rectangular plate, in.

r_o = outer radius of circular plate, in.

r_i = inner radius of plate with a concentric circular hole, in.

r = radius of circle over which the load is distributed, in.

t = plate thickness, in.

y = maximum deflection in the plate, in.

m = $1/\mu$, reciprocal of the Poisson's ratio given in Table 2-1

No plate can be supported with absolute rigidity, and the values of s and y obtained by the use of these equations are subject to modification according to the experience and judgment of the designer in determining the degree of rigidity. A plate or cylinder head held in place by evenly spaced bolts near its outer edge is usually considered to be freely supported, since the bolts are generally small enough to be considerably distorted when the load is applied. Heads cast integral with heavy cylinder walls approach the condition of rigid or fixed supports. These cast heads should be provided with generous fillets at the juncture with the walls to prevent the formation of crystalline cleavage planes at the corners when the casting is cooling. Large corner radii will prevent this action. Semispherical heads are the strongest forms to resist internal pressure; however, flat heads may be desirable and when used should be provided with generous corner radii and should be free from abrupt changes in thickness.

21

Seals, Packing, Gaskets, and Shields

21-1 Definitions. A *gasket* is a relatively elastic or plastic member placed between two relatively rigid, static machine members to act as a barrier to the flow of any substance, usually a fluid. The flat material between the two flanges of a pipe connection is a gasket, as is the member which prevents fluid flow between the block and cylinder head of an engine.

A *seal* is designed to prevent passage of a substance between two machine members which have relative motion, although it is also usually expected to be effective when there is no such motion. An example is the member employed to confine grease inside a ball bearing, or to prevent leakage of oil from around a rotating shaft.

Packing prevents the passage of a substance between machine members that are occasionally but not usually in relative motion to each other; for example, the material placed around the stem of a valve to prevent fluid leakage.

A *shield* is a protective machine element employed to prevent the direct impingement of a substance onto a machine element, but often not completely preventing the substance from contacting the machine

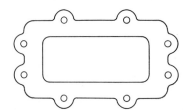

Fig. 21-1 Flat gasket made from sheet material.

element. An example of a shield is the enclosure around a chain or belt drive and the element which prevents dirt from directly entering a rolling-element bearing and yet allows oiling.

21-2 Gaskets. The major types of gaskets in common use are flat gaskets manufactured from sheet material, formed gaskets, ring gaskets (both circular and noncircular in cross section), specially shaped gaskets to fit into grooves or similar recesses in flanged members, and hardening and nonhardening fluids which when smeared over the surfaces to be sealed act as a gasket. Gaskets are made of asbestos, cork, vegetable fiber, rubber, neoprene, Teflon, other plastics, leather, copper, lead, other metals, and other materials as well as a combination of two or more materials. Occasionally powdered graphite, wax, and other materials are impregnated into the gasket material, mainly for the purpose of permitting easier and better initial tightening of the two members touching the gasket.

Figure 21-1 depicts a *flat gasket* made from sheet material. It is suitable for flat flanged surfaces bolted together. Particularly if the gasket is of soft material, care must be exercised in the design to be sure that the flanges have sufficient rigidity and employ enough bolts to prevent bulging of the flanges; the result of bulging is that the gasket lacks backing and may leak. In general, five or more bolts should be used. It usually is better to use a larger number of small bolts rather than only a few larger bolts.

A corrugated metal-asbestos *shaped gasket* is illustrated in Fig. 21-2. The concentric ridges of such a gasket are readily compressed when the flanges are bolted together, causing rather high stresses at these ridges and thus good sealing action. Adequate thickness flanges and a sufficient number of bolts should be employed. Various materials are used in shaped gaskets.

The use of an O ring as a *ring gasket* is shown in Fig. 21-3. In general, when an O ring is used as a gasket, the O-ring groove should be in only one of the members to be sealed, the other member remaining flat. The depth of the groove should be about three-quarters to seven-eighths the diameter of the torus of the O ring. For sealing against very high pressures (1,000 psi and above), shallower grooves and thus greater compres-

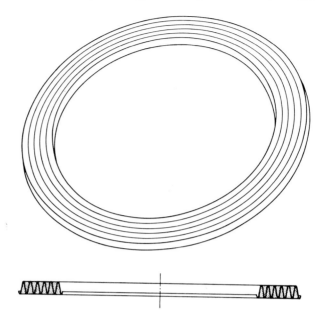

Fig. 21-2   Corrugated metal–asbestos gasket.

sion of the O ring are desirable. Usually the joint is tightened until metallic contact exists between the mating flat surfaces of the two sealed flanges. Although in general the O-ring groove is circular, it may be any desired shape of correct length into which a standard O ring may be deformed.

Permatex, Form-a-Gasket, and shellac are examples of fluids which when spread on the mating surfaces of two adjacent machine members can effect a sealing action, often referred to as a *hardening-type fluid gasket.*

Gaskets formed by curing a material into a shaped groove in a machine member are known by the trade names *Gask-o-Seal, Strip-o-Seal,* and others. In general their sealing action is excellent against a surface

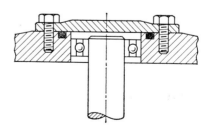

Fig. 21-3   O-ring gasket for cover plate.

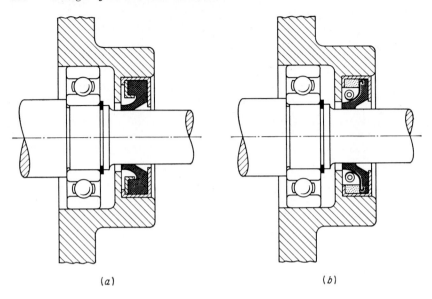

(a)   (b)

Fig. 21-4   Elastic ring shaft seals. (a) Without spring. (b) With spring.

on a mating member, behaving in much the same manner as an O ring
in a shaped groove. The sealed surfaces need not be flat so long as they
correctly mate. Such gaskets are also excellent seals against vacuum,
as in space capsules.

21-3   Seals.   With very few exceptions, almost all seals are used in
conjunction with circular cross-sectioned members. In rotating machinery, *shaft seals* are used to retain oil or grease, to exclude dirt and abrasive
particles, to prevent entrance of moisture and other fluids, to permit a
pressure differential between the exterior and the interior of the sealed
volume, to permit a controlled atmosphere in a sealed volume such as in
refrigeration systems and turbines, and for various other reasons. A ring
of an elastic material with a lip, with or without spring loading, is perhaps
the most common type of shaft seal. These *elastic ring seals* are depicted
in Fig. 21-4. Usually such seals come in a metallic case which is easily
pressed into a cylindrical housing, generally with no means of retention
other than the press fit. Since the bore of the hole in the uninstalled seal
is smaller than the diameter of the shaft, the seal must be deformed
slightly when assembled onto the shaft. For longest seal life, it is best
that a lubricant be able to contact at least one side of the seal. The lip
construction of most elastic ring seals permits sealing against fluid flow
in one direction along the shafts but not in the reversed direction; hence
care must be exercised in installing the seal to be sure that the desired

direction of sealing is obtained. Fluid flow to the right is prevented in the shaft seal of Fig. 21-4.

Felt shaft seals operate in essentially the same manner as elastic ring shaft seals but are without lips, relying upon compression of the felt for sealing action. Such seals usually are inadequate for sealing against pressure differential; however they find wide application as dirt excluders (Fig. 21-5). Where it is desired to seal a lubricant within a housing against a differential pressure, both elastic ring seals and felt seals may be used; the felt seal is usually on the outside to provide a dirt barrier and so protect the inner elastic ring seal from abrasion.

O rings and *G-T rings* can be used as shaft seals as indicated in Figs. 21-6 and 21-7, particularly for rotating shafts. For translational shaft motion, neither type of ring wears well if adequate lubrication of the seal is not provided. When used as shaft seals, the G-T rings appear to wear slightly less than O rings, presumably due to the greater contact surface between the G-T ring and the shaft. Proper dimensions of the recess into which the O or G-T rings fit (either with or without backing rings) is of great importance for proper sealing action. Manufacturers provide specific information concerning the proper recess dimensions for each ring.

Molded cups of leather, rubber, plastic, and other materials form good shaft seals for pistons and other male members reciprocating inside female bores. O rings and G-T rings also may be used as piston seals. Figures 21-8 and 21-9 respectively depict molded cup and O-ring piston seals.

Shaped rings can be used for seals on pistons, V and U cross sections being rather common. Figure 21-10 illustrates a V-ring piston seal con-

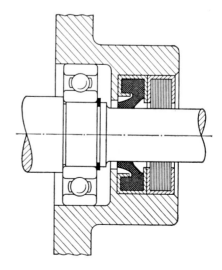

Fig. 21-5 Shaft with inner elastic ring seal and outer felt seal.

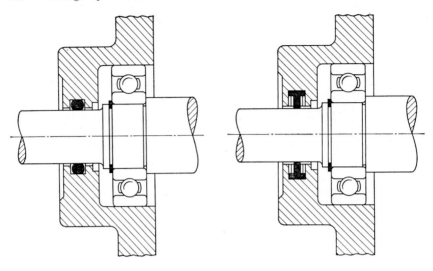

Fig. 21-6   O-ring shaft seal.   **Fig. 21-7**   G-T ring shaft seal.

sisting of several V rings with two backing rings. Such piston seals frequently are referred to as piston packing. Usually some means is provided
for applying compressive loading onto the seal material rather than
simply relying upon the tight fit of the seal in a fixed-dimension recess.

Molded cups and shaped rings are good seals for reciprocating piston
action, although under static conditions or under piston rotational conditions the sealing action is not so good as that provided by some of the
previously mentioned seals.

Metallic seals consisting of a spring-loaded nonrotating member
pressing against the shoulder of a rotating shaft form a shaft seal useful
for sealing refrigerant compressors, water pumps, and similar rotating
shafts, the fluid being sealed usually acting as the lubricant for the

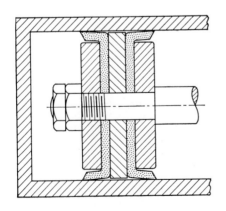

Fig. 21-8   Molded cup piston seal.

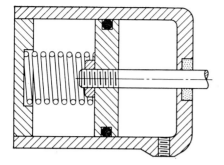

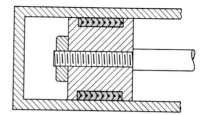

Fig. 21-9 Piston with O-ring seal.   **Fig. 21-10** Piston with V-ring seal.

rubbing surfaces. Usually the spring loading results from the elastic deformation of the sealing member. With adequate force between the sealing member and the shaft and with proper lubrication, essentially no fluid leaks past this seal until the seal is worn out.

Labyrinth seals may be employed where it is desired to minimize friction and some leakage of fluid past the seal can be tolerated. As indicated in Fig. 21-11, no actual contact exists between the shaft and the labyrinth seal; diametral clearances of 0.001 to 0.005 in. are common. Often the outside groove of the seal has a fluid drain to permit removal of most or all of the liquid which may have collected at this point.

21-4 Packing. Various types of packing can be employed to create a seal between two machine members which usually have no relative motion between them but upon occasion may have translational and/or

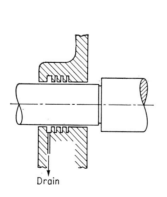
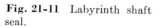

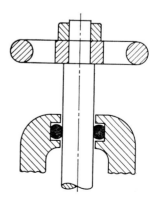

Fig. 21-11 Labyrinth shaft seal.   **Fig. 21-12** O-ring packing for valve stem.

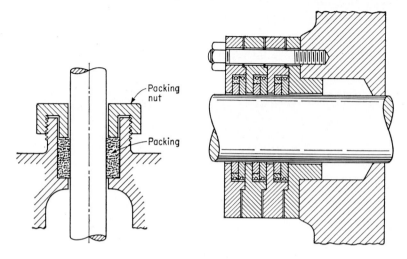

Fig. 21-13 Bulk packing for centrifugal pump shaft.

Fig. 21-14 Spring-loaded floating metal packing for shaft.

rotational motion. Asbestos, leather, cork, vegetable fiber, rubber, neoprene, plastics, and other materials are used, often impregnated with graphite, wax, or other substances to reduce friction when motion occurs. Figure 21-12 shows the use of an O ring as packing for the valve stem of a water valve. No means need be provided for compressing the O ring; the compression that occurs during installation is sufficient. Shaped rings (V, U, rectangular, and other cross sections) as well as *bulk packing* material (spiral cord of flax, asbestos, and other materials), which are placed in enclosed volumes around male members, serve well so long as there is some means of supplying compressive loading. This type of packing is common for the rotating shafts of jet and centrifugal pumps as illustrated in Fig. 21-13. *Metallic packing* consists of spring-loaded split metallic rings that are fitted into circular grooves in the housing of a shaft. These or a combination of metal and nonmetallic materials, and still other arrangements employing metal components, find usage as packing for the shafts of compressors, turbines, and other rotating equipment. Spring-loaded floating metal packing for a shaft is illustrated in Fig. 21-14.

21-5 Shields.   Shields are protective devices provided to reduce the possibility of damage to machine elements from outside causes (such as dirt or water spray) or to protect humans from harm. One of the most common shields is the *ball-bearing shield* as illustrated in Fig. 21-15. Similar type shields are employed for other machine elements and must

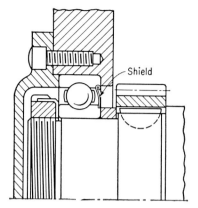

Fig. 21-15   Ball bearing with shield.

be designed for each specific application. Although metal is the most common shield material, neoprene, rubber, plastics, and other materials are used. *Safety shields* are cages partially or completely enclosing moving and/or heated machine elements to prevent or reduce the possibility of human contact with these elements. The safety shield for V-belt drives is a common example.

Problems for Assigned Work

A variety of problems has been assembled to bring out specific points and methods. It is impossible, in the space permitted for this portion of the book, to provide problems covering every phase of design, and problems from other sources may be used to advantage. A design problem may have many correct answers and in some cases no acceptable answer. The solution of a design problem depends upon many factors, some known, some unknown; some definite in value, some quite variable; some dependent upon judgment, some upon experience, some upon appearance, and some upon restrictions that leave the designer with no choice. The solution of a design problem contains the thought, judgment, and integrity of the designer. It is his creation.

Chapter 2.   Materials

2-1   A beam of SAE 3240 oil-quenched steel has a Brinell hardness of 229. It is subjected to a harmonically varying flexural stress which varies between 25,000 psi tension and 15,000 psi compression. Determine the probable value of the flexural endurance for this steel beam under the loading conditions indicated. Assume a rough-finished surface on the beam.

2-2   An annealed SAE 1095 steel is shaped into a shaft with ground surfaces. It is subjected to rapidly varying torque between the limits of 30,000 psi shear in one direction to 15,000 psi shear in the opposite direction. Determine the probable shear endurance limit for this shaft loaded under the specified conditions

2-3   A steel machine member is subjected to axial loading varying from 50,000 psi tension to 30,000 psi compression. The tensile yield strength of the material is 100,000 psi, and the flexural endurance limit for fully reversed harmonic loading is 70,000 psi, including the effect of the surface finish of the machine member. Determine the permissible working stress for a design factor of 1.3.

2-4   An extension spring of oil-tempered SAE 9260 steel wire is subjected to a periodic loading resulting in shear stresses varying between limits with a 3:1 ratio. A design factor of 1.75 is considered applicable. Graphically determine the maximum permissible shear stress in the spring.

2-5   A piece of steel is kept at a constant temperature of 1000°F and under a constant static tensile load. Experimentally determined data of tensile elongation versus time are as follows:

Time, hr:	0	100	200	300	400	500	600	700	800	900	1000
Elongation, %:		0.055	0.086	0.094	0.102	0.110	0.126	0.134	0.142	0.150	0.157	0.165

Under the above conditions, predict the time when the stress in the steel is 75,000 psi.

2-6   By careful inspection of tabular values of the physical properties of ferrous materials, determine the approximate value of the ratio of the tensile yield stress to the tensile ultimate stress for most steels. What is the approximate value of this ratio for the relationship between the endurance limit in reversed bending to the tensile ultimate stress?

2-7   By careful inspection of tabular values of the physical properties of ferrous materials, derive an expression relating the flexural endurance limit in reversed bending for steels as a function of the Brinell hardness number of the material. This is expressible as an equation. Indicate what limits, if any, are applicable to the use of your derived equation.

2-8   A steel shaft has a core hardness of 30.5 on the Rockwell C scale. What do you believe are the tensile ultimate strength, flexural endurance limit, and shear endurance limit for this material?

2-9   After consulting three references sources, tabulate the tensile yield strength and the shear yield strength of water-quenched SAE 2340 steel, drawn to 800°F. How do you justify the differences in your tabulated values?

Chapter 3.   Stresses and Strains

3-1   Two 0.90-in.-thick sheets of aluminum 10 in. wide are bonded together face to face. What shear stress is developed on the bond when the sheets are bent around a mandrel 60 in. in diameter?

3-2   A steel shaft $1\frac{1}{2}$ in. in diameter and 2 ft long is held rigidly at one end and has a handwheel 18 in. in diameter keyed to the other end. The modulus of rigidity of steel is 12,000,000 psi. (*a*) What load applied tangent to the rim of the wheel will produce a torsional shear stress of 10,000 psi? (*b*) How many degrees will the wheel turn when this load is applied? (*c*) Will there be any bending stress in this shaft? If so, what will be its magnitude?

3-3 A car weighing 4,000 lb is equipped with a full-floating axle, which prevents bending stress on the axle. The maximum load on a rear wheel is 1,250 lb. The tires are 26 in. in diameter and the coefficient of friction between the tires and the road is 0.7. The engine is capable of developing a torque of 250 ft-lb. The total gear reduction is 3.89 in high, 12.17 in low, and 16.7 in reverse. The minimum diameter of the axle is $1\frac{1}{16}$ in. (*a*) Determine the torsional stress in the axle when the car is running in high, low, and reverse gears if the tires do not slip. (*b*) Determine the torsional stress in the axle when the tires slip. (*c*) Since there are high local stresses at the sharp corners at the bottom of the splines, the stress at the root of the splines should not exceed 50 per cent of the stress in the straight portion of the axle. Determine the minimum root diameter of the splines.

3-4 A pair of spur gears having diameters of 36 in. and 12 in., respectively, revolve 50 and 150 times per minute, respectively, and transmit 20 hp without shock. Find the diameters of the shafts for a maximum torsional stress of 6,000 psi.

3-5 A steel weighing 480 lb/ft³ has an ultimate strength of 80,000 psi. What is the maximum length of steel rod that could be hung vertically from its upper end without rupturing?

3-6 A copper tube of $1\frac{1}{2}$-in.² cross section is slipped over a bolt of 1-in.² area. The length of the tube is 20 in., and the pitch of the threads on the bolt is $\frac{1}{8}$ in. The modulus of elasticity of the steel is 30,000,000 psi and of the copper 15,000,000 psi. If the tube is compressed between the head and nut of the bolt, what will be the unit stress produced in both parts by one-fourth turn of the nut?

3-7 A short prism has a rectangular cross section 2 in. by 4 in. A compressive load of 2,400 lb is applied 1 in. from the short side and $\frac{1}{2}$ in. from the long side. Determine the stress at each corner.

3-8 A flange coupling on a line shaft is held together by four $\frac{3}{4}$-in. UNF bolts, arranged on a bolt circle 6 in. in diameter. Each bolt transmits 5,000 lb in shear, and the tensile load on the bolt due to tightening is 12,000 lb. Determine the maximum shear stress and maximum tensile stress.

3-9 How much work in foot-pounds is required to stress a tension member 4 in.² in cross section and 5 ft in length to the elastic limit if E is 30,000,000 psi and the elastic limit stress is 34,000 psi? How does the energy stored in the bar vary with the stress imposed?

3-10 Three beams of equal size and shape are made from cast iron, soft steel, and nickel steel respectively. Which of these beams will be the stiffest? State your line of reasoning in determining your answer.

3-11 A beam is to be rectangular in section with depth twice the width. The distance between supports is 6 ft and there are two concentrated downward loads: 6,000 lb 18 in. from the left support, and 5,000 lb 40 in. from the left support. (*a*) Draw shear and bending-moment diagrams. (*b*) Using steel with E equal to 30,000,000 psi and ultimate strength of 75,000 psi, determine the beam dimensions if the maximum stress is to be 15,000 psi. (*c*) Using cast iron with E equal to 10,000,000 psi and ultimate strength of 30,000 psi, determine the beam dimensions if the maximum stress is 5,000 psi.

3-12 A beam of I section is 2 in. wide and 4 in. deep with all sections $\frac{1}{2}$ in. thick. It is supported at points 5 ft apart and carries a concentrated load of 400 lb at a distance of 2 ft from the left support. (*a*) Determine the horizontal shear in the vertical section just to the left of the load and at distances of 0, 1, and 2 in.

from the neutral axis. (*b*) Determine the horizontal shear in the vertical section just to the right of the left support and at distances 0, 1, and $1\frac{1}{2}$ in. from the neutral axis.

3-13   Draw to scale the shear and bending-moment diagrams for the beam in Fig. P3-1. What should be the dimensions of a rectangular cross section for this beam if the bending stress is limited to 15,000 psi and the beam is of uniform cross section with the depth three times the width?

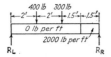

Fig. P3-1

3-14   A beam of SAE 1020 steel 10 ft long has an I section 4 in. deep and 2 in. wide, with the web and flanges $\frac{1}{4}$ in. thick. This beam is to be replaced by one made of AA 2017 aluminum alloy. Assume that in each case the stress is to be one-half the yield stress. (*a*) Determine the dimensions of the aluminum beam to support the same load. (*b*) Determine the dimensions if the aluminum beam is to have the same stiffness as the steel beam. (*c*) Determine the relative weights of these three beams.

3-15   A column of steel has a cross section of $2\frac{1}{2}$ by 6 in. and is loaded through ball joints at both ends. Determine the load that will cause a stress of 15,000 psi, assuming the yield stress to be 33,000 psi and the length to be 42 in.

3-16   A steel column of circular cross section is to be made of SAE 1045 steel having a yield strength of 60,000 psi. The length of the column is 6.25 ft and the end fixity may be assumed to be 2.75. (*a*) Determine the required diameter if the failure load is 346,000 lb. (*b*) Determine the required diameter if the failure load is 40,000 lb.

3-17   A rectangular steel tube has a cross section as shown in Fig. P3-2 with $t = t_1 = t_2 = 0.105$ in., $b = \frac{7}{8}$ in., and $h = 1\frac{1}{4}$ in. The end fixity when bending about the long axis is 1.5, and 1.0 when bending about the short axis. The yield strength of the material in compression is 80,000 psi. Determine the failure load if the column is 33 in. long and if the column is 50 in. long.

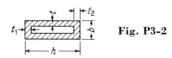

Fig. P3-2

3-18   The piston rod of a reciprocating water plunger pump may be considered to be a column with one fixed and one guided end. Determine the size of rod necessary for an 8-in. cylinder using 200-psi maximum water pressure. The rod is 20 in. in length and is made of steel having an ultimate strength of 70,000 psi and an elastic limit of 38,000 psi. The permissible working stress is 8,000 psi.

3-19   A 1.00-in.-diameter steel ball is pressed onto a horizontal steel fixed plane with a normal force of 15,000 lb between the ball and the plane. Determine the maximum compressive contact stress. Is this stress reasonable?

3-20 A roller bearing has 0.096-in.-diameter rollers each 0.875 in. long. The inner diameter of the outer race is 4.0823 in. A 5-ton load is supported by this bearing. Load distribution analysis for this particular bearing indicates that the maximum radial load between any one roller and the outer race is 0.649 times the total radial load supported by the entire bearing. Determine the maximum contact stress resulting between the rollers and the outer race.

3-21 Determine the maximum stress at section *A-A* for the link shown in Fig. P3-3.

Fig. P3-3

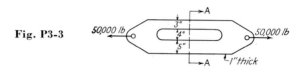

3-22 A block of cast iron 2 by 3 by 4 in. in size has a compressive load of 10,000 lb applied to the 3- by 4-in. face and a compressive load of 15,000 lb applied to the 2- by 3-in. face. (*a*) What is the maximum equivalent stress according to the maximum-strain theory? (*b*) What will be the maximum equivalent stress if the 10,000-lb load is made tension instead of compression?

3-23 A machine member of bronze is subjected to external forces producing a direct shear stress of 8,000 psi and a direct tensile stress of 12,000 psi. Determine the maximum principal stress (tension) and the maximum shear stress induced in this machine member.

3-24 An aluminum plate $\frac{1}{2}$ by 4 by 6 in. in size has a tensile load of 3,000 lb applied parallel to the 6-in. side and a tensile load of 2,000 lb applied parallel to the 4-in. side. (*a*) What is the maximum normal stress? (*b*) What is the maximum shear stress? (*c*) If an additional load of 10,000 lb in compression is applied to the flat faces, what will be the maximum shear stress?

3-25 A punch press has a frame similar to Fig. P3-4 with a cross section at *A-A* as shown in Fig. P3-2. The dimensions are *a* equals 4 in., *b* equals 6 in., and t_1 equals t_2 equals 1 in. Determine the dimension *h* so that the maximum tensile or compressive stress will not exceed 6,000 psi. $F = 10$ tons; $t = 1$ in.

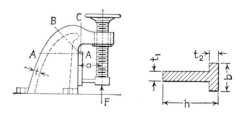

Fig. P3-4   **Fig. P3-5**

3-26 A small punch has a capacity sufficient to punch a $\frac{5}{8}$-in. hole in a $\frac{1}{2}$-in. steel plate whose ultimate strength in tension is 90,000 psi. The punch frame is shaped like a C clamp with a T cross section (Fig. P3-5). The dimensions are t_2 equals $1\frac{1}{4}$ in., t_1 equals 1 in., and *b* equals 6 in. The center line of the punch is $6\frac{1}{2}$ in. from the inner surface of the flange. Determine the depth of leg h if the stress is not to exceed 14,000 psi.

3-27   A block of pure cast copper 1 by 2 by 3 in. in size is subjected to a compressive load of 1,000 lb on the 1- by 3-in. face. The maximum allowable direct tensile stress is 2,000 psi. (*a*) Using the maximum-shear theory, calculate the maximum tensile load that can be applied to the 1- by 2-in. face. (*b*) Using the maximum-strain theory, calculate the maximum tensile load that can be applied to the 1- by 2-in. face.

3-28   A cube of annealed SAE 1095 steel is loaded with a 10,000-psi tensile stress on two opposite faces and 22,500-psi compressive stresses on two additional opposite faces. By use of the maximum distortion energy theory, determine the maximum permissible stress that can be applied normal to the third pair of faces of the cube.

3-29   The bus bar in a power-generating station consists of a $\frac{1}{4}$- by 2-in. copper bar rigidly anchored at 3-ft intervals. Power is transmitted over this bus bar at 2,300 volts, and the resistance causes the temperature to rise to 160°F when the room temperature is 80°F. The ultimate strength of the copper is 36,000 psi, and the coefficient of thermal expansion is 0.0000109 in./(in.)(°F). What change in stress is caused by the operation of the plant?

3-30   A steel bar 1 in. in diameter is inserted inside a copper tube 2 in. in outside diameter and rigidly attached to it. If the completed bar is 12 in. long, what will be the unit stress in each material when the temperature has been increased by 200°F?

3-31   A $\frac{3}{4}$-in. rivet is driven hot, and the average temperature of the metal is 900°F. The length after driving is $1\frac{1}{4}$ in. What will be the tensile stress in the rivet after cooling to 70°F?

3-32   A solid steel shaft circular in cross section has a change in diameter from 3.75 in. to 3.24 in. with a fillet radius of 0.15 in. at the place of change. The torsional moment in the shaft at that spot is 425 ft-lb with no shock. Determine the maximum torsional shearing stress in the shaft at the cross-section change.

3-33   A long solid steel bar of uniform 1.250-in. diameter has a 0.375-in.-diam hole at its center with its axis perpendicular to the length of the bar.   A static tensile load of 100,000 lb is applied longitudinally to the bar. Determine the probable maximum stress in the bar near or at the hole.

3-34   The shaft of Prob. 3-32 is subjected simultaneously to a 625 ft-lb torsional moment and a 300 ft-lb flexural moment at the change in cross section. Considering combined stress concentration, determine the probable maximum combined stress in the shaft.

Chapter 4.   Design Stresses

4-1   The piston of an automobile engine is held to the connecting rod by a wrist pin having an external diameter of $1\frac{3}{16}$ in. and an internal diameter of $\frac{7}{16}$ in. The length of the bearing in the rod end is 1 in. and the bearing in each boss of the piston is $\frac{5}{8}$ in. The maximum load transmitted is 3,200 lb. (*a*) Determine the bearing pressure between the rod end and the wrist pin. (*b*) Compute the maximum bending stress in the pin, assuming that it is a simple beam with a uniform load. (*c*) Compute the maximum deflection of the pin. (*d*) If the pin is made of SAE 3240 steel, what is the apparent factor of safety?

4-2 A diesel engine developing 1,850 hp at 105 rpm is used on a ship to drive the propeller. The shaft of the propeller is to be made of $3\frac{1}{2}$ per cent nickel steel, heat-treated, with an ultimate strength in tension of 130,000 psi, and an ultimate in shear about 70 per cent of the ultimate in tension. Angular twist is limited to 1 deg in 20 diameters. (*a*) Using a design factor of 2, find the diameter of solid shaft required. (*b*) Using the same data, find the diameter of hollow shaft required if the outside diameter is twice the inside diameter. (*c*) What is the percentage of saving in weight by using the hollow shaft?

4-3 An automobile has a rear axle $1\frac{1}{8}$ in. in diameter at the smallest section where bending is negligible. The engine develops 75 hp at 3,000 rpm. The gear reduction between the engine and the rear axle is 12.4 in low gear and 4.3 in high gear. The transmission efficiency is 88 per cent. This axle is made of heat-treated nickel steel (SAE 2340) having an ultimate strength of 150,000 psi and a yield stress of 125,000 psi in tension. (*a*) Determine the working stress when in low gear and in high gear assuming that the engine is developing its full power. (*b*) What is the apparent factor of safety? (*c*) Estimate the probable design stress used by assuming the various factors and giving reasons for such assumptions.

4-4 A 12- by 14-in. by 350-rpm single-acting oil engine has a compression pressure of 325 psi and a maximum explosion pressure of 600 psi. The connecting rod for this engine is to be made of SAE 1045 annealed steel. What should be the design stress?

4-5 For the rectangular cross-section beam shown in Fig. P4-1, the load *F* varies from minus 30 lb to plus 50 lb and will undergo 1,200,000,000 repetitions in the service life of the machine. The load may be considered as suddenly applied. All surfaces are finish ground. The beam is made of SAE 4140 treated to a Brinell hardness of 380. What design factor was used?

Fig. P4-1

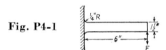

4-6 A spring subjected to torsional stress is made of hard-drawn brass wire. The ultimate strength in tension is 75,000 psi and the yield stress 26,000 psi. The modulus of elasticity is 14,000,000 psi. The load on the spring varies rapidly from 75 to 100 lb. What is the permissible design stress if the load is applied with minor shock?

4-7 A spring subjected to bending stresses is made of hard-drawn brass. The ultimate strength in tension is 84,000 psi, and the yield stress is 64,000 psi. The modulus of elasticity is 14,000,000 psi. The load carried by this spring varies rapidly from 300 to 700 lb. Determine the permissible design stress, considering the shock to be negligible. Determine the design stress if the load is applied with moderate shock.

4-8 A tension member subjected to a load of 30,000 lb has a cross section $\frac{1}{2}$ in. by 6 in. and is pierced by a central hole $\frac{1}{2}$ in. in diameter. (*a*) What is the average stress at a section away from the hole? (*b*) What is the probable maximum stress at the hole? (*c*) Assuming this member to have a design factor of 1.5, specify the material to be used.

4-9   An SAE1045 annealed steel shaft subjected to torsion transmits 25 hp at 300 rpm. At one end this shaft is increased in diameter by $\frac{1}{2}$ in. and the two sections are joined by a fillet of $\frac{1}{4}$-in. radius. Determine the shaft diameters for a design factor of 2 and moderate shock.

4-10   A rectangular beam of cast iron has a cross section $1\frac{1}{4}$ in. by 3 in. The supports are 15 in. apart, and there is a central load of 1,000 lb. At a distance of 4 in. from one support there is a semicircular groove in the lower surface. The groove radius is $\frac{1}{16}$ in. (*a*) What is the bending stress at the load? (*b*) What is the probable stress at the groove? (*c*) What design stress should have been used?

4-11   A solid cylindrical machine member of 2-in. diameter is subjected to a light shock load producing completely reversed stresses in bending. The member is rough finished and has a diameter decrease to $1\frac{1}{4}$ in. by use of a $\frac{1}{4}$-in.-radius fillet. The life of the member is estimated to be 100,000,000 repetitions. The member is made of a material equivalent to 0.5C normalized steel. For a design factor of 1.5, find the design stress for the member.

4-12   Same as Prob. 4-11 except the estimated life is 1,000,000 repetitions.

4-13   Same as Prob. 4-11 except the load varies from a maximum in tension to a minimum in tension with a ratio of 3 to 2.

4-14   Same as Prob. 4-11 except the 2-in. diameter is constant for its entire length and a 0.20-in. diameter transverse hole is drilled in the member.

Chapter 5.   Keys, Cotters, Retainers, and Fasteners

5-1   A belt pulley is fastened to a $2\frac{15}{16}$-in. shaft, running at 200 rpm, by means of a key $\frac{3}{4}$ in. wide by 5 in. long. The permissible stresses in the key are 8,000 psi in shear and 14,000 psi in compression. (*a*) Determine the horsepower that can be transmitted. (*b*) What depth of key is required?

5-2   A 48-in. cast-iron pulley is fastened to a $4\frac{1}{2}$-in. shaft by means of a $1\frac{1}{8}$-in. square key 7 in. long. The key and shaft are SAE 1025 steel, annealed. (*a*) What force acting at the pulley rim will shear this key? (*b*) What force acting at the pulley rim will crush the cast-iron keyway if the strength of cast iron is 24,000 psi in tension and 96,000 psi in compression? (*c*) What should be the maximum force applied at the rim of the pulley if the load is applied with moderate shock?

5-3   A key $\frac{15}{16}$ in. wide, $\frac{5}{8}$ in. deep, and 12 in. long is to be used on a 200-hp, 1160-rpm squirrel-cage induction motor. The shaft diameter is $3\frac{7}{8}$ in. The maximum running torque is 200 per cent of the full-load torque. Determine the maximum shear and compressive stresses on the key and the maximum direct shear stress on the shaft, considering the effect of the keyway.

5-4   A shaft and key are made of the same material and the key width is one-fourth the shaft diameter. (*a*) Considering shear only, determine the minimum key length in terms of the shaft diameter. (*b*) The shear strength of the key material is 75 per cent of its crushing strength. Determine the thickness of the key to make the key equally strong in shear and crushing. (*c*) How do the dimensions determined disagree with standard practice? Give reasons why standard practice gives better key design.

5-5 A gear with a hub 3 in. long is fastened to a $1\frac{15}{16}$-in. commercial steel shaft by means of a $\frac{1}{2}$-in.-square feather key. What force is required to move the gear along the shaft when transmitting the full torque capacity of the shaft? Assume the coefficient of friction to be 0.15.

5-6 A standard key fixes the hub of a gear to a shaft which transmits 325 hp at 200 rpm. Determine the key dimensions for a factor of safety of 5 if the key is made of SAE 1025 water-quenched steel with a BHN of 183.

5-7 The transmission gears of an automobile are carried on a $2\frac{1}{4}$-in., SAE 10-spline shaft and slide when under load. The hub length of each gear is $1\frac{5}{8}$ in. Determine the total horsepower that can be transmitted at 3,000 rpm with 800 psi permissible pressure on the splines.

5-8 A 12-in. gear transmitting 50 hp at 120 rpm is to be fastened to a shaft of SAE 1020 steel with a permanent fit by means of an SAE six-spline fitting. The hub is 1.5 times the shaft diameter. Determine the spline dimensions, using a factor of safety of 5.

5-9 A 12-in. lever is fixed to a $1\frac{1}{2}$-in. shaft by means of a taper pin passed through its hub perpendicular to the axis, the mean diameter of the pin being $\frac{3}{8}$ in. What pull on the end of this lever will cause a shear stress on the pin of 9,000 psi, and what torsion stress will this produce in the shaft?

5-10 A tensile load is transmitted through a $1\frac{3}{4}$-in. rod fitted with a flat cotter key $\frac{7}{16}$ by $1\frac{5}{8}$ in. in cross section. The edge of the opening for the cotter is $1\frac{3}{8}$ in. from the rod end. The opposite end of the rod is threaded, and the permissible stress in the threaded portion is 9,000 psi. Determine the weakest part of the joint if all parts are made of SAE 1020 steel.

5-11 A single $\frac{1}{4}$-in. hexagonal socket setscrew with cup point is used to fasten a helical gear onto a $1\frac{1}{8}$-in.-diameter steel shaft. The seating torque of the setscrew is approximately 90 in.-lb. Estimate the maximum torque and axial thrust load the gear can apply to the shaft.

5-12 A $2\frac{5}{8}$-in. steel shaft has a chain sprocket fastened onto it by use of setscrews. An 85 in.-lb torque is to be transmitted. Determine the number and size of setscrews to be used.

5-13 Select a basic external retaining ring to transmit a 6,000-lb thrust load from a shaft onto a gear with a bore of 2.875 in. Give the dimension of the ring and the groove in shaft.

5-14 A thrust plate of a multiple disk clutch is to be located within a bored housing with an inside diameter of $4\frac{1}{4}$ in. by means of an internal Y-formed retaining ring. Considering shock conditions, select an internal wire-formed retaining ring which will permit maximum thrust to be transmitted from the thrust plate on the housing. What are the dimensions of the ring and groove into which it fits?

Chapter 6. Threaded Members

6-1 When a regular hexagonal nut is tightened on its bolt, the axial pull produced on the bolt causes a direct tensile stress across the root section of the bolt and a shear stress across the root of the threads. Using a standard $1\frac{1}{2}$-in.

finished nut and UNC threads, determine if the bolt will rupture by tension before the threads shear off. Assume s_s equals $0.75s_t$.

6-2 A generator weighing 3,000 lb is furnished with an eyebolt in the housing for lifting purposes. Assume the bolt to be made of SAE 1020 steel with UNC threads. (*a*) Determine the required bolt diameter. (*b*) Determine the depth to which the bolt should extend into the cast-iron motor housing.

6-3 A casting weighing 3 tons is lifted by means of a $1\frac{1}{4}$-in. eyebolt having UNC threads. The bolt extends $1\frac{5}{8}$ in. into the casting. (*a*) Determine the direct tensile and the direct shear stresses in the threaded portion of the bolt. (*b*) If the bolt is made of SAE 1025 steel, what is the apparent factor of safety?

6-4 An electric motor weighs 1,800 lb and is provided with an eyebolt screwed into the cast-iron frame for lifting purposes. (*a*) What size bolt with UNC threads should be used, assuming ordinary bolt steel? (*b*) How far should the bolt extend into the casting?

6-5 The head of a cylinder 24 in. in diameter is subjected to a pressure of 200 psi. The head is held in place by 16 UNC bolts $1\frac{1}{4}$ in. in diameter. A copper gasket is used to make the joint steamtight. Determine the probable stress in the bolts.

6-6 A 12-in. cylinder has its head bolted on by twelve 1-in.-8UNC bolts arranged on a $13\frac{1}{2}$-in. bolt circle. The pressure is 125 psi, and the joint is made with a hard gasket and through bolts. Determine the probable unit stress in the bolts. Are these bolts large enough if made of steel having an ultimate strength of 70,000 and a yield stress of 38,000 psi?

6-7 The cylinder of an engine is $4\frac{3}{4}$ in. in diameter. The maximum gas pressure in the cylinder is 500 psi. Determine the number of $\frac{1}{2}$-in.-13UNC nickel-steel bolts required to hold this cylinder to the crankcase, assuming the ultimate tensile strength of the nickel steel to be 110,000 psi.

6-8 The head of a cylinder 24 in. in diameter is subjected to a pressure of 200 psi. The head is held on by means of sixteen $1\frac{1}{4}$-in.-7UNC bolts arranged on a 27-in. bolt circle. No gasket is used. Determine the probable stress in the bolts.

6-9 A steel bolt having a cross-sectional area of 0.20 in.² is tightened so that the initial stress in the body of the bolt is 20,000 psi. The effective spring constant of the gasket is 10,000 lb/in. The effective length of bolt is 5 in. If a tensile load of 400 lb is applied to the bolt, what will be the stress in the bolt?

6-10 A $\frac{5}{8}$-in.-18UNF bolt is to be tightened so that the average tensile stress at the stress area is 30,000 psi. (*a*) Determine the torque wrench setting from Eq. (6-1). (*b*) Assuming $f = f_c = 0.15$ and employing Eq. (6-7), determine the torque wrench setting. (*c*) Same as part (*b*) except assume $f = f_c = 0.10$. (*d*) Determine the average stresses at the thread-stress area when the torque in part (*a*) is applied.

6-11 Determine the stretch that should be given a $\frac{5}{8}$-in.-18UNF steel bolt (Fig. 6-10) when tightened to produce an average direct-tension stress at the stress area of 30,000 psi.

6-12 Determine the rotation of the nut for tightening a $\frac{5}{8}$-in.-18UNF stud with lengths as shown in Fig. 6-11 so that the tightening load will be 10,000 lb.

6-13 A $\frac{1}{16}$-in.-thick copper gasket is to be used in making a tight joint with a 1-in.-8UNC bolt. The outside diameter of the gasket is 2 in., and the inside diameter is $1\frac{1}{16}$ in. The effective length of the bolt is 4 in. Determine the value of K.

6-14 Design the through bolts for fastening the head of a pressure vessel to the flange of the vessel. The inside diameter of the vessel is 12 in. A hard-copper gasket ($K = 0.25$) 1 in. wide and $\frac{1}{16}$ in. thick is used. The inside of the gasket is flush with the inner wall of the vessel. The pressure in the chamber varies from 500 psi to 1,200 psi 50 times/min. The bolts are made of SAE 2340 steel, water-quenched and drawn to 800°F.

6-15 Standard 10-in. pipe flanges for pressures up to 125 psi have an out-side diameter of 16 in. and are provided with twelve $\frac{7}{8}$-in.-14UNF bolts arranged on a $14\frac{1}{4}$-in. bolt circle. A hard asbestos gasket is used. (*a*) Determine the stress in the bolts due to pressure. (*b*) Determine the stress due to tightening the bolts. (*c*) What is the factor of safety? (*d*) What is the factor of safety as a ratio between the pressure that would cause failure and the actual pressure?

6-16 A $2\frac{15}{16}$-in. line shaft is supported by a wall bracket so that the shaft center is $7\frac{1}{4}$ in. from the wall. The bracket is supported by three $\frac{3}{4}$-in.-10UNC bolts. Two bolts are $5\frac{1}{2}$ in. below the shaft center, and the third bolt is 17 in. below the shaft center. The lower edge of the bracket is 2 in. below the lower bolt. The shaft puts a vertical load of 4,500 lb on the bracket. (*a*) What is the unit shear stress on the bolts, assuming the shear to come on the body of the bolt? (*b*) What is the unit tensile stress in each bolt? (*c*) What is the maximum combined shear stress on the bolts? (*d*) What is the maximum combined tensile stress on the bolts?

6-17 The inside diameter of the stator of a 100-kw, 950-rpm motor is 38 in. Starting torque may be assumed to be 200 per cent of the running torque. The maximum belt pull on the motor shaft is 2,000 lb. The shaft center is 24 in. above the base line. There are four foundation bolts spaced 24 in. on centers axially, and 42 in. on centers normal to the axis. The width of the base is 48 in. Bolt steel has an ultimate strength of 60,000 psi with an elastic limit of 30,000 psi. Determine the load on each bolt and the proper diameter of these bolts.

6-18 A 2-in. square-thread steel screw is used in a common screw jack. There are $2\frac{1}{4}$ threads per inch. The body of the jack is made of cast iron. The coefficient of friction on the threads is 0.15. The thrust collar at the top has an inside diameter of $1\frac{1}{2}$ in. and an outside diameter of 3 in. The coefficient of friction for the collar is 0.25. The load to be lifted by the jack is 2 tons. (*a*) Determine the efficiency of the screw alone. (*b*) Determine the efficiency of the screw and thrust collar. (*c*) Determine the pull necessary on the end of a 2-ft lever to operate this screw jack.

6-19 In the machine frame shown in Fig. P6-1 the guide post is bolted to the base by means of eight bolts, four on each side. (*a*) Determine the proper size of bolts using ordinary bolt steel, considering that the load is applied with

Fig. P6-1

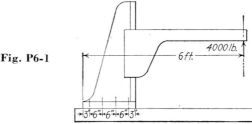

4000 lb.

6 ft.

$\leftarrow$ 3″ $\leftarrow$ 6″ $\leftarrow$ 6″ $\leftarrow$ 6″ $\leftarrow$ 3″ $\rightarrow$

heavy shock. (b) Determine the proper size of bolts using steel having an ultimate strength in tension of 110,000 psi.

6-20   The lead screw of a lathe has a 2-in. Acme thread, $2\frac{1}{2}$ threads/in.
To drive the tool carriage this screw must exert an axial force of 600 lb. The
thrust is carried on a collar $4\frac{1}{2}$ in. in outside and $2\frac{1}{4}$ in. in inside diameter. The
lead screw revolves 30 rpm. (a) Determine the efficiency of the screw and collar,
assuming a coefficient of friction of 0.15 for the threads and 0.10 for the thrust
collar. (b) Determine the horsepower required to drive this screw.

6-21   In a large gate valve used in a high-pressure water line the gate weighs
1,000 lb and the friction, due to water pressure and causing resistance to opening,
is 500 lb. The valve stem is $1\frac{1}{2}$ in. in diameter and fitted with three square threads
to the inch. The valve stem is nonrotating and is raised by a rotating wheel with
internal threads acting as a rotating nut on the valve stem. This wheel presses
against a supporting collar of $1\frac{5}{8}$ in. inside diameter and 3 in. outside diameter.
Assume the coefficient of friction for the threads to be 0.15 and for the collar 0.25.
(a) Determine the efficiency of the screw and collar. (b) Determine the torque or
turning moment that must be applied to the wheel to raise the valve gate.

6-22   (a) Determine the force necessary at the end of a 10-in. wrench used
on the nut of a $\frac{3}{4}$-in.-10UNC bolt in order to produce a tensile stress of 25,000 psi
in the bolt. Assume the coefficients of friction of the threads and the nut to be
0.15. (b) Determine the efficiency of the nut and bolt. (c) Assuming the bolt to
be made of SAE 1015 steel, find the allowable stress by Seaton and Routhwaite's
equation and explain the difference between this stress and that in part (a).

6-23   A $2\frac{1}{2}$-in.-4UNC bolt is 18 in. long. The threaded portion is 5 in. long.
The maximum stress in the bolt is limited to 10,000 psi. This bolt is subjected to
severe shock loads. (a) How much energy, in foot-pounds, can this bolt absorb?
(b) What will be the total elongation produced? (c) If the unthreaded portion is
turned down to the root diameter of the thread, how much energy can be absorbed? (d) What will be the total elongation produced in part (c)? (e) What
conclusions would you draw concerning bolts subjected to shock loading?

6-24   A split nut is held from rotating but is propelled along a 3-in. Acme
screw having a single thread of two threads per inch. The nut advances against
a load of 12,000 lb. A thrust collar of 4 in. outside diameter and 3 in. inside
diameter is used on the screw. Assume the coefficient of friction in the thread to
be 0.12 and the collar 0.08. (a) Determine the horsepower required to drive the
screw. (b) Same as (a) using a double-thread screw.

6-25   A load of W lb falling 0.10 in. creates an impact tensile load on a
$1\frac{1}{2}$-in.-6UNC bolt. The bolt is 12 in. long between the head and nut, and $1\frac{1}{2}$ in. is
threaded. A maximum direct tensile stress of 18,000 psi is permissible. (a) When
a full-diameter bolt is used, what is the permissible load W? (b) When the bolt
shank is reduced to the root diameter of the threads, what is the permissible
load W?

Chapter 7.   Welded and Riveted Connections

7-1   Two $\frac{1}{2}$-in. plates are placed so they overlap 2 in. and are joined by a
single lap shielded weld placed normal to the load. The weld is a $\frac{1}{2}$-in. fillet weld
6 in. long. Determine the total static load that this joint will sustain.

7-2 Two $\frac{5}{8}$-in. plates are placed so they overlap and are welded with two $\frac{5}{8}$-in. unshielded fillet welds each 8 in. long and placed normal to the load. Determine the total static load that this joint can sustain.

7-3 What load can be sustained by the joint in Prob. 7-2 if the welds are placed parallel to the load line?

7-4 A $\frac{3}{8}$- by 4-in. steel plate is welded to a steel gusset plate by means of parallel fillet welds. Determine the length of welds required if a load of 15,000 lb is to be transmitted.

7-5 A $3\frac{1}{2}$- by 4- by $\frac{3}{8}$-in. structural-steel angle is used as a tension member to support a moderate shock load of 40,000 lb. The 4-in. leg of this angle is welded to a steel plate by means of two $\frac{3}{8}$-in. parallel welds. Determine the length of each weld.

7-6 A 6- by 4- by $\frac{1}{2}$-in. angle is welded to its support by two $\frac{1}{2}$-in. fillet welds along the back and edge of the 6-in. leg. A load of 4,500 lb is applied normal to the gravity axis of the angle at a distance of 15 in. from the center of gravity of the welds. Assume each weld to be 3 in. long and determine the maximum shear stress in the welds.

7-7 The bracket shown in Fig. P7-1 is welded to the column by $\frac{1}{2}$-in. fillet welds along the top and bottom of the plates rather than by use of rivets. Determine the load that can be applied at F if the maximum shear stress in the welds is to be 4,000 psi.

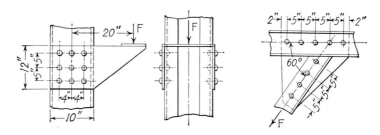

Fig. P7-1   **Fig. P7-2**

7-8 The riveted joint shown in Fig. P7-2 is changed to a welded joint (a) When $\frac{3}{8}$-in. fillet welds are used along both edges of the channel, what load may be applied at F so that the shear stress in the welds will be 8,000 psi? (b) The tension member is a 5- by 5- by $\frac{3}{8}$-in. angle. What length welds should be used to join the angle and gusset plate under the conditions mentioned?

7-9 A gear is built up of a forged-steel rim and hub and a web of steel plate joining them. The web is riveted to the rim by means of 36 rivets arranged in two concentric circles of 16- and 20-in. diameters with 16 rivets in the 16-in. circle and 20 rivets in the 20-in. circle. All rivets are $\frac{3}{4}$ in. in diameter. The pitch diameter of the gear is 26 in., and the turning force applied at the pitch line is 8,500 lb. Determine the maximum shear stress assuming that all rivets in each row are equally loaded and in single shear.

7-10 The operating arm of a brake lever is attached to the brake shoe by six rivets arranged in two parallel rows. The distance between the rows is 3 in., and the pitch in each row is $2\frac{1}{2}$ in. A load of 500 lb is applied at a distance of 24 in. from the center of gravity of the rivets. Determine the rivet diameter required if the shear stress is not to exceed 6,000 psi.

7-11   Two steel plates $\frac{1}{2}$ by 6 in. in size are connected by a double-strap triple-riveted butt joint with $\frac{5}{8}$-in. rivets. The inner row of rivets contains three rivets with a pitch of 2 in. The second row contains two rivets with a pitch of 2 in. The third row contains one rivet. The back pitch is $1\frac{1}{2}$ in., and the distance from the outer and inner rivets to the plate edges is 1 in. The cover plates are $\frac{5}{16}$ in. thick and the corners are cut away at 45 deg, starting at the middle or second row of rivets. When the total load transmitted by the plates is 40,000 lb, determine the following unit stresses: (a) Compression in main plate. (b) Compression in cover plates. (c) Tension in main plate at outer row. (d) Tension in main plate at middle row. (e) Tension in main plate at inner row. (f) Tension in cover plates at outer row. (g) Tension in cover plates at middle row. (h) Tension in cover plates at inner row. (i) Shear in rivets. (j) Shear in main plate between inner row of rivets and plate edge. (k) Shear in cover plate between outer rivet and plate edge.

7-12   A double-riveted lap joint is made of steel having an ultimate strength of 55,000 psi in tension and 95,000 psi in compression. The rivets have a strength of 44,000 psi in shear. The plates are $\frac{3}{4}$ in. thick. Rivets 1 in. in diameter are used in $1\frac{1}{16}$-in. holes spaced $3\frac{1}{4}$ in. on centers. Determine the manner in which this joint will fail and at what load.

7-13   The bracket in Fig. P7-1 is to carry a load F equal to 10,000 lb. If the rivets are $\frac{5}{8}$ in. in diameter, what is the maximum shear stress developed in the rivets?

7-14   The load F in Fig. P7-1 is 7,500 lb. If the center rivet of each plate is omitted, what size rivets are required if the shear stress is not to exceed 4,000 psi?

7-15   In Fig. P7-2 the rivets are $\frac{3}{4}$ in. in diameter. What load may be applied at F if the permissible shear stress is 10,000 psi?

Chapter 8.   Shafts

8-1   A machinery shaft is subject to torsion only. The bearings are 8 ft apart. The shaft transmits 250 hp at 200 rpm. Allow a shear stress of 6,000 psi after an allowance for keyways. (a) Determine the shaft diameter for steady loading. (b) Determine the shaft diameter if the load is suddenly applied with minor shock.

8-2   The shaft of a 50-hp 850-rpm electric motor is 31 in. from center to center of bearings and is $2\frac{1}{2}$ in. in diameter. (a) If the magnetic pull on the armature is 1,500 lb concentrated midway between the bearings, determine the maximum shear and the maximum tensile stress in the shaft. (b) If the shaft extends beyond the bearings and carries a 6-in. gear $7\frac{1}{2}$ in. from the bearing center, determine the maximum shear and the maximum tensile stress in the shaft.

8-3   A punch press has sufficient capacity to punch six holes 1 in. in diameter in a $\frac{3}{4}$-in. boiler plate. The maximum force exerted acts on the driving crank with a moment arm of $1\frac{1}{4}$ in. Considering torsion only, determine the required size of shaft, using SAE 2340 steel heat-treated to a hardness of 250 Brinell.

8-4   A 15-hp, 1,725-rpm motor drives a centrifugal pump through a single set of 5:1 reduction gears. Determine the diameters of the shafts on the motor and pump if ordinary steel shafting with standard keys is used.

8-5 The cross shaft of the braking system of an automobile is shown in Fig. P8-1. The brakes are identical. Assume that $A = 15$ in., $B = 33$ in., $C = 44$ in., and $d_1 = 1$ in. Determine d_2 so that the forces applied to the brakes will be equal.

Fig. P8-1

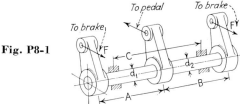

8-6 A 75-hp motor operating at 90 rpm is to drive a lathe in which the loads may be considered to be suddenly applied and the shock of a medium intensity. The shaft is fitted with a standard key. Considering torsion only, what shaft diameter is required if the shaft material has a yield stress of 35,000 psi in tension?

8-7 A machinery shaft supported on bearings 8 ft apart is to transmit 250 hp at 200 rpm while subjected to a bending load of 1,000 lb located at a distance of 2 ft from one bearing. Allow a shear stress of 6,000 psi and a bending stress of 12,000 psi. (*a*) Determine the shaft diameter for steady loading. (*b*) Determine the shaft size if the transverse load is steady and the torsional load is suddenly applied.

8-8 A steel shaft 2 in. in diameter is 6 ft between supports. A vertical load of 500 lb is applied 18 in. from one bearing, and a load of 400 lb at 30 deg with the vertical and perpendicular to the shaft is applied at a distance of 30 in. from the same bearing. What is the maximum tensile stress in the shaft?

8-9 The shaft joining a turbine and generator is 6 in. in diameter, and is made of SAE 1045 steel water-quenched and drawn to 1000°F. Use a design factor of 1.5. A key is used to hold the coupling flange to the generator shaft. This key is fitted on all four sides, and the allowable stresses may be taken as 6,000 psi in shear and 14,000 psi in bearing. (*a*) Determine the horsepower that can be transmitted by the shaft at 1,000 rpm. (*b*) Determine the required width, depth, and length of the key.

8-10 A shaft 24 in. between bearings supports a 20-in. pulley 10 in. to the right of the left-hand bearing, and the belt drives a pulley directly below. Another pulley 15 in. in diameter is located 5 in. to the right of the right-hand bearing, and the belt is driven from a pulley horizontally to the right. The coefficient of friction for the belts is 0.30 and the angle of contact 180 deg. The maximum tension in the belt on the small pulley is 800 lb. Find shaft diameter, allowing $s_t = 8,000$ psi and $s_s = 6,000$ psi.

8-11 A mild carbon-steel shaft transmitting 15 hp at 210 rpm is supported on two bearings 27 in. apart and has keyed to it two gears. An 18-tooth, $14\frac{1}{2}$-deg involute three-pitch gear is located 5 in. to the right of the right-hand bearing and delivers power to a gear directly below the shaft. An 80-tooth, four-pitch gear is located 6 in. to the right of the left-hand bearing and receives power from a gear directly over it. Calculate the diameter of the shaft, assuming working stresses of $s_t = 12,000$ psi and $s_s = 10,000$ psi.

8-12   A steel shaft is supported by bearings 4 ft apart. A gear B, 12 in. in diameter, is located 10 in. to the left of the right-hand bearing and is driven by a gear A, directly behind it. A belt pulley C, 24 in. in diameter, is located 18 in. to the right of the left-hand bearing and drives a 24-in. pulley directly behind it. The ratio of belt tensions is 2:1, with the tight side on top. The gear teeth are $14\frac{1}{2}$-deg involute. The stresses in the shaft are not to exceed 16,000 psi in tension and 8,000 psi in shear. The deflection at the gear is not to exceed 0.008 in., and the torsional deformation is not to exceed 1 deg in 20 diameters. The gears transmit 50 hp, and the shaft rotates at 200 rpm. Determine the required shaft diameter. Assume the bending load to be gradually applied and the torsion load to be applied with minor shock.

8-13   The maximum torque delivered by a certain truck engine is 600 ft-lb. The overall efficiency of the gearing and drive shafts is 85 per cent. The weight on each rear wheel is 3,000 lb. The total speed-reduction ratio between the engine and rear axle is 12.5:1. The tires are 29 in. in diameter, and the coefficient of friction of rubber on pavement is 0.8. (*a*) Can the tires be made to slip? (*b*) What diameter of axle must be used if the torsional stress is not to exceed 10,000 psi? The axle is made of nickel steel having an ultimate tensile strength of 115,000 psi.

8-14   A shaft 48 in. long is supported at the ends by simple bearings. A vertical load of 2,000 lb is applied 10 in. from the left end, a load of 3,000 lb acting down and forward at an angle of 60 deg with the vertical is applied 24 in. from the left end, and a load of 2,500 lb acting down and forward at an angle of 30 deg with the vertical is applied 40 in. from the left end. A torque of 10,000 in.-lb is applied at the first load, a torque of 6,000 in.-lb is taken off at the second load, and a torque of 4,000 in.-lb is taken off at the third load. (*a*) Determine the maximum bending moments, graphically. (*b*) Determine the required shaft diameter if the allowable stress is 10,000 psi in tension and 6,000 psi in shear, and the maximum deflection at any point of loading is 0.01 in.

8-15   During the design of a ship it is found that the propeller shaft will be required to transmit 2,000 hp at 110 rpm. The designer has the choice of three steels, A, B, and C, having the properties and prices shown in the table. Prices are per pound after heat-treating and complete machining, i.e., prices per pound of completed shaft.

Steel	Cost per lb, dollars	Ult. strength in tension, psi	Ult. strength in shear, psi
A	0.15	70,000	50,000
B	0.195	90,000	62,500
C	0.24	135,000	95,000

The factor of safety to be used is 12 based upon the ultimate strengths, and the angular twist is to be limited to 1 deg in 20 diameters. (*a*) If weight is the chief consideration, what material and what shaft diameter would you use? (*b*) What diameter shaft would you use if price is the chief consideration?

(c) Same as (a) using a hollow shaft whose inside diameter is three-quarters of the outside diameter. The hollow shaft costs 2 cents/lb more than the solid shaft. (d) Same as (b) using the hollow shaft. (e) Comparing (b) and (d), what percentage of weight and of cost could be saved by using the hollow shaft?

8-16   The shaft of a certain engine is 24 in. long between the bearings. For 8 in. from the left-hand bearing, the diameter is 2 in. For 6 in. from the right-hand bearing, the diameter is $2\frac{1}{4}$ in. The center portion of the shaft is 3 in. in diameter. A load of 1,000 lb is concentrated at a point 10 in. from the left-hand bearing, and a load of 2,000 lb at a point 15 in. from the left-hand bearing. (a) Determine the maximum bending stress in the shaft. (b) By graphical solution, determine the maximum deflection in the shaft.

8-17   A 2-in. steel shaft 40 in. long is simply supported at the ends. It carries a disk A, weighing 100 lb, 15 in. from the left-hand bearing and a second disk B, weighing 75 lb, 25 in. from the left-hand bearing. (a) Determine the critical speed of the shaft alone. (b) Determine the critical speed of the shaft and disk A. (c) Determine the critical speed of the shaft and disk B. (d) Determine the critical speed of the shaft and both disks. (e) What error in critical speed would there be if the weight of the shaft is neglected?

8-18   Determine the critical speed of a small electric motor designed to run at 1,200 rpm if the shaft is $\frac{3}{4}$ in. in diameter and 18 in. between supports. The rotating element may be considered as being a single disk, its weight of 50 lb being concentrated at the center. Is the speed of this motor satisfactory?

8-19   A 3-in. shaft supported on bearings 5 ft apart carries a 3,000-lb disk $1\frac{1}{2}$ ft from the left-hand bearing and a 4,000-lb disk $2\frac{1}{2}$ ft from the left-hand bearing. (a) Determine the critical speed when self-aligning ball bearings are used. (b) Will an operating speed of 1,800 rpm be satisfactory for this shaft? (c) Determine the critical speed if sleeve bearings are used and fitted tight enough so that they may be considered rigid supports.

Chapter 9.   **Rolling-element Bearings**

9-1   Determine the force necessary to move a weight of 5,000 lb supported on four steel wheels of 2 ft diameter, assuming the coefficient of rolling friction to be 0.008.

9-2   The dome of an observatory, weighing 8,000 lb, rests on 12 cast-iron balls of 3.75 in. diameter. The balls run in cast-iron upper and lower races of 15-ft pitch diameter. The coefficient of rolling friction may be taken as 0.01. The dome is rotated at $\frac{1}{4}$ rpm by means of a gear of 6-in. pitch diameter keyed to a vertical shaft at the axis of the dome. Determine the horsepower of the motor required.

9-3   The weight on each wheel of a motor-driven traveling crane is 6,000 lb. The wheels are 2 ft in diameter and are carried on journals 4 in. in diameter and 8 in. long. The coefficient of rolling friction is 0.02 and of sliding friction 0.05. Determine the horsepower of the motor required to drive a four-wheel crane traveling 500 fpm.

9-4   A locomotive, weighing 60 tons, has four pairs of driving wheels. The coefficient of friction between the wheels and rails is 0.20. The drivers are 48 in. in diameter with journals 6 in. in diameter by 9 in. long. The coefficient of rolling friction is 0.01. The coefficient of journal friction is 0.03. (a) Determine the maximum pull that the locomotive can exert. (*b*) What force is required to move the locomotive forward, neglecting wind resistance? (*c*) What is the resistance per ton of weight? (*d*) Assuming the same resistance ratio for the train, what total load can the locomotive pull on level tracks? (*e*) What total train weight could the locomotive pull on a 10 per cent grade?

9-5   A rotating tube mill in a cement mill consists of a horizontal cylinder 7 ft 6 in. in diameter. This mill weighs 50 tons when loaded and is supported on 16-in. rollers, two at each end, carried on 5-in. journals. The rollers are located 30 deg on either side of the vertical center line. The coefficient of rolling friction is $0.001/R$ and the coefficient of journal friction is 0.06. Determine the horsepower friction loss when the mill revolves at 20 rpm.

9-6   Select a ball bearing to carry satisfactorily a 16,000-lb radial load together with a 3,000-lb thrust load. The journal supported by the bearing rotates at a constant 1,600 rpm for an estimated 100,000-hr life, while the outer race of the bearing is pressed into a stationary housing. Although prime consideration should be given to minimize cost of the bearing, do not unjustifiably use a bearing bore larger than necessary. Assume there is a 30 per cent probability that the bearing will fail before its expected design life.

9-7   Select a ball bearing to support a radial load of 4,200 lb with moderate shock and a thrust load of 2,100 lb for 18 hr per day for 10 years. The supported shaft is approximately 3 in. in diameter at the bearing as determined by strength considerations. The bearing inner race is pressed directly onto the shaft without the use of a sleeve on the shaft. The outer race of the bearing rotates at 400 rpm while the shaft remains stationary. Since it is not imperative that this bearing not fail in service, assume a 50 per cent probability that this bearing will not fail before its expected design life.

9-8   Select a single-row angular contact ball bearing to successfully support a 2,500-rpm continuously rotating idler gear which applies a 3,650-lb radial load and a 3,650-lb thrust load with heavy shock onto the stationary shaft. Since whether or not the bearing fails is not of extreme importance, assume there is a 40 per cent chance that the bearing will fail before its expected life. The limits of the bore of the gear are $2\frac{3}{4}$ in. and $3\frac{1}{2}$ in.

9-9   Select a deep-groove ball bearing to support a 285-lb radial load and a 228-lb thrust load with light shock. This bearing is supported by a stationary shaft and in turn supports an idler helical gear which rotates 3,000 rpm for an expected life of 7 years. Strength calculations for the shaft indicate a minimum theoretical diameter of 2.92 in. at the location of the gear.

9-10   Select a straight roller bearing to satisfactorily support a 4,000-lb radial load without any thrust. The journal rotates at a constant 2,600 rpm for an estimated 70,000 hr with an acceptable 70 per cent chance that this bearing will not fail before its expected design life. The outer race is held fixed. Consider that moderate shock conditions exist. Strength considerations for the shaft diameter indicate a minimum of 3.875 in. at the bearing.

Chapter 10.   Sliding-element Bearings

10-1   The specific gravity of a lubricating oil was found to be 0.875 at 100°F. What is the probable specific gravity at 160°F?

10-2 A lubricating oil has a specific gravity of 0.857 at 60°F and a viscosity of 76 Saybolt standard seconds at 200°F. Determine the viscosity of this oil at 200°F in both centipoises and reyns.

10-3 A journal bearing operates under the conditions resulting in 89,718 ft-lb/min of heat generated when the operating temperature of the oil is 160°F. The natural heat-dissipating capacity of the bearing is 26,428 ft-lb/min. Determine the minimum required quantity in gallons per minute of SAE 20 Gulf Coast oil if the lubricant enters the bearing at 110°F.

10-4 The respective viscosities of a zero VI Gulf Coast oil, an unknown oil, and a 100 VI Pennsylvania oil at 100°F are 13, 11.62, and 10 μreyns. What is the viscosity index of the unknown oil?

10-5 A 2.000-in.-diameter lathe-turned tin-babbitt bearing 1.600 in. long supports a 1.998-in.-diameter hardened and cylindrically ground steel journal rotating at a constant 1,200 rpm. The radial load on the bearing is 2,600 lb. An SAE 20 Gulf Coast oil with an oil-film operating temperature of 160°F exists as the lubricant for the bearing. Does the bearing operate under hydrodynamic or marginal lubrication conditions? What is the power loss in the bearing due to shear action in the oil film?

10-6 A stationary high-speed lathe-turned bronze bearing of 2.000-in. bore and 1.600-in. length supports an 1,800-rpm cylindrically ground steel journal of 1.996-in. diameter. A radial load of 10,500 lb is supported by the bearing. The operating temperature of the well-ventilated bearing lubricated with SAE 20 oil is 155°F. Does the bearing operate marginally or hydrodynamically? What is the probable coefficient of bearing friction f_b? Is some form of auxiliary cooling (such as pressurized lubrication) necessary?

10-7 A bearing 2.085 in. in diameter and 1.762 in. long supports a journal running at 1,200 rpm. It operates satisfactorily with a diametral clearance of 0.0028 in. and a total radial load of 1,400 lb. At the 160°F operating temperature of the oil film, the bearing modulus ZN/p was found to be 16.48. Select an oil suitable for use in this bearing.

10-8 A stationary journal bearing 1.875 in. in diameter and 1.500 in. long is bored and lathe turned from a bronze casting. The 1,600-rpm journal is 1.864 in. in diameter and is of hardened and ground steel. The SAE 20 lubricant has an operating temperature of 170°F. The radial load on the bearing is 11,250 lb. Does the bearing operate under hydrodynamic or marginal lubrication conditions? What is the probable power loss in the bearing due to shear of the oil film?

10-9 An SAE 20 oil at an oil-film temperature of 120°F is employed to lubricate a 1.996-in.-diameter cylindrically ground steel journal rotating at 1,200 rpm. The lathe-turned bronze bearing is stationary, has a 2.000-in. bore, and has a length of 1.000 in. Under constant loading conditions, determine the probable maximum radial load the bearing can carry and still maintain hydrodynamic lubrication.

10-10 A 1.875-in.-diameter steel shaft rotating at 1,000 rpm is to carry a radial load of 800 lb. The bearing is 2.5625 in. long and has a 0.0022-in. diametral clearance. An operating minimum oil-film thickness to diameter ratio of 0.00015 in./in. is assumed for the SAE 20 lubrication oil in the bearing. The natural heat-dissipating capacity of the bearing is to be based upon a 110°F ambient air temperature. Determine the probable operating oil-film temperature in this bearing.

10-11   A suspension bearing is to support 70,000 lb, rotate at 400 rpm, and is lubricated with SAE 30 lubricating oil. A coefficient of friction of 0.022 is assumed. The inside diameter of the collar thrust area is 14.0 in. and a maximum unit loading of 1,200 psi is allowed. Determine the outside diameter to be used for the bearing area if this diameter is to be an integral multiple of $\frac{1}{4}$ in.

10-12   Design a horizontal-shaft multiple-collar bearing to carry a 20,000-lb thrust if the inside diameter of each thrust area is 6.00 in., the ratio of the outside to inside diameter of each thrust area is approximately 1.75, the coefficient of friction is 0.08, and the maximum allowable unit loading is 500 psi. All thrust collars are to be identical in size, and the outside diameter of each collar is to be an integral multiple of $\frac{1}{8}$ in.

10-13   A ship traveling 18 mph requires 2,500 hp at the propeller, which turns at 80 rpm. The thrust collars on the propeller shaft are 18 in. in outside diameter and 12 in. in inside diameter. Assume the coefficient of friction to be 0.08 and the permissible pressure on the bearing surfaces to be 50 psi. (*a*) Determine the number of thrust collars required. (*b*) Determine the power lost in the bearing.

10-14   A ship traveling 46 knots (1 knot = 6080.2 ft/hr) requires 65,872 hp at a single propeller turning at 91 rpm. The entire propeller thrust is carried by a Kingsbury thrust bearing having an 18 in. inside diameter and a 46 in. outside diameter. The coefficient of friction is 0.0013. Determine the probable power loss in this bearing due to friction.

Chapter 11.   Spur Gears

11-1   Two shafts, connected by full-height $14\frac{1}{2}$-deg true-involute-tooth gears, are to have a 5:1 velocity ratio. Determine the minimum number of teeth required on each gear.

11-2   A full-depth 20-deg involute rack is to drive a pinion. If both have standard addendums and true-involute-tooth curves, what will be the least number of teeth that can be used on the pinion?

11-3   A 64-tooth full-depth 20-deg involute gear rotates at 100 rpm and drives a second gear. (*a*) What is the highest speed at which the second gear can run? (*b*) How many teeth are on the second gear? (*c*) If 20-deg stub teeth (AGMA) are used, at what speed can the second gear run, and how many teeth will it have?

11-4   A standard $14\frac{1}{2}$-deg involute gear has 60 teeth. (*a*) If this is the larger gear of a pair, what is the largest speed reduction possible? (*b*) If this is the smaller gear of a pair, what is the largest speed reduction possible?

11-5   A cast-iron gear with $14\frac{1}{2}$-deg cast teeth has a pitch diameter of 14 in., 21 teeth, and a 4-in. face width. The permissible working stress is 5,000 psi. (*a*) What is the diametral pitch? (*b*) What is the circular pitch? (*c*) What is the tooth thickness? (*d*) What is the outside diameter? (*e*) What is the root diameter? (*f*) What is the clearance? (*g*) What is the backlash? (*h*) What is the permissible pitch-line load?

11-6   A pair of ordinary cast-iron $14\frac{1}{2}$-deg gears with cast teeth transmit 15 hp at 400 rpm of the pinion. The pinion diameter is approximately 6 in. and the velocity ratio is 4. Determine the circular pitch, face, number of teeth, and pitch diameter of each gear.

11-7 A 28-tooth $1\frac{1}{4}$-in.-pitch $14\frac{1}{2}$-deg cast-tooth cast-iron pinion has a face width of $3\frac{1}{4}$ in. Determine the horsepower which it can transmit at 72 rpm and at 144 rpm.

11-8 A cast-iron cast-tooth $14\frac{1}{2}$-deg pinion approximately 8 in. in diameter transmits 35 hp at 100 rpm. The speed of the gear is 20 rpm. Find the circular pitch, face, exact pitch diameter, and number of teeth of each gear.

11-9 A 1-pitch $14\frac{1}{2}$-deg involute gear has 14 teeth. Draw one tooth three times the actual size, and determine graphically the value of the Lewis factor. Use radial flanks below the base line and a root fillet equal to $1\frac{1}{3}$ times the clearance.

11-10 A 24-tooth $14\frac{1}{2}$-deg involute gear of 8-in. diameter and 3-in. face width has teeth cut with milling cutters. The gear transmits 16 hp at 175 rpm. (a) Determine the apparent working stress in the teeth. (b) Determine the equivalent static stress. (c) Select a material suitable for this gear.

11-11 A pair of carefully cut gears with 20-deg stub teeth is to transmit 30 hp at 300 rpm of the gear with a speed reduction of 5:1. The 3-in. pinion is made of SAE 1035 steel with a hardness of 250 Brinell and drives a cast-iron gear. Determine the diametral pitch and face width.

11-12 A pair of $14\frac{1}{2}$-deg involute cut-teeth gears, of ordinary cast iron, is to transmit 10 hp at 1,120 rpm of the pinion. The velocity ratio is 4, the maximum center distance is $7\frac{1}{2}$ in. Assume steady load. Find the diametral pitch, the face width, and the number of teeth on each gear.

11-13 A pair of ordinary cast-iron 20-deg Fellows gears transmit 20 hp at 1,100 rpm of the pinion. The velocity ratio is 4 and the center distance is 10 in. (a) Determine the pitch, number of teeth, and face width of each gear. (b) Check the dynamic and wear loads for the gears designed in part (a) and state if satisfactory. If the gears are not satisfactory state what change or changes are necessary in order for the gears to be satisfactory.

11-14 A pair of 4-pitch $14\frac{1}{2}$-deg involute spur gears with a face width of $2\frac{1}{2}$ in. is made of SAE 1045 steel with a Brinell number of 400. The pinion has 16 teeth and turns at 1,200 rpm. The speed reduction is 4. What horsepower can be safely transmitted with continuous service and pulsating loads?

11-15 A 4-in. good-grade cast-iron $14\frac{1}{2}$-deg cut-teeth pinion transmits 10 hp at 1,720 rpm to a gear of similar material. The speed ratio is 3. By use of the Lewis equation determine the tentative pitch, face number of teeth, and pitch diameter of each gear. Check these gears by the Buckingham procedure and make the necessary modifications for dynamic and wear qualities.

11-16 Design a pair of gears with AGMA stub teeth to transmit 75 hp from a 7-in. pinion running at 2,500 rpm to a gear running at 1,500 rpm. Both gears are to be made of SAE 3245 steel with a hardness of 260 Brinell. Approximate the pitch by means of the Lewis equations. Then adjust the dimensions to keep within the limits set by the dynamic load and wear equations.

11-17 A pair of 7-pitch cut-teeth spur gears with 20-deg AGMA stub teeth is made of SAE 3245 heat-treated steel. The pinion has 21 teeth and the gear 25. The pinion speed is 3,000 rpm. The face is $1\frac{1}{2}$ in. (a) Basing the strength upon the Lewis equation, what horsepower can be transmitted? (b) If the Brinell number of each gear is 600 and the stress-concentration factor K is 1.6, determine the dynamic and wear loads and state if satisfactory. If the gears are not satisfactory, suggest the necessary change.

11-18   A pair of 4-pitch, $14\frac{1}{2}$-deg composite spur gears with a face width of $2\frac{1}{2}$ in. is made of SAE 1045 steel with a Brinell number of 400. The pinion has 28 teeth and turns at 1,200 rpm. The speed reduction is 4. What horsepower can be transmitted as indicated by the Lewis equation? Assuming this horsepower to be transmitted, check the dynamic and wear loads and find the allowable stress-concentration factor K.

11-19   A pair of 20-deg full-depth spur gears must transmit 25 hp at a speed of 5,000 rpm of the pinion. The velocity ratio is 10 and the diameter of the pinion is 3 in. There should be between 15 and 24 teeth in the pinion. The pinion is made of SAE 1045 steel with a Brinell number of 250, and the gear is made of ordinary cast iron. Determine the pitch, the number of teeth, and the face width of these gears.

11-20   An 18-tooth Bakelite pinion running at 1,750 rpm drives a 145-tooth cast-iron gear. Teeth of 3-pitch AGMA stub form are used, and the face width is 4 in. Determine the horsepower that can be transmitted.

11-21   Determine the pitch, face, number of teeth, and diameters of a pair of $14\frac{1}{2}$-deg gears which transmit 20 hp at 720 rpm of the 7-in. pinion. The pinion is made of Micarta and the gear is made of good-grade cast iron. The velocity is 3 and the load steady.

11-22   A 5-in. Fellows pinion delivers 50 hp at 1,750 rpm to a gear rotating at 500 rpm. Both gears are to be SAE 3245 with a BHN of 300. Determine the pitch, face width, diameter, addendum, and number of teeth for each gear by using the Lewis method. Then check these dimensions by the Buckingham procedure and if not satisfactory make the necessary changes.

Chapter 12.   Helical, Worm, and Bevel Gears

12-1   A 20-tooth helical gear has a pitch diameter of 10 in. The helix angle is 23 deg, and the pressure angle measured in a plane perpendicular to the axis of rotation is 20 deg. The addendum is 0.8 divided by the diametral pitch. (*a*) Find the diametral pitch. (*b*) Find the circular pitch in a plane normal to the teeth. (*c*) Find the pressure angle in a plane normal to the teeth.

12-2   A pair of precision, 12-pitch Gleason helical gears, enclosed and well lubricated, have 20-degree stub teeth and 22.5 degree helix angles. They are of equal diameter and are on parallel shafts 13 in. apart. Each gear has a face width of 1.625 in. and rotates at 1,800 rpm for 24 hr/day for a period of several years; 90 hp with moderate shock is to be transmitted with unidirectional tooth loading. One gear is of cast iron and the other is 1045 steel with a core hardness of 210 Brinell and a surface hardness of 260 Brinell. Make a complete check of these gears and state your reasons why you believe these gears to be properly, or improperly, designed. Do not change any dimensions, materials, or hardnesses.

12-3   A 75-hp motor, running at 450 rpm, is geared to a pump by means of double-helical gearing. The SAE 1035 forged-steel pinion on the motor shaft is 8 in. in diameter and drives the good-grade cast-iron gear on the pump shaft at 120 rpm. Determine the diametral pitch and the face width.

12-4 A single-stage turbine running at 30,000 rpm is used to drive a reduction gear delivering 3 hp at 3,000 rpm. The gears are 20-deg involute herringbone gears of 28 pitch and $2\frac{1}{8}$-in. effective width. The pinion has 20 teeth with a helix angle of 23 deg. (*a*) Determine the tangential pressure between the gear teeth at the pitch line. (*b*) Determine the load normal to the tooth surface. (*c*) Determine the apparent stress in the teeth, by the Lewis equation. (*d*) Determine the pitch-line velocity, in feet per minute. (*e*) Determine the equivalent static stress in the teeth. (*f*) What material would you use for each of these gears?

12-5 Design a pair of equal-diameter 20-deg-stub-tooth helical gears to transmit 50 hp with moderate shock at 1,200 rpm. The two shafts are parallel and 18 in. apart. Each gear is to be steel.

12-6 Two helical gears A and B are on nonparallel shafts 10.0 in. apart and their diametral ratio $D_a : D_b = 2 : 1$. The helix angles of gears A and B respectively are 22.75 deg and 42.85 deg. Determine the pitch diameters of each gear. What is the angular speed ratio of the two gears?

12-7 Design the teeth for two herringbone gears for a single-reduction speed reducer to have a velocity ratio of 3.80. The speed reducer is to transmit 36 hp, and the pinion is to have a speed of 3,000 rpm. The helix angle should be 30 deg and the teeth are to be 20-deg stub teeth in the plane of rotation. The length of the face of the pinion should not exceed twice the pitch diameter. The material of the gears is a high-carbon steel, heat-treated to have a yield point of approximately 60,000 psi and a Brinell hardness of 450.

12-8 A speed reducer for oil-field use is to have herringbone gears and a total reduction of 40:1. No gear is to have less than 18 teeth and all gears are to have the same pitch. The low-speed shaft is to run at 25 rpm and is to deliver 240,000 in.-lb torque. Determine the pitch and face width of all gears when AGMA stub-tooth herringbone gears are used. Use the Schmitter gear equation.

12-9 A pair of 4-pitch $14\frac{1}{2}$-deg involute machine-cut bevel gears of SAE 3245 steel have a 2:1 reduction. The pitch diameter of the driver is 10 in., and the width of face 2 in. Determine (*a*) the pitch angle of the pinion; (*b*) the pitch angle of the gear; (*c*) the face angle of the pinion; (*d*) the face angle of the gear; (*e*) the cutting angle of the pinion; (*f*) the cutting angle of the gear; (*g*) the maximum diameter of the pinion; (*h*) the virtual number of teeth on the pinion; (*i*) the equivalent tangential pressure at the large end of the teeth that may be transmitted at 200 rpm of the driver; (*j*) the pitch diameter at the center of pressure on the pinion; (*k*) the resultant tooth pressure at the center of pressure; (*l*) the horsepower transmitted.

12-10 A pair of machine-cut bevel gears made of SAE 1035 steel have 20 and 30 teeth. The $14\frac{1}{2}$-deg full-height tooth form is used. The smaller gear has a 5-in. pitch diameter, a $1\frac{1}{4}$-in. face width, and runs at 600 rpm. Using the Lewis equation, determine the horsepower that can be transmitted under steady load and continuous service.

12-11 A pair of bevel gears have their shafts at right angles. The larger gear has 50 teeth of 4–5 pitch and is made of SAE 1045 steel. The pinion has 20 teeth and is made of alloy steel hardened to 300 Brinell. The pinion runs at 850 rpm and transmits 30 hp. Determine the required face width.

12-12 The ring gear of a truck differential has 56 teeth of 4-pitch and is made of SAE 2345 steel hardened to 240 Brinell. The pinion has 13 teeth with

$1\frac{5}{8}$-in. face width and is made of SAE 3245 steel hardened to 300 Brinell. The teeth are of the AGMA straight-tooth form. Determine the following: (*a*) the horsepower that may be transmitted at 1,000 rpm of the pinion; (*b*) the resultant tooth pressure and the radius at which this pressure acts on both gears; (*c*) the magnitude of the thrusts along the shafts.

12-13 A pair of bevel gears have a 1:1 velocity ratio, a pitch diameter of 8 in., a face of $1\frac{1}{2}$ in., and they rotate at 250 rpm. The teeth are 5-pitch $14\frac{1}{2}$-deg involute and accurately cut. These gears transmit 8 hp. Determine: (*a*) outside diameter of gears; (*b*) equivalent static stress in the teeth; (*c*) resultant tooth load, tangent to the pitch cone; (*d*) resultant radial load on the bearings; (*e*) resultant thrust load on the gear shafts.

12-14 A 20-deg, 5-pitch, 15-tooth, and $1\frac{1}{2}$-in.-face Gleason bevel gear of steel hardened before cutting to a BHN of 300 is on a shaft rotating at 3,000 rpm and drives a shaft at right angles with a speed ratio of 2. A 20-deg, 4-in.-diameter stub-tooth pinion on the latter shaft drives a gear on a shaft parallel to the shaft of this pinion. The speed ratio is 3. Design the spur gears using a good grade of cast iron.

12-15 A centrifugal pump submerged in a well is driven at 1,250 rpm by a 25-hp, 900-rpm motor through a pair of bevel gears. The gear on the motor shaft is cast steel and has a pitch diameter of 8 in. with a face width of $1\frac{1}{2}$ in. The pinion is made of SAE 1040 steel. The teeth are of the $14\frac{1}{2}$-deg full-height form. (*a*) Determine the diametral pitch. (*b*) Determine the radial load on the pinion. (*c*) Determine the end thrust on the pinion. (*d*) Determine the end thrust on the gear.

12-16 The straight-tooth bevel pinion driving the differential of an automobile has 15 teeth and a $1\frac{1}{4}$-in. face width. The ring gear has 56 teeth. AGMA teeth of 5-pitch are used. The pinion transmits 40 hp at 2,500 rpm. The pinion is supported on two bearings placed $1\frac{1}{2}$ and $5\frac{3}{4}$ in. behind the large pitch circle. (*a*) Determine the probable stress in the teeth. (*b*) Determine the resultant tooth load and the radius at which this pressure acts. (*c*) Determine the thrust along the axis. (*d*) Determine the radial load on each bearing.

12-17 A three-thread worm, rotating at 1,000 rpm, drives a 31-tooth worm gear and transmits 15 hp. The worm has $14\frac{1}{2}$-deg teeth with $\frac{3}{4}$-in. pitch, 2-in. pitch diameter, and an included face angle of 60 deg. The coefficient of friction is 0.05. (*a*) Determine the helix angle of the worm. (*b*) Determine the speed ratio. (*c*) Determine the center distance. (*d*) Determine the apparent stress in the worm-wheel teeth. (*e*) Determine the probable efficiency of this worm-gear set.

12-18 A triple-thread cast-iron worm running at 225 rpm receives $7\frac{1}{2}$ hp through its shaft. The total efficiency of the worm and its bearings is 92 per cent. The velocity ratio is to be 10:1, and the distance between shafts is to be 8 in. The worm is to have a $14\frac{1}{2}$-deg pressure angle and a lead angle greater than 15 deg. Supporting bearings of both worm and worm gear are placed on 6-in. centers. (*a*) Determine the worm-shaft diameter with a permissible stress of 7,500 psi in torsion. (*b*) Using a standard circumferential pitch, determine the number of teeth and pitch diameters of the worm and worm gear. (*c*) Determine the radial, thrust, and tangential forces on both the worm and worm gear. (*d*) Determine all bearing loads. (*e*) Determine the heat that must be dissipated per minute.

12-19 An elevator is driven by a 50-hp 1,200-rpm motor through a worm drive. The worm has four threads of $14\frac{1}{2}$-deg pressure angle and a pitch diameter of $4\frac{1}{4}$ in. The worm gear has 52 teeth of $1\frac{3}{4}$-in. pitch and 4-in. face. The worm bearings are 18 in. center to center, and the worm-gear bearings are 9 in. center to center. Assume the coefficient of friction to be 0.03. (*a*) Determine center distance of the shafts. (*b*) Determine the efficiency of the drive. (*c*) Determine the loads on each bearing. (*d*) Determine the heat that must be dissipated, in Btu per minute.

Chapter 13. Springs

13-1 A coiled compression spring of oil-tempered steel wire has seven active coils of $\frac{7}{16}$-in. wire wound in a coil of $3\frac{1}{4}$-in. outside diameter. The spring is used to produce axial pressure on a clutch. The free length is $7\frac{1}{2}$ in. With the clutch engaged, the length is $5\frac{5}{8}$ in. Determine the stress in the wire and the force exerted against the clutch plate.

13-2 A coiled spring having $8\frac{1}{2}$ active coils of $\frac{3}{8}$-in. steel wire has an outside diameter of $3\frac{1}{4}$ in., a free length of 8 in., and an operating length of 5 in. Determine the stress in the spring and the force exerted. Will this spring buckle?

13-3 The plunger of an oil pump is held against the operating cam by a spring made of No. 10 W&M gauge steel wire coiled with an outside diameter of $\frac{7}{8}$ in. There are five active coils. The open length is $1\frac{3}{16}$ in. What are the length and stress when a load of 30 lb is applied?

13-4 The valve springs of an automobile engine are made of No. 8 W&M gauge wire with an outside diameter of $1\frac{1}{8}$ in. The spring load is 43 lb at a spring length of $2\frac{1}{4}$ in. (valve closed) and 96 lb at a length of $1\frac{29}{32}$ in. (valve open). There are seven effective coils. (*a*) Determine the maximum stress in the wire. (*b*) Determine the modulus of rigidity of the wire material. (*c*) Determine the open length of the spring.

13-5 A rod of 0.12-in. diameter is coiled into a tension spring of 20 effective turns with a mean radius of 1 in. The modulus of rigidity is 12,000,000 psi. (*a*) Determine the stress due to a load of 3 lb. (*b*) Determine the corresponding spring elongation. (*c*) If the initial compression in the coils were 1 lb, what would the stress and elongation be when the 3-lb load was applied? (*d*) Determine the natural frequency.

13-6 The valve spring of a gasoline engine is $1\frac{19}{32}$ in. long when the valve is open, and $1\frac{29}{32}$ in. when the valve is closed. The spring loads are 50 lb with the valve closed and 80 lb with the valve open. The inside diameter of this spring cannot be less than 1 in. (*a*) Determine the required wire size for a maximum operating stress of 60,000 psi. (*b*) Determine the number of active coils required. (*c*) Determine the spring length when completely closed, assuming squared and ground ends. (*d*) Determine the pitch to which this spring should be wound.

13-7 A gas-engine valve spring is loaded to 75 lb when the valve is closed and 115 lb when the valve is open. The valve lift is $\frac{5}{16}$ in. The outside diameter is to be from $1\frac{1}{2}$ to $1\frac{3}{4}$ in., and the permissible stress is 60,000 psi. (*a*) Determine the wire diameter, outside diameter, and number of effective coils. (*b*) Determine the open or free length of the spring, assuming the ends to be squared and ground.

13-8 An engine valve spring exerts a force of 65 lb when the valve is open and 40 lb when the valve is closed. The spring has an outside diameter of $1\frac{1}{2}$ in. The valve lift is $\frac{3}{8}$ in. The permissible stress is 65,000 psi. Determine the wire size and the number of effective coils required.

13-9 A weight of 600 lb falls a distance of 3 ft and strikes a coil compression spring which absorbs the blow with a deflection of 12 in. The mean diameter of the coils is to be eight times the wire diameter. If the allowable stress is 60,000 psi and the modulus of rigidity is 11,500,000 psi, find the diameter of the wire and the number of effective coils required.

13-10 A $\frac{1}{2}$- by 1-in. rectangular wire forms a spring of six effective coils with a 3-in. outside diameter. A 3,000-lb load is supported. (*a*) Determine the maximum stress if the long side of the wire is parallel to the spring axis. (*b*) Determine the deflection at this load. (*c*) What load will produce the same stress if the wire is wound with the short side parallel to the spring axis?

13-11 The spring of a discharge valve for a reciprocating plunger pump is made of phosphor bronze (SAE 81). It has five effective coils and $1\frac{1}{2}$ noneffective coils. When the maximum allowable deflection of the spring is 0.5 in., the total axial load on the spring is 3 lb. The mean coil diameter is $1\frac{1}{2}$ in. When the spring is at its maximum lift (deflection) it could be compressed $\frac{1}{4}$ in. before it is completely closed. Find the diameter of the wire and the free length of the spring.

13-12 A coil spring is made of eight effective coils of $\frac{1}{4}$- by $\frac{1}{2}$-in. wire and has an outside diameter of $2\frac{1}{2}$ in. The permissible stress is 50,000 psi. (*a*) Determine the load that can be supported if the wire is coiled with the long side parallel to the axis. (*b*) Determine the load if the wire is coiled with the short side parallel to the axis. (*c*) Determine the deflection in each case. (*d*) Check the buckle in each case.

13-13 A diesel engine weighing 160,000 lb is mounted on 12 springs in order to protect the building from vibration. The springs have $2\frac{1}{2}$ effective coils and are made of SAE 3245 steel, oil-quenched and drawn to 1000°F. (*a*) What size of square wire and what outside diameter would you use for these springs? (*b*) What will the spring deflection be when the engine is not running?

13-14 A $\frac{1}{4}$-in. round wire is coiled to form a conical spring with 10 effective coils and an outside diameter of 3 in. at one end and 2 in. at the other. (*a*) Determine the load that will produce a stress of 50,000 psi. (*b*) Determine the deflection at this load.

13-15 A clutch spring is made of $\frac{3}{16}$- by $\frac{5}{16}$-in. rectangular wire wound to form a conical spring with outside diameters of $5\frac{3}{4}$ and $3\frac{3}{8}$ in. at the ends. There are $3\frac{1}{2}$ effective coils. The long side of the wire is parallel to the axis. (*a*) What clutch force is required to produce a spring stress of 50,000 psi? (*b*) Determine the free length of the spring if the operating length is $1\frac{1}{8}$ in. (*c*) Determine the spring stress if the clutch plate moves $\frac{1}{4}$ in. in disengaging.

13-16 Determine the natural frequency of the spring in Prob. 13-1.

13-17 Determine the natural frequency of the spring in Prob. 13-2.

13-18 A laminated spring is made of six graduated and two full-length leaves, each 0.134 in. thick and $\frac{3}{4}$ in. wide. The effective length of the spring is 30 in. The permissible stress is 40,000 psi. (*a*) Determine the central load that may be carried, assuming no initial stress in the spring. (*b*) Determine the deflection.

13-19   A semielliptic laminated spring is made of No. 10 BWG steel 2 in. wide. The length between supports is $26\frac{1}{2}$ in. and the band is $2\frac{1}{2}$ in. wide. The spring has two full-length and five graduated leaves. A central load of 350 lb is carried. (*a*) Determine the maximum stress in each set of leaves for an initial condition of no stress in the leaves. (*b*) Determine the maximum stress if initial stress is provided to cause equal stresses when loaded. (*c*) Determine the deflection in part (*a*). (*d*) Determine the deflection in part (*b*).

13-20   A cantilever spring has an effective length of 21 in. and has two full-length and eight graduated leaves, each $2\frac{1}{4}$ in. wide. The spring is to sustain a load of 900 lb with a stress of 50,000 psi in all leaves. (*a*) Determine the thickness of the spring leaves and give the BWG number. (*b*) Determine the deflection at full load.

13-21   A truck spring has 10 leaves of graduated length. The spring supports are $42\frac{1}{2}$ in. apart and the central band is $3\frac{1}{2}$ in. wide. The central load is to be 1,200 lb with a permissible stress of 40,000 psi. Determine the width and thickness of the steel spring material and the deflection when loaded. The spring should have a ratio of total depth to width of about $2\frac{1}{2}$.

13-22   A spring has an overall length of 44 in. and sustains a load of 16,000 lb at its center. The spring has three full-length leaves and 15 graduated leaves with a central band 4 in. wide. All leaves are to be stressed to 60,000 psi when fully loaded. The ratio of total spring depth to width is to be approximately 2. (*a*) Determine the width and thickness of the leaves. (*b*) Determine the initial space that must be provided between the full-length and graduated leaves before the band is applied. (*c*) What load is exerted on the band after the spring is assembled?

13-23   A disk spring is made of $\frac{1}{8}$-in. sheet steel with an outside diameter of 5 in. and an inside diameter of 2 in. The spring is dished $\frac{3}{16}$ in. The maximum stress is to be 80,000 psi. (*a*) Determine the load that may be safely carried. (*b*) Determine the deflection at this load.

Chapter 14.   Couplings and Clutches

14-1   A plain flange coupling for a 3-in. shaft has the following dimensions: bore, 3 in.; hub diameter, $5\frac{3}{8}$ in.; hub length, $3\frac{3}{8}$ in.; flange diameter, 10 in.; flange thickness, $1\frac{1}{16}$ in.; bolt diameter, $\frac{3}{4}$ in.; bolt-circle diameter, $8\frac{1}{4}$ in.; number of bolts, six; and key size, $\frac{3}{4}$ in. square. All parts are made of SAE 1020 steel, annealed. This coupling is rated at 50 hp at 100 rpm. (*a*) Determine the bearing, shear, and tensile stresses in all parts of the coupling. (*b*) What factor of safety does this coupling have?

14-2   Two $1\frac{7}{16}$-in. shafts are connected by a flange coupling. The flanges are fitted with six bolts of SAE 1020 steel on a 5-in. bolt circle. The shafts run at 350 rpm and transmit a torque of 8,000 in.-lb. Assume a factor of safety of 5. (*a*) What diameter bolts should be used? (*b*) How thick should the flanges be? (*c*) Determine the key dimensions. (*d*) Determine the hub length. (*e*) What horsepower is transmitted?

14-3   A jaw clutch for a $4\frac{3}{16}$-in. shaft has three jaws with radial faces. The dimensions are: inside diameter of jaws, $4\frac{3}{8}$ in.; outside diameter, $11\frac{1}{2}$ in.; axial height of jaws, 2 in.; and key size, 1 by 1 by $6\frac{3}{8}$ in. Assume $\frac{1}{8}$-in. clearance between the jaws and a working stress in the shaft of 6,000 psi. (*a*) What horsepower can be transmitted at 100 rpm? (*b*) Determine the shear and bearing stresses in the key and the bearing stress on the jaw faces.

14-4   A cone clutch has a face angle of 15 deg with a maximum diameter of 24 in. and a face width of 3 in. The coefficient of friction is 0.2 and the permissible pressure on the cone surface is 12 psi. (*a*) What torque may be transmitted? (*b*) What horsepower may be transmitted at 800 rpm? (*c*) What axial force must be exerted at this power?

14-5   An engine developing 40 hp at 1,250 rpm is fitted with a cone clutch built into the flywheel. The cone has a face angle of $12\frac{1}{2}$ deg and a maximum diameter of 14 in. The coefficient of friction is 0.2. The normal pressure on the clutch face is not to exceed 12 psi. (*a*) Determine the face width required. (*b*) Determine the spring force required to engage this clutch.

14-6   A flat-rim clutch 10 in. in diameter uses four friction blocks, 2 in. wide and 6 in. along the circumference. Coefficient of friction = 0.55. If this clutch rotates at 200 rpm, what force is required on each block if 6 hp is transmitted?

14-7   A block clutch has four wooden shoes each contacting with 75 deg of the inside of a 12-in. drum. The coefficient of friction is 0.3 and the maximum contact pressure is not to exceed 35 psi. This clutch is to transmit 15 hp at 250 rpm. Determine the required width of shoes.

14-8   A disk clutch consists of two steel disks in contact with one asbestos-fabric-faced disk having an outside diameter of 10 in. and an inside diameter of 8 in. Determine the horsepower that can be transmitted at 1,000 rpm if the coefficient of friction is 0.35 and the disks are pressed together by an axial force of 2,000 lb.

14-9   A single-plate clutch is to have a maximum capacity of 75 hp at 1,800 rpm. The clutch facing has a coefficient of friction of 0.4 and a permissible pressure of 30 psi. The clutch is engaged through 12 springs of $1\frac{1}{4}$ in. mean diameter. The springs compress $\frac{1}{16}$ in. for disengagement, with an increase in pressure of 10 per cent. Determine the diameters of the clutch facing if the inner diameter is 0.7 of the outer diameter.

14-10   A multiple-disk clutch is to be used on machine tools. There are eight driven disks having an outside diameter of 3 in. and an inside diameter of $2\frac{1}{4}$ in. The disks are metal and run in an oil spray. The coefficient of friction may be taken as 0.02 and the permissible unit pressure as 100 psi. (*a*) Determine the axial force required. (*b*) Determine the horsepower that can be transmitted at 600 rpm.

14-11   A six-cylinder engine is rated at 160 hp at 3,600 rpm. The maximum torque is developed at 2,800 rpm and 120 hp. The multiple-disk clutch consists of fabric-faced disks of $8\frac{1}{2}$ in. outside diameter and $6\frac{1}{4}$ in. inside diameter in contact with five driven disks. The coefficient of friction is 0.20. Determine the axial force necessary to engage this clutch.

14-12   A gasoline-engine-driven tractor is equipped with a multiple-disk clutch with six driven disks faced with asbestos fabric whose coefficient of friction is 0.25. The disks are 8 in. in inside diameter and 12 in. in outside diameter. The construction of the clutch requires five springs and limits the spring diameter to $1\frac{1}{2}$ in., and the length when the clutch is engaged to $2\frac{3}{4}$ in. When the clutch is disengaged the clutch springs are $2\frac{1}{2}$ in. long, and the pressure is 45 per cent higher than when engaged. The maximum stress in the springs is not to exceed

70,000 psi. Find the horsepower that can be transmitted at 1,200 rpm, allowing 10 psi pressure on the disks.

14-13   A multiple-disk clutch has 22 pairs of identical mating friction surfaces each with 10.25 in. outside diameter and 3.125 in. inside diameter annular contact area. The disks are cast iron and steel operating in an oil bath with a maximum pressure of 1,200 psi. After this clutch has been in operation for a period of time and has become worn, determine the probable power rating of this clutch at 1,800 rpm.

14-14   A 22.5-hp, 3,600-rpm motor is to be fitted with a centrifugal starting clutch having eight equally spaced shoes inside a 12.0-in.-inside-diameter drum. Assuming that the torque at centrifugal engagement is 75 per cent of the torque at rated horsepower and speed, and that the engagement is to occur at a motor speed of 3,000 rpm, determine the weight of each shoe. Assume a coefficient of friction of 0.325, a 35-deg arc of contact between each shoe and the drum, and a 5.12-in. radius from the center of the drum to the mass centroid of each shoe.

14-15   A hydraulic coupling having a mean diameter of outer passageway equal to 14 in. has been successfully transmitting a torque of 200 ft-lb at 1,850 rpm of the engine with a slip of 3.5 per cent. The design of a geometrically similar coupling is proposed to deliver 850-hp output to a shaft turning at 450 rpm with the same coupling efficiency and using the same oil. Estimate the required mean diameter of the outer passageway. What is the horsepower required of the engine delivering power into the coupling?

14-16   A hydraulic torque converter has an outer vane diameter of 14 in. and successfully delivers 200 hp at 2,200 rpm output speed. Assuming the same design coefficient, what would you estimate to be the power rating of a geometrically similar torque converter having a 20-in. outer vane diameter and a 1,200-rpm output speed?

Chapter 15.   Brakes

15-1   Determine the force with which the brake shoe of a diesel-electric train must be pressed against the wheel to absorb 250,000 ft-lb of energy in 20 sec if the mean velocity of the wheel relative to the brake shoe is 45 fps and the coefficient of friction is 0.2.

15-2   A simple brake band (Fig. 15-1) has a 30-in. drum fitted with a steel band $\frac{5}{32}$ in. thick lined with brake lining having a coefficient of friction of 0.25 when not sliding. The arc of contact is 245 deg. This brake drum is attached to a 24-in. hoisting drum that sustains a rope load of 1,800 lb. The operating force has a moment arm of 60 in., and the band is attached 5 in. from the pivot point. (*a*) Find the force required just to support the load. (*b*) What force will be required if the direction of rotation is reversed? (*c*) What width of steel band is required if the tensile stress is limited to 7,500 psi?

15-3   A simple band brake operates on a drum 24 in. in diameter that is running at 200 rpm. The coefficient of friction is 0.25. The brake band has a contact of 270 deg, and one end is fastened to a fixed pin and the other end to the brake arm 5 in. from the fixed pin. The straight brake arm is 30 in. long and is placed perpendicular to the diameter that bisects the angle of contact. (*a*) What is the minimum pull necessary on the end of the brake arm to stop the wheel if

45 hp is being absorbed? What is the direction of rotation for this minimum pull? (b) What width steel band $\frac{3}{32}$ in. thick is required for this brake if the maximum tensile stress is not to exceed 8,000 psi?

15-4   The brake shown in Fig. P15-1 is fitted with a cast-iron brake shoe. The coefficient of friction is 0.3. The braking torque is to be 3,000 in.-lb. (a) Determine the force F required with counterclockwise rotation. (b) Determine the force F required with clockwise rotation.

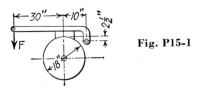

Fig. P15-1

15-5   In the brake of Prob. 15-4, where must the pivot point be placed to make the brake self-energizing with counterclockwise rotation?

15-6   A differential brake band has an operating lever 9 in. long. The ends of the brake band are attached so that their operating arms are $1\frac{1}{2}$ and 6 in. long. The brake-drum diameter is 24 in., the arc of contact 300 deg, the brake band $\frac{1}{8}$ by 4 in., and the coefficient of friction 0.22. (a) Find the least force required at the end of the operating lever to subject this band to a stress of 8,000 psi. (b) What is the torque applied to the brake-drum shaft? (c) Is this brake self-locking? Prove your answer.

15-7   An elevator brake is constructed as shown in Fig. P15-2. Each brake shoe is 5 in. long. The coefficient of friction is 0.3, and the permissible pressure 50 psi average. The operating force F is 100 lb. (a) Determine the braking torque for clockwise rotation. (b) Determine the braking torque for counterclockwise rotation. (c) Determine the required width of the brake shoes.

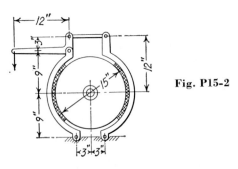

Fig. P15-2

15-8   The rope drum of an elevator hoist is 2 ft in diameter, and the speed of the elevator is 600 fpm. This drum is fitted with a brake drum 3 ft in diameter having four cast-iron brake shoes each subtending an arc of 45 deg on the brake drum. This elevator weighs 4,000 lb loaded, and the brake is to have sufficient capacity to stop the elevator in 12 ft. The coefficient of friction of cast iron on cast iron may be taken as 0.2. (a) Determine the radial force required on each brake shoe. (*b*) If the allowable pressure on the brake shoe is 50 psi, determine the width of shoes required. (*c*) How much heat is generated in stopping this elevator?

15-9   Determine the torque capacity of the automotive-type hydraulically actuated two-shoe brake as indicated in Fig. P15-3 if the applied force *F* onto each shoe is 300 lb, the coefficient of friction *f* is 0.32, brake constant *K* is 12.5 lb/in.^{3}, width of each shoe is 1.5 in., *a* is 4 in., *b* is 7.5 in., *r* is 5.0 in., θ_1 is 7 deg, θ_2 is 144 deg, θ_3 is 203 deg, and θ_4 is 340 deg.

Fig. P15-3

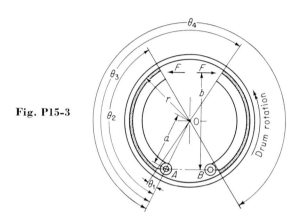

15-10   An automotive brake consists of two identical shoes, each of which is as indicated in Fig. 15-6; θ_1 is 10 deg and θ_2 is 110 deg. The maximum pressure between the shoes and the inside diameter of the drum is 200 psi for each shoe. The inside diameter of the drum is 10.0 in. The width of each shoe is 1.75 in., and the effective coefficient of friction is 0.25. The distance *a* is 3.875 in., and *b* is 8.5 in. Assuming clockwise rotation of the drum, what is the total braking torque on the drum? If the direction of drum rotation is reversed and the actuating forces remain the same, will the resulting braking torque be larger, equal to, or smaller than that for clockwise rotation?

15-11   A spot brake of the type indicated in Fig. 15-8 has two hydraulic cylinders each with 2.0 in.^{2} actuating piston area, and there is a 12-in. distance between the axes of these cylinders and the wheel shaft. What braking torque can be applied for a 1,000-psi hydraulic pressure within the cylinders and a coefficient of friction of 0.53? How much energy is absorbed by this brake if this braking torque is held constant over a period of 10 sec?

Chapter 16.   Belts

16-1   A 36-in. driving pulley and a 48-in. driven pulley are arranged on 10-ft centers. The output of the driven shaft is 150 hp. Assume a belt speed of 4,200 fpm, a coefficient of friction of 0.3, a slip of 1.5 per cent at each pulley, and 5 per cent friction loss at each shaft. (*a*) Determine the revolutions per minute of each shaft. (*b*) Determine the difference in belt tensions. (*c*) Determine the size of machine-wire-laced leather belt required. (*d*) Determine the required shaft sizes, assuming pure torsion and an allowable stress of 8,000 psi. (*e*) Determine the overall efficiency of this transmission.

16-2   A 25-hp, 870-rpm motor with a 12-in. pulley drives a centrifugal pump at 290 rpm by means of a medium double leather belt laced with wire by hand. Determine the width and length of the belt required, allowing for 20 per cent overload and 4-ft center distance.

16-3   A 50-hp, 1,200-rpm high-torque squirrel-cage motor is used to drive a punch press. The motor pulley is cast iron 14 in. in diameter. The driven pulley is cast iron 42 in. in diameter. The center distance is 8 ft and is inclined at 55 deg with the horizontal. Select a suitable leather belt.

16-4   A 5-hp, 900-rpm high-torque squirrel-cage motor drives a plate shear by means of a light double-ply leather belt. A 9-in. motor pulley is used and a 6-in. driven pulley, both of cast iron. The drive is horizontal with a 36-in. center distance. (*a*) Determine the effective tension required for starting. (*b*) Determine the probable unit stress in the belt.

16-5   A ventilating fan having an 18-in. cast-iron pulley is driven from a 25-hp at 870 rpm normal-torque motor placed directly below it. The steel pulley on the motor is 10 in. in diameter, and the center distance is 5 ft. Select a leather belt for this drive.

16-6   Using the data in Prob. 16-5, select a suitable rubber belt.

16-7   Two shafts, 6 ft apart, are connected by an open leather belt $\frac{1}{4}$ in. thick and 4 in. wide. The driving shaft rotates at 1,750 rpm and carries an 8-in. pulley. The driven shaft rotates at 438 rpm. (*a*) Determine the horsepower that may be transmitted under steady-load conditions. (*b*) Determine the additional horsepower that may be transmitted when a 6-in. idler is placed 9 in. from the center of the driver on a line parallel to the original belt.

16-8   A mine ventilating fan running at 120 rpm is driven from a 12-in. pulley on a 50-hp, 720-rpm motor. The center distance is 18 ft. Determine the width of six-ply rubber belt required.

16-9   The main drive in a tile plant is from a 250-hp, 550-rpm motor. A 30-in. pulley is used on the motor and a 130-in. pulley on the driven shaft. An endless belt with 25 per cent overload capacity is to be used. The center distance is 20 ft. (*a*) What width of 36-oz duck eight-ply rubber belt is required? (*b*) If the coefficient of friction is 0.3, determine the probable working stress.

16-10   A ventilating fan fitted with a 16-in. cast-iron pulley is driven from a 10-in. pulley on a 25-hp, 1,200-rpm motor placed 5 ft directly below it. Select a rubber belt for this drive.

16-11   A 10-hp, 1,200-rpm motor is arranged for a Rockwood drive as shown in Fig. 16-6. The pulley diameter is 8 in., and the motor weight is 350 lb. The distance *b* is 16 in., and the distance *c* is 18 in. The starting torque on the motor is 250 per cent of rating. (*a*) Determine the effective tension required for starting the drive and the values of F_1 and F_2. (*b*) Determine the distance *a* from the hinge point to the motor center line. (*c*) Determine the tensions F_1 and F_2 in the belt at rated load. (*d*) Determine the tensions at zero load. (*e*) Compare the results of (*a*), (*c*), and (*d*) with the tensions required with a simple open belt drive.

16-12   The drive from a motor to a centrifugal pump consists of three size B V belts. The motor pulley has a 4.3-in. pitch diameter, and the pump pulley has a pitch diameter of 15.4 in. The motor runs at 1,200 rpm. What power can be transmitted if the center distance is 3 ft?

16-13   An oil-field pumping jack is fitted with a gear-reduction unit delivering 26,000 in.-lb torque at 25 rpm. The total reduction in the gears is 40:1. The gear unit is fitted with a 13-in. pitch diameter V-belt pulley and is driven by a 1,200-rpm motor. The center distance is 4 ft. Determine the number of size C V belts required.

16-14   A 15-hp, 1750-rpm squirrel-cage normal-torque electric motor is to drive an air compressor at 435 rpm. The distance from shaft center to shaft center should be about 30 in. The motor has a $1\frac{3}{8}$-in. shaft and the compressor shaft is $2\frac{1}{4}$ in. Both shafts have standard keys. Select the belts and pulleys for a narrow V-belt drive.

Chapter 17.   Power Chains

17-1   A roller chain operating under steady-load conditions transmits 5 hp from a shaft rotating at 600 rpm to one operating at 750 rpm. (*a*) Determine the chain required using at least 15 teeth in the sprockets. (*b*) Determine the sprocket pitch diameters. (*c*) Determine the shortest advisable center distance. (*d*) Determine the number of links of chain required.

17-2   A 10-hp, 1,200-rpm motor drives a line shaft at 250 rpm. The shaft center distance is to be approximately 2 ft. The motor shaft has a diameter of $1\frac{1}{4}$ in. The starting torque of the motor is from 1.75 to 2.00 times the running torque. The load is applied with moderate shock. (*a*) Select a roller chain for this drive. (*b*) Determine the sprocket pitch diameters. (*c*) How many chain links are required, and what is the exact center distance?

17-3   A roller chain is used to drive the camshaft of an internal-combustion engine. Both shafts rotate at 325 rpm and the center distance is approximately 21 in. The crankshaft is 5 in. in diameter, and the root spaces of the sprockets must clear the shaft by at least $\frac{5}{8}$ in. Three horsepower is required to drive the shaft. Determine all necessary dimensions for the chain and sprockets and show by calculations that it is safe in every way.

17-4   A truck equipped with a 50-hp engine uses a roller chain as the final drive to the rear axle. The driving sprocket runs at 225 rpm and the driven sprocket at 100 rpm with a center distance of approximately 3 ft. The chain speed is to be approximately 500 fpm. The transmission efficiency between the engine and the driving sprocket is 85 per cent. (*a*) Determine the pitch and width of chain to be used. (*b*) Determine the number of teeth in each sprocket and the pitch diameters.

17-5   A rotary engine driving a wire-line reel on an oil-well rig develops 250 hp at 1,000 rpm. The reel runs at 50 rpm maximum. The engine may be slowed down to 200 rpm, the torque remaining constant. This is severe service for the chain drive between the engine and reel. (*a*) Select a suitable roller chain. (*b*) Determine the number of teeth and the pitch diameter of each sprocket.

17-6   A silent chain operating under good service conditions transmits 100 hp from a 600-rpm motor to a shaft running at 167 rpm. (*a*) Determine a suitable number of teeth for each sprocket. (*b*) Determine the pitch and width of chain required for this drive.

Chapter 18.   Wire Ropes

18-1   Determine the probable bending stress and equivalent bending load in a $1\frac{1}{2}$-in. 6 by 19 steel rope made from a 0.095-in. wire, when it is used on a 92-in. sheave. Also check fatigue strength.

18-2   A $\frac{3}{4}$-in. 6 by 37 alloy-steel rope is made of 0.034-in. wire. (a) Determine the bending stress when used with a 24-in. sheave. (b) Determine the equivalent bending load. (c) Estimate the probable number of bends the rope will withstand.

18-3   A $1\frac{1}{2}$-in. 6 by 19 plow-steel rope is used to lift the cage of a mine hoist. The mine is 600 ft deep, the rope sheave is 12 ft in diameter, and the cage acceleration is 5 ft/sec.² (a) Determine the equivalent bending load. (b) Determine the load due to the rope. (c) Determine the acceleration load. (d) Determine the useful load with a factor of safety of 5.

18-4   A mine hoist is to lift ore from a depth of 2,000 ft. The skip weighs 5,000 lb and 4 tons are to be raised each trip. The maximum speed of load is 1,800 fpm and this speed is reached in 15 sec; $1\frac{3}{8}$-in. alloy-steel 6 by 37 rope is used. The sheave diameter is $2\frac{1}{2}$ ft. Find the factor of safety.

18-5   In an office building the elevator rises 1,200 ft with an operating speed of 1,000 ft/min and reaches full speed in 35 ft. Assume a drive similar to Fig. 18-5, 6 by 19 cast-steel ropes, and 36-in. sheaves. The loaded elevator weighs 2 tons. Determine the number of $\frac{5}{8}$-in. ropes required if the factor of safety is 6.

18-6   An oil well using $4\frac{1}{2}$-in. drill pipe (same dimensions as 4-in. standard pipe) is drilled to a depth of 4,000 ft. Assume 50 lb every 40 ft for pipe joints. The rope sheaves are 30 in. in diameter, and the acceleration is 10 ft/sec.² Determine the number of strands of 1-in. 6 by 37 plow-steel rope required for lifting the string of pipes using a factor of safety of $2\frac{1}{2}$.

18-7   Consider the following information on a traction-type elevator: lift height, 600 ft; maximum speed, 600 fpm; distance between stops, 15 ft; maximum acceleration, 2 ft/sec²; weight of car, 2,000 lb; rated capacity, 3,000 lb; counterweight, 3,900 lb; number of ropes, 4; size of ropes (6 by 37), $\frac{7}{16}$ in. in diameter; ultimate strength, 5.5 tons; weight per foot, 0.3 lb; diameter of sheaves, 24 in.; and coefficient of friction, 0.135. (a) Determine the actual load on each rope and the factor of safety. (b) If the angle of contact is 180 deg, will the friction between the rope and the grooves be sufficient to prevent slipping under the worst conditions?

18-8   A 70-story building is provided with 75 elevators, 24 of which operate at 1,200 fpm. The highest-rise cars have a total travel of 779 ft and carry a rated load of 3,500 lb, the elevator weighing approximately 3,000 lb. It requires 120 ft to accelerate to full speed. The ropes used on these elevators are special $\frac{11}{16}$-in. 8 by 19 ropes of 32,400-lb rated strength. (a) Assuming the head sheave to be 30 in. in diameter, determine the number of ropes required per elevator, with a factor of safety of 8, including the bending stress. (b) If the empty elevator rests with its platform level with the lowest floor, how much will it drop when fully loaded?

18-9   Select a wire rope for a vertical mine hoist to lift 1,500 tons of ore in each 8-hr shift from a depth of 3,000 ft. Assume a two-compartment shaft with the hoisting skips in balance. Use a maximum velocity of 2,500 ft/min with acceleration and deceleration periods of 15 sec each and a rest period of 10 sec for discharging and loading the skips. A hoisting skip weighs approximately 0.6 of its load capacity. The factor of safety should be approximately 5.

Chapter 19.   Fits, Tolerances, Limits, and Surface Roughnesses

19-1   The journal of a generator is $1\frac{7}{8}$ in. in diameter. The bearing is to have a medium fit. Determine the allowance, the tolerance on the shaft, and the tolerance on the bearing bore.

19-2   The shaft of an 800-rpm turbine is $4\frac{1}{2}$ in. in diameter at the bearing. Give the machining limits for the shaft and the bearing bore to provide a free fit.

19-3   The outer race of a No. 312 ball bearing is to be given a snug fit, and the inner race is to be given a tight fit. Determine the machining dimensions for each of the following: bearing bore, shaft diameter, bearing outside diameter, and housing bore.

19-4   A No. 220 ball bearing is to be used with the inner race stationary. Determine the machining dimensions of the bearing bore, the shaft, the bearing outside diameter, and the housing bore.

19-5   The crank disk on a gear speed-reducing unit is to be given a medium press fit. Determine the machining dimensions for the $5\frac{7}{8}$-in. shaft and for the crank-disk bore.

19-6   A cast-iron coupling is to be pressed onto a $3\frac{15}{16}$-in. steel shaft. The coupling hub has a length of 6 in. and an outside diameter of $7\frac{1}{2}$ in. The coupling is bored 0.003 in. smaller than the shaft. (*a*) Determine the pressure between the shaft and coupling. (*b*) Determine the maximum tensile stress in the coupling hub. (*c*) Determine the force required to force the coupling on the shaft.

19-7   A cast-steel crank is to be shrunk onto a 10-in. steel shaft. The outside diameter of the crank hub is $17\frac{1}{2}$ in. The maximum tangential stress in the hub is to be 20,000 psi. The coefficient of friction between the hub and shaft is assumed to be 0.15. (*a*) Determine the required bore of the crank. (*b*) Determine the probable value of the normal pressure between the shaft and hub. (*c*) Determine the torque that may be transmitted without using a key, if hub length is 10 in. (*d*) Determine the stress in the shaft due to torsion.

19-8   The cast-iron crank of a gas engine is shrunk on its shaft. The steel shaft is $4\frac{1}{2}$ in. in diameter, and the crank hub is $8\frac{1}{2}$ in. in diameter. The maximum torque to be transmitted is 100,000 in.-lb. Assume the coefficient of friction to be 0.1. (*a*) Determine the shrinkage allowance to give a tangential stress of 7,500 psi in the hub. (*b*) Determine the normal pressure between the hub and shaft after assembly. (*c*) Determine the required hub length if no key is used. (*d*) To what temperature must the hub be raised if 0.003-in. clearance is required for assembly?

19-9   A steel ring with a bore of 3.495 in. and an outside diameter of $5\frac{3}{4}$ in. has been shrunk onto a steel shaft of 3.500-in. diameter. Determine the maximum stresses set up in the shaft and in the ring.

19-10   An aluminum-alloy ring is shrunk onto a steel shaft having a diameter of 2.500 in. The ring has a bore of 2.497 in. and an outside diameter of 3.000 in. Assume the modulus of elasticity of aluminum alloy to be 10,000,000 psi. Determine the probable tangential stress in the ring.

19-11   The cylinder of a high-pressure pump is made in the form of a compound cylinder with outer part shrunk on. Both parts are of steel. The bore of the pump cylinder is 5 in., the inner cylinder has an outside diameter of $9\frac{1}{2}$ in., and the outer cylinder has an outside diameter of 13 in. (*a*) Determine all machining dimensions so that the tangential stress at the cylinder bore will vary between the limits of 5,000 and 7,500 psi in compression, after assembly. (*a*) What will be the maximum tensile stress in the outer cylinder? (*b*) What will be the maximum tangential stress in the inner and outer cylinders when the pump pressure is 5,000 psi?

19-12   Determine typical average values of the rms μin. and the predominant-peak surface roughness of each of the following surfaces: cylindrically ground steel shaft, ground and lapped bronze bearing, and lathe-turned cast-iron shaft.

19-13   Microscopic examination of the cross section of a metallic surface indicates the following deviation (in microinches) of the surface from the mean surface: $+62$, $+127$, -13, $+23$, -69, -42, $+23$, -91, $+4$, $+38$. On the basis of these 10 readings, taken at equal spacings along the same line on the surface, determine the approximate root-mean-square surface roughness of the metallic surface. Assuming the surface to have been ground to an average peak-to-valley roughness/rms roughness ratio of 4.5, determine the average peak-to-valley roughness of the metallic surface.

Chapter 20.   Cylinders, Pipes, Tubes, and Plates

20-1   (*a*) What is the tangential stress in a steel cylinder of 6 in. inside diameter and $7\frac{1}{2}$ in. outside diameter with an internal pressure of 750 psi? (*b*) What is the longitudinal stress? (*c*) What is the maximum shear stress? (*d*) Assuming the maximum-normal-stress theory to be applicable, determine the equivalent stress.

20-2   A cast-iron pipe is to deliver water at the rate of 31,000 gpm and a flow rate of $1\frac{1}{2}$ ft/sec. The maximum pressure in the pipe is 125 psi. The permissible stress in the cast iron is 3,000 psi. Determine the pipe diameter and the wall thickness.

20-3   A cast-iron pipe is 10 in. in inside diameter and the metal is $\frac{3}{8}$ in. thick. The pipe contains water under a head of 250 ft. (*a*) What is the apparent factor of safety considering water pressure only? (*b*) Determine the stress caused by bending if the pipe is full of water, horizontal, 24 ft long, and simply supported at the ends. (*c*) What is the maximum combined tensile stress?

20-4   A 4-in. steel water pipe is subjected to an internal pressure of 175 psi. The permissible stress is 10,000 psi. (*a*) Determine the required wall thickness. (*b*) Compare this thickness with that of a standard-weight steel pipe. (*c*) Can you give any reason for differences between the thickness determined in (*a*) and that of a standard pipe?

20-5   Determine the wall thickness required for a brass condenser tube of 1 in. inside diameter. The vacuum in the condenser is to be 26 in. of Hg and the water pressure is to be 35 psi on the inside of the tube. Allow $\frac{1}{16}$ in. on the wall thickness for corrosion.

20-6   A steel lap-welded tube 10 ft long with an outside diameter of $6\frac{5}{8}$ in.

is subjected to an external pressure of 150 psi. Determine the required wall thickness.

20-7 A brass condenser tube 1 in. in diameter and 4 ft long has a wall thickness equivalent to No. 16 BWG. (*a*) Determine the collapsing pressure. (*b*) Determine the permissible working pressure with an apparent factor of safety of 10.

20-8 A seamless cold-drawn steel tube is 6 in. in outside diameter and 12 ft long. The tube is to be subjected to an external pressure of 110 psi with an apparent factor of safety of 8. Determine the required tube thickness.

20-9 A cylinder 10 in. in diameter is to be filled with fluid at a pressure of 3,000 psi. The cylinder is made of steel with an ultimate strength of 60,000 psi. Determine the outside diameter of the cylinder using an apparent factor of safety of 5.

20-10 A closed-end cast-iron cylinder of 8 in. inside diameter is to carry an internal pressure of 2,000 psi with a permissible stress of 3,000 psi. (*a*) Determine the wall thickness by means of the Lamé, Clavarino, and maximum-shear equations. (*b*) Which result would you use? Give reasons for your conclusion.

20-11 A steel tank for shipping liquefied gas is to have an inside diameter of 8 in. and a length of 40 in. The gas pressure is 1,500 psi. The permissible stress is to be 8,000 psi. (*a*) Determine the required wall thickness, using the thin-cylinder equations. (*b*) Determine the thickness, using Clavarino's equations.

20-12 A single-acting triplex pump delivers 120 gpm at 275 psi when operating at 29 rpm. The cylinders have a bore of $5\frac{1}{2}$ in. and a stroke of 8 in. (*a*) Allowing $\frac{1}{8}$ in. for reboring and variation in casting thickness, determine the outside diameter of the cylinder when made of cast iron having an ultimate strength of 30,000 psi. Use an apparent factor of safety of 8. (*b*) Same as (*a*) using cast iron with an ultimate strength of 18,000 psi. (*c*) Same as (*a*) using cast bronze. (*d*) If the metals used in (*a*), (*b*), and (*c*) cost 7, 4, and 20 cents/lb, respectively, compare the relative costs of the three cylinders.

20-13 A cylinder of 8-in. bore has an outside diameter of 14 in. Using Lamé's equations, determine the maximum tangential and radial stresses under the following conditions. (*a*) Internal pressure of 2,000 psi. (*b*) External pressure of 2,000 psi. (*c*) Internal pressure of 2,000 psi, and an external pressure of 1,000 psi.

20-14 The cylinder of a hydraulic press has a bore of 18 in., and the working pressure is $2\frac{1}{2}$ tons/in.^{2} The working stress is limited to 4 tons/in.^{2} Assume Lamé's equations to apply. (*a*) Determine the required wall thickness. (*b*) Plot curves showing the variation of the radial and tangential stresses throughout the wall thickness.

20-15 Same as Prob. 20-14, assuming Clavarino's equation to apply and $\mu = 0.28$.

20-16 A steel ram of a hydraulic press has a 6-in. inside diameter and a 12-in. outside diameter and is subjected to an external pressure of 6,000 psi. Plot curves showing the variation of the stress throughout the wall thickness, computing points for each $\frac{1}{2}$-in. change in radius. How does the maximum stress compare with that obtained by the thin-cylinder formulas?

20-17 A flat circular plate of 12-in. diameter is supported around the edge and is subjected to a uniform pressure of 150 psi. The stress is to be limited to

10,000 psi. (*a*) Determine the thickness of the steel plate required. (*b*) Determine the maximum deflection and state where it occurs.

20-18 Same as Prob. 20-17 except that the load is concentrated on a small central area. The total load is 5,000 lb.

20-19 A flat circular plate of 10-in. diameter is rigidly supported around the edge and supports a load of 3,000 lb uniformly distributed around the circumference of a circle of 4-in. diameter. (*a*) Determine the thickness of the steel plate required if the stress is not to exceed 12,000 psi. (*b*) Determine the maximum deflection in the plate.

20-20 A cast-iron cylinder is to have a flat head cast integral with the cylinder walls. The internal working pressure is 2,000 psi and the permissible working stress 3,500 psi. Determine the wall thickness if the inside diameter is 6 in.

20-21 A cylinder of 30-in. inside diameter is to contain liquid at a pressure of 400 psi. The cylinder head is made of cast iron and is attached by bolts arranged on a 34-in. circle. (*a*) Determine the required head thickness. (*b*) Determine the required number and size of bolts if made of material having an ultimate strength of 70,000 psi.

Chapter 21.   Seals, Packing, Gaskets, and Shields

21-1 A helical speed reducer employed to drive an oil pumping rig has input and output shafts each as close to 1.875 in. in diameter as possible. In order to keep the oil inside the reducer and at the same time prevent the entrance of sand and other dirt into the reducer, shaft seals are desirable. Select shaft seals suitable for this purpose from those listed in manufacturers' catalogs. The maximum rotational speed of each shaft is 1,200 rpm; the shafts are ground to size.

21-2 A single-cylinder plunger-type hydraulic pump employs a ground-and-polished bronze plunger of approximately 2.75-in. diameter. The plunger has a stroke of 8 in. with a maximum of 325 strokes/min. Select a seal suitable for this plunger. Give all dimensions pertaining to the plunger diameter and the groove containing the seal.

21-3 An automotive-type hydraulic jack has a cylinder of approximately 2.375-in. diameter and operates at a maximum pressure of 4,200 psi. Select an O ring suitable for sealing the hydraulic oil inside the cylinder. Give the recommended diameter for the plunger and the dimensions for the groove into which the O ring fits.

21-4 A 6-in. pipe is used to carry hot wax of approximately 300°F temperature. Sections of the pipe are fastened together by use of flat circular flanges of 10-in. outside diameter with a gasket between them. Select a flat gasket material suitable for this application.

Index